MOLECULAR MACHINES

MOLECULAR MACHINES

editor

Benoît Roux

University of Chicago, USA

W⊃ **World Scientific**

NEW JERSEY · LONDON · SINGAPORE · BEIJING · SHANGHAI · HONG KONG · TAIPEI · CHENNAI

Published by

World Scientific Publishing Co. Pte. Ltd.

5 Toh Tuck Link, Singapore 596224

USA office: 27 Warren Street, Suite 401-402, Hackensack, NJ 07601

UK office: 57 Shelton Street, Covent Garden, London WC2H 9HE

British Library Cataloguing-in-Publication Data
A catalogue record for this book is available from the British Library.

Cover photo adapted from McGuffee SR, Elcock AH (2010) Diffusion, Crowding & Protein Stability in a Dynamic Molecular Model of the Bacterial Cytoplasm. *PLoS Comput Biol* **6(3)**: e1000694.doi:10.1371/journal.pcbi.1000694.

MOLECULAR MACHINES

ISBN-13 978-981-4343-44-2
ISBN-10 981-4343-44-7

Typeset by Stallion Press
Email: enquiries@stallionpress.com

Printed in Singapore by Mainland Press Pte Ltd.

Preface

In everyday language, a machine is generally an intricate device composed of different moving parts that is able to utilize energy to perform some useful work. Complicated machines can also be found in biology where they are complex macromolecular assemblies of proteins, nucleic acids, and carbohydrates that consume energy in order to perform specific functions. The concerted action of all those tiny "molecular machines" underlies all the activities of the living cell. At the present time, we know the structure of several biological macromolecular systems in great atomic detail. Yet this is generally insufficient to fully comprehend how they are able to accomplish specific tasks. For this, it is necessary to identify the different moving parts and understand how they work together. Despite some great progress, our understanding of the function of the biological molecular machines is clearly still in its infancy. Breaking new ground with these difficult problems is likely to require novel paradigms, permitting a seamless integration of structural, dynamical and functional data from experiments and theories. To some extent, what will be needed is not yet completely established. The goal of this volume is to provide an introduction to the world of complex biological molecular machines to a broad audience of students and researchers in the biosciences. Each chapter is written by leading experts to cover the results from cutting-edge research, while remaining broadly accessible. Although it is not possible to offer a comprehensive review of this vast subject, the volume covers a wide range of biological systems from both theoretical and experimental fronts. The volume starts with a review of the structure of bacterial cytoplasm in Chapter 1, thus setting the context for the function of any biological machine. Biological macromolecular systems must perform in the crowded and confusing intracellular environment. Chapter 2 follows by describing how the energy from sunlight is captured and stored by the chromatophore in photosynthethic bacteria. Plans and blueprints are required to know how to make complicated molecules. This information is encoded into the long DNA molecules and Chapter 3 reviews the function of polymerases, which serve to read and copy the DNA. Chapter 4 describes how the genetic information is decoded and translated by the ribosome into an amino acid sequence to form a protein. While the polypeptide chain emerging from the ribosome can sometimes fold spontaneously into a complicated shape, special help is required in some cases. Chapter 5 reviews how chaperonins assist the proper folding of proteins. Chapter 6 reviews the function of myosin and actin, and their role in the mechanism of muscle contraction. Chapter 7 reviews the function of protein kinases, which are part of a critical communication and signaling network inside the cell. The last six chapters are focused on the proteins residing in the cell membrane. Membrane proteins, such as transporters and ion channels (and the chromophore reviewed in Chapter 2), provide great examples of vectorial molecular machines exploiting concentration gradients of ions or the electrical potential difference across the membrane to perform useful work. Chapter 9 presents a comparative review of five membrane transporters, the leucine transporter (LeuT), the glycerol-3-phosphate transporter (GlpT), the glutamate transporter (GlT), an ATP-binding cassette (ABC) transporter, and the Na^+-coupled galactose transporter (vSGLT), while Chapter 8 is focused on the Na^+/H^+ antiporter, Chapter 10 is focused on the ABC transporter family, and Chapter 11 is focused on Na^+-coupled secondary transporters. Chapter 12 describes the structure and function of the voltage-gated K^+-selective channel, and Chapter 13 provides a broad view of the role of ion channels in cardiac muscle contraction.

Lastly I would like to personally thank Jin-Yun Liang who first suggested the creation of this volume, and Shelley Chow and Jihan Abdat from World Scientific for their great help and guidance during the editorial process.

Contents

About the Authors

Chapter 1: Molecular Behavior in Biological Cells: The Bacterial Cytoplasm as a Model System

Adrian H. Elcock[*] and Andrew S. Thomas
Department of Biochemistry
University of Iowa
Iowa City, IA 52242
[*]Corresponding author: adrian-elcock@uiowa.edu

Chapter 2: The Light-Harvesting Apparatus in Purple Photosynthetic Bacteria
Johan Strümpfer,[1,3] Jen Hsin,[1,2] Melih Şener,[1,2] Danielle Chandler,[1,2] and Klaus Schulten[*,1,2,3]
[1]Beckman Institute, Urbana-Champaign
[2]Department of Physics
[3]Center for Biophysics and Computational Biology
University of Illinois at Urbana-Champaign, Urbana, IL 61801
[*]Corresponding author: kschulte@ks.uiuc.edu

Chapter 3: DNA Polymerases: Structure, Function and Modeling
Tamar Schlick[*]
Department of Chemistry and Courant Institute of Mathematical Sciences
New York University
251 Mercer Street, New York, NY 10012
[*]Corresponding author: schlick@nyu.edu

Chapter 4: Information Processing by Nanomachines: Decoding by the Ribosome
Karissa Y. Sanbonmatsu,[*,1] Scott C. Blanchard,[2] and Paul C. Whitford[1]
[1]Theoretical Biology and Biophysics
Theoretical Division, Los Alamos National Laboratory
MS K710, Los Alamos NM 87545
[2]Department of Physiology and Biophysics
Weill Cornell Medical College
1300 York Avenue, New York, NY 10065
[*]Corresponding author: kys@lanl.gov

Chapter 5: Chaperonins: The Machines Which Fold Proteins
Del Lucent,[1] Martin C. Stumpe,[2] and Vijay S. Pande[*,3]
[1]OCE Postdoctoral Fellow
Center for Materials Science and Engineering
343 Royal Parade, Parkville VIC, 3052, Australia
[2]Simbios
Stanford University
Stanford, CA 94305
[3]Department of Chemistry
Stanford University
Stanford, CA 94305
[*]Corresponding author: pande@stanford.edu

Chapter 6: Muscle and Myosin
Ronald S. Rock, Jr.[*]
Department of Biochemistry and Molecular Biology
The University of Chicago
929 E. 57th Street, Chicago, IL 60637 USA
[*]Corresponding author: rrock@uchicago.edu

Chapter 7: Protein Kinases: Phosphorylation Machines
[*]Elaine E. Thompson,[1] Susan S. Taylor,[1,2,3] and J. Andrew McCammon[1,2,3,4]
[1]Department of Chemistry and Biochemistry
[2]Department of Pharmacology
[3]Howard Hughes Medical Institute
[4]Center for Theoretical Biological Physics
University of California at San Diego
9500 Gilman Drive, La Jolla, CA 92093, USA
[*]Corresponding author: elainet@mccammon.ucsd.edu

Chapter 8: Computational Studies of Na[+]/H[+] Antiporter: Structure, Dynamics and Function
Assaf Ganoth, Raphael Alhadeff, Isaiah T Arkin[*]
The Hebrew University of Jerusalem,
Institute of Life Sciences, Department of Biological Chemistry,
Edmund J Safra Campus, Givat-Ram, Jerusalem, 91904, Israel
[*]Corresponding author: arkin@savion.huji.ac.il

Chapter 9: Membrane Transporters — Molecular Machines Coupling Cellular Energy to Vectorial Transport Across the Membrane
Zhijian Huang,[1,2] Saher A. Shaikh,[1,2] Po-Chao Wen,[1,2,3] Giray Enkavi,[1,2,3] Jing Li,[1,2,3] and Emad Tajkhorshid[*,1,2,3]
[1]Department of Biochemistry
[2]Beckman Institute
[3]Center for Biophysics and Computational Biology,
University of Illinois at Urbana-Champaign, Urbana, IL 61801
[*]Corresponding author: emad@life.illinois.edu

Chapter 10: ABC Transporters
E. P. Coll and D.P. Tieleman[*]
[1]Dept. of Biological Sciences, BIOL 415
University of Calgary
2500 University Dr. NW, Calgary, Alberta, Canada, T2N 1N4
[*]Corresponding author: tieleman@ucalgary.ca

Chapter 11: Sodium-coupled Secondary Transporters: Insights from Structure-based Computations
Elia Zomot, Ahmet Bakan, Indira H. Shrivastava, Jason DeChancie, Timothy R. Lezon and Ivet Bahar[*]
Department of Computational & Systems Biology
School of Medicine, University of Pittsburgh
3064 BST3, 3501 Fifth Ave, Pittsburgh, PA 15213
[*]Corresponding author: bahar@pitt.edu

Chapter 12: Voltage-gated Ion Channels: The Machines Responsible for the Nerve Impulse
Benoit Roux[*] and Francisco Bezanilla
Department of Biochemistry and Molecular Biology
The University of Chicago
929 E. 57th Street, Chicago, IL 60637, USA
[*]Corresponding author: roux@uchicago.edu

Chapter 13: Voltage-gated Channels and the Heart
Jonathan R. Silva[*,1] and Yoram Rudy[2]
[1]Department of Cell Biology and Physiology
Washington University School of Medicine
[2]Washington University in St. Louis
The Fred Saigh Distinguished Professor of Engineering,
Professor of Biomedical Engineering, Medicine, Cell Biology & Physiology, Radiology, and Pediatrics

Director, Cardiac Bioelectricity and Arrhythmia Center (CBAC)
Washington University in St. Louis
Cardiac Bioelectricity Center
290 Whitaker Hall, Campus Box 1097, One Brookings Drive, St. Louis, MO 63130-4899, USA
*Corresponding author: jonsilva@gmail.com

Molecular Behavior in Biological Cells

1

The Bacterial Cytoplasm as a Model System

Adrian H. Elcock and Andrew S. Thomas

---

1. Introduction

Given a choice, most, if not all of the molecular machines covered in this book would probably prefer to perform their functions in their intended physiological environment, rather than in a test-tube. This is in contrast to the wishes of most experimental and computational biophysicists, however, for whom there are compelling reasons for choosing to study molecules *in vitro*, not least of which is the fact that reconstituting interesting behavior with purified components (proteins, DNA etc.) can unambiguously establish the identities of the factors responsible for the observed behavior. But the *in vitro* environment, while being very well suited to providing clean, unequivocal insights, is in many respects very different from the actual environment encountered by most molecular machines, and this is especially true for those that function *inside* biological cells.[1] First, there is clearly a vast increase in environmental complexity: a protein *in vitro* may only be accompanied by a few other types of molecules (water molecules, dissolved salts, and buffer molecules that help to maintain pH), but the same protein inside, for example, the cytoplasm of a bacterial cell may encounter hundreds to thousands of different kinds of protein molecules, RNA molecules, metabolites (glucose, ATP etc.), lipid molecules, and even the bacterial chromosome (DNA). A number of these molecules may be required partners for the protein, e.g. cofactors or substrates of enzymes, or other proteins with which it may form important functional complexes. But the vast majority of them will not be, and each protein must therefore be able to find its intended partners from a very wide array of unsuitable alternatives.

Not only is the compositional complexity vastly greater *in vivo*, but so is the total concentration of macromolecules. The macromolecular concentration inside a human red blood cell (erythrocyte), for example, is around 350 mg/mL,[2,3] while that in the cytoplasm of the bacterium *Escherichia coli* is around 300 mg/mL, of which ~220 mg/mL is protein.[4] For comparison, the protein concentration inside typical crystals (of the type used to solve structures by X-ray crystallography) is of the order of 150 mg/mL; it is therefore interesting that the cellular interior can remain sufficiently like a fluid that macromolecules can still diffuse (albeit somewhat slowly) from one end of the cell to another. This is especially so given that most proteins, when concentrated to such high levels will "crash out", i.e. precipitate from solution. As we shall see, this very high concentration of "stuff" can, in principle, have consequences not only for the kinetics of macromolecular processes but also for their thermodynamics.

2. A Simple Analogy for Molecular Behavior *in vivo*

This chapter will be concerned primarily with outlining the ways in which molecular behavior *in vivo* might be different from that observed *in vitro*. As a starting point, we will illustrate potential differences by examining a concrete example: the diffusional behavior of the protein CheY from *E. coli*; most of the same basic principles should apply, however, to more or less any protein. CheY is a small 128-residue (14 kDa) protein that shuttles back and forth between chemoreceptors (which sense chemical signals present in the exterior environment) and the flagellar motor (which turns the bacterium's flagellum either counterclockwise, driving the bacterium forward, or clockwise, causing it to tumble).[5]

How are we to think about the diffusional aspect of CheY's function at a molecular level? To answer this question we will try to construct a real world analogy starting from the moment that the protein becomes phosphorylated and released by the histidine kinase CheA (which is attached to the membrane-bound chemoreceptors) to the moment that it modulates the organism's swimming behavior by directly binding to the flagellar motor control protein, FliM. One possibility is to think of CheY as a courier, bearing a message that is important, perhaps vital, to the survival of the nation (i.e. organism). We will make life somewhat simpler for ourselves by restricting ourselves to a 2D setting: we will consider the cell as analogous to a large airport terminal. CheY's job, therefore, is simply to carry its message (phosphate group) from one end of the terminal to the other, where another person — more important than our courier — awaits to receive the message.

Before we even consider adding the rest of the intracellular milieu, it is useful to imagine how things might work if, aside from the chemoreceptors and the flagellar motor, an *E. coli* cell were to be filled only with water. In terms of our analogy, this would correspond to a situation in which the entire terminal is deserted. We might naively imagine that passing the message to the receiver requires only that the courier walk the length of the terminal. But since our courier and recipient represent molecules, they have neither eyes, ears, noses nor any other senses that might enable them to identify each other at long distances. Furthermore, it is important to remember that since they also do not have brains, they have no conception of what it is that they are "supposed" to do: the courier in our scenario, for example, doesn't know that he/she is supposed to deliver a message anymore than the recipient knows that he/she is supposed to receive a message. Because of these issues, it is more realistic to assume that our courier will not recognize our recipient until they literally bump — face to face — into each other. We can mimic this in our real world model by applying a blind-fold to *both* the courier and the recipient, and by assuming that they have no other way of communicating with one another other than by a sense of touch.

It is also important for us to recognize that our courier will not simply move across the platform in a straight line: molecules in aqueous solution do not move as if shot from a cannon (i.e. ballistically); instead, on the kinds of timescales and lengthscales that we are interested in here (nanoseconds and nanometers and beyond), they move *diffusively*, that is, they are subject to repeated collisions with solvent molecules that continually reorient and redirect them, leading them to exhibit Brownian motion. A simple way to add this to our real world situation is to imagine that our courier is continually buffeted by sudden, short-lived gusts of wind that come from completely unpredictable directions (clearly, it's not a very well designed airport terminal). Note that while this "drunken sailor" method of movement may not seem particularly efficient, it may, depending on the time- and lengthscales over which movement is required, be completely adequate for the purposes of the cell. But in other cases — e.g. when cargo has to be transported from one end of a eukaryotic cell to the other (as, for example, happens in nerve cells in the human body) — diffusion is not efficient enough and motor proteins, moving along pseudo-static "tracks", are therefore needed (see other chapters in this book). In passing, it is worth noting that we could extend the analogy to account for such motor-driven transport by imagining that the courier, still blind-folded, occasionally stumbles on to the terminal's moving walkways; note that there is no guarantee that he/she will end up on a walkway moving in the right direction!

Having struggled to set up a crude analogy for how a message might be passed from one protein to another in a biological cell filled only with water, we can now rerun the entire scenario in a way that more realistically mimics intracellular conditions. As noted above, the principal difference is that there are very high concentrations of other macromolecules present; we can account for this by imagining that the platform is now extremely crowded with other people, some the size of small children, some the size of Shrek, but *all* of them blind-folded. Clearly, the situation will be chaotic: our courier, who was already having a difficult enough time navigating the deserted terminal is now also repeatedly jostled and blocked by the crowd. We can probably imagine that this will slow the courier considerably although the extent to which he/she is slowed may perhaps depend on his/her size: small children, for example, are often quite adept at finding their ways through crowded rooms, whereas we could imagine that Shrek might have a great deal more difficulty. A schematic illustration of these differences is shown in Figure 1. But we can also imagine a more complicated situation. What if the courier encounters a member of the crowd who happens to share some of the physical features of the message's recipient? In such cases, we must expect that the courier will tend to remain close to the "impostor" for some period of time (perhaps a very long time), with the result that his/her diffusive exploration of the environment might be significantly slowed. In molecular terms, such a situation could occur if CheY encounters macromolecules that have similar electrostatic potentials or similar patches of exposed hydrophobic residues, to those of the flagellar motor protein FliM. In passing, therefore, we can speculate that evolution might have acted to decrease the extent to which such non-specific (and unintended) interactions are likely to occur.

It is hopefully apparent from the above description that we *can* construct a real world analogy for the situation encountered by CheY in the *E. coli* cytoplasm, but we have to be prepared to alter it in some fundamental ways to make it work. While it is probably useful at an illustrative level — since it allows us to begin to frame our thinking about what life might be like inside a cell — it should not be thought that it describes all of the possible differences between the *in vitro* and *in vivo* environments (for example, Section 3).

3. Macromolecular Crowding Effects

One of the more subtle effects of the highly crowded environment faced by molecules *in vivo* can only be revealed by using some ideas from statistical thermodynamics.[6] This is the effect of

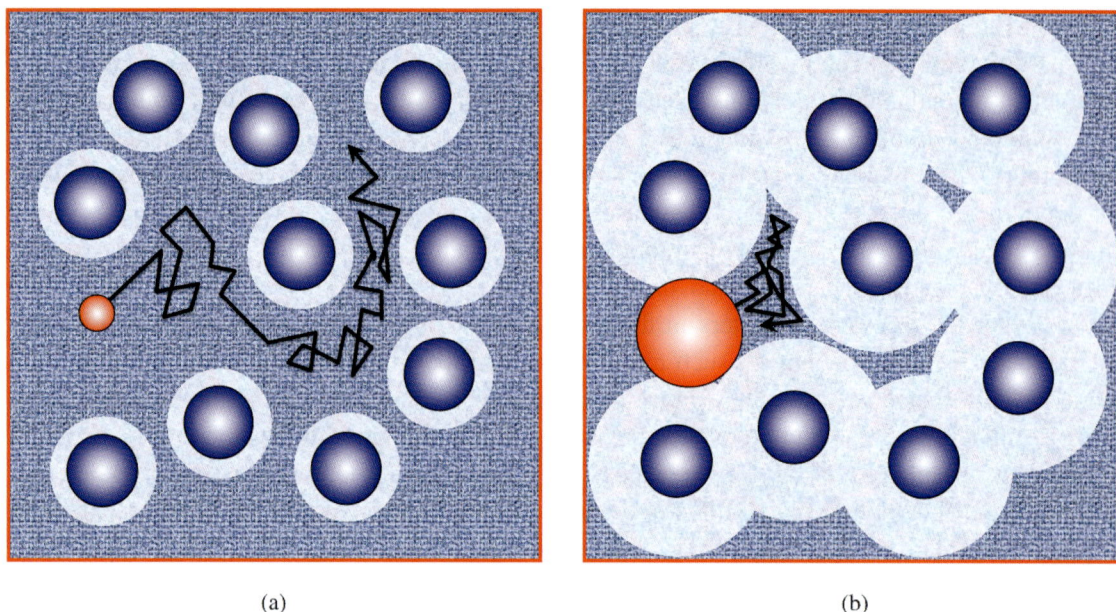

(a)   (b)

Figure 1.   A schematic illustration of how diffusion might be affected by highly crowded conditions for (a) a small protein, and (b) a large protein. The diffusing protein of interest is shown as a red circle, the other macromolecules
comprising the crowded environment are shown as blue circles. The light blue "haloes" around each crowder molecule
indicate the volume that is inaccessible to the diffusing protein due to steric interactions: note that the excluded volume
experienced by the large protein (b) is considerably larger than that experienced by the small protein (a). The black lines
indicate potential diffusional trajectories for the two proteins.

steric ("excluded-volume") interactions between a protein of interest (again, e.g. CheY) and its
environment of surrounding "crowder molecules". As we will see, such interactions can, in principle, have very significant thermodynamic effects, and this realization has led, in recent years,
to the development of an entire sub-field of biophysics devoted to understanding and predicting
effects due to "macromolecular crowding".[7,8] An illustrative example of the crowding effect is
shown in Figure 2 in which we consider an idealized protein folding equilibrium. In what follows, we will make a number of simplifications in order to ease our calculations, but it is
important to bear in mind that the basic argument and conclusions are unaffected by these
simplifications.

    We consider a simplified "protein" that can exist in one of only two conformations: one, a
highly compact native state, and another, an extended unfolded state. These are shown in Figures
2a and 2b respectively. As shown in these figures, we will assume that the protein exists within the
confines of a 2D box consisting of 12×12 squares. We will further assume that the energies of
both protein conformations are equal. This assumption makes our life considerably easier since, in
the language of statistical thermodynamics,[6] it means that the two conformations have identical
"Boltzmann weights"; all we have to do to estimate the equilibrium constant for the folding reaction, therefore, is take the ratio of the number of possible conformations of the folded protein and
the unfolded protein. In the present case, therefore, $K_{eq} = 1/1 = 1$, which in turn means that the free
energy of folding of the protein, which is given by $\Delta G = -RT \ln K_{eq} = -RT \ln (1) = 0$.

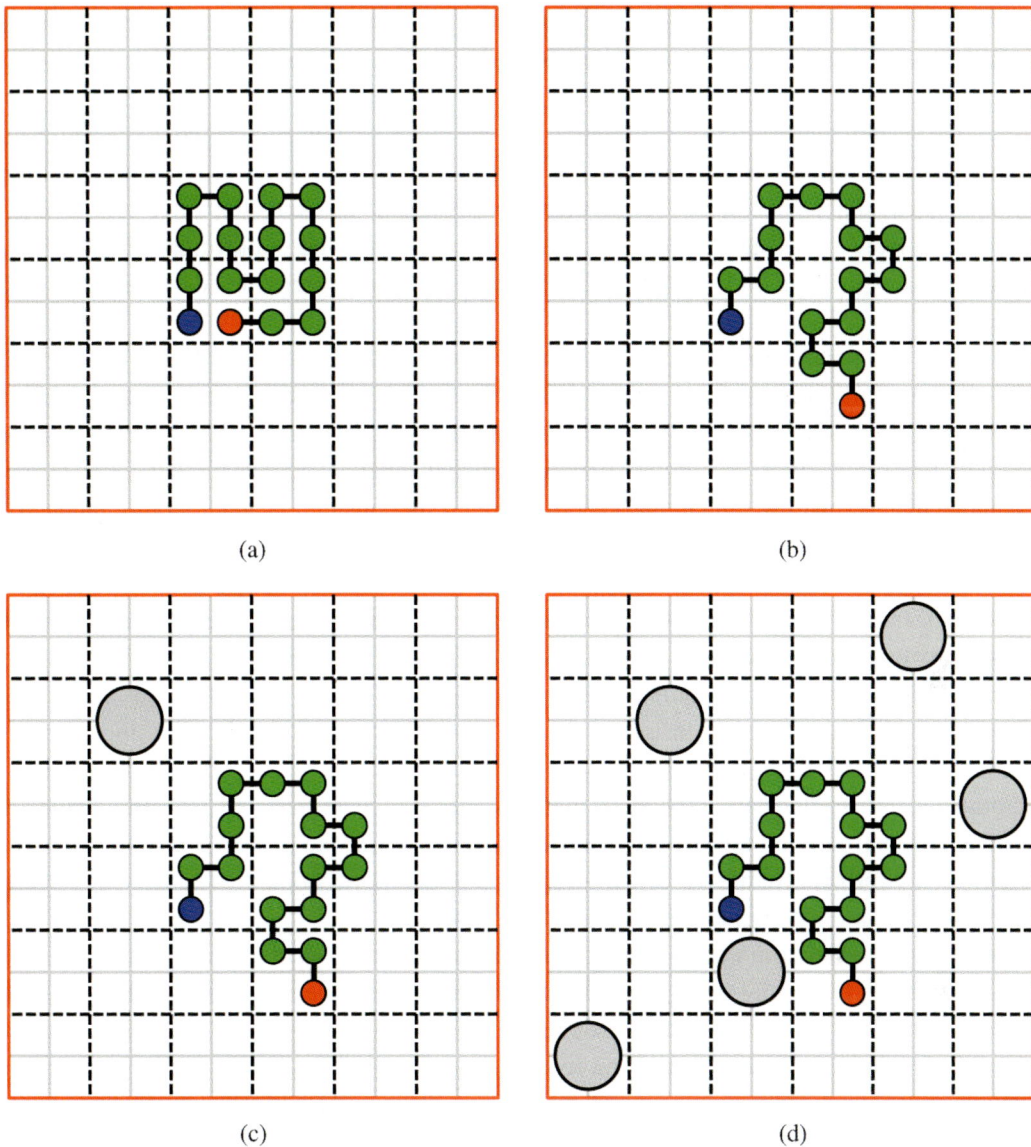

Figure 2. A simple 2D model of crowding effects on a protein Folding reaction. (a) Folded state structure of a model protein consisting of 16 "amino acids" (circles) connected by peptide bonds (thick lines); the blue and red circles represent, respectively, the N- and C- termini of the protein. (b) The same protein in a putative unfolded conformation. (c) One possible configuration of the same system with a single added crowder molecule (large grey circle); note that for simplicity we assume that the crowder can only be placed within one of the large dashed squares, of which there are 36 in total. (d) One possible configuration of the same system with five added crowder molecules.

Next we consider what happens when we add an idealized large macromolecule to the system (represented by the large filled circle in Figure 2c). We will assume, again just for the purposes of making the calculations simpler, that the "crowder molecule" can only occupy the larger, dashed squares on the grid (i.e. in the absence of the protein, it would be restricted to 6×6 possible positions).

We can now assess the effect of this single crowder on the protein's folding equilibrium simply by asking how many positions can be occupied by the crowder molecule without clashing with the protein; in the language of statistical thermodynamics this means that we assume that any configuration of the system that has a steric interaction between the protein and crowder has an infinitely positive energy, and therefore has a Boltzmann weight of zero, and so can be ignored. Counting up the number of available positions for the crowder molecule, we find that there are 32 and 29, respectively, for the folded and unfolded states of the protein. The equilibrium constant for the folding reaction is now given by 32/29, which in terms of free energy, is $\Delta G = -RT \ln (1.10) = -0.05$ kcal/mol.

What does this mean? It means that the mere presence of the crowder molecule has shifted the protein folding equilibrium in favor of the folded form. This, in essence, is the macromolecular crowding effect: excluded-volume interactions of the crowder molecules with the protein of interest tend to favor the latter, assuming a more compact state.[9] The reason that it is termed a "macromolecular" crowding effect can be demonstrated by considering what would happen if the crowder molecule was a much smaller molecule, e.g. one that occupies only a single square of the system and which therefore could occupy each of the 12×12 small, grey squares in the absence of the protein. In this case, we find that the number of possible positions for placing this "micromolecular" crowder molecule is the same for both the folded and unfolded conformations of the protein (128 in both cases). The equilibrium constant for the folding reaction in this case is therefore given by 128/128, which again is 1, the same result that we obtained in the micromolecular crowder molecule's absence. In other words, there is no formal "crowding effect" induced by very small molecules. It is important to be clear that this does *not* mean that the "real life" inclusion of high concentrations of small molecules will cause no changes in the observed thermodynamics of processes such as protein folding (they certainly can); it means only that the changes they elicit are *not* likely to be due to the excluded-volume effect outlined above.

Returning to our macromolecular crowder molecule, we can consider what happens as its concentration increases by adding multiple copies of the molecule to the system. Now the calculations get more complicated because we have to enumerate all of the possible positions that could be occupied *simultaneously* by the crowder molecules (this is why we made the earlier assumption of only allowing the crowder to occupy 6×6 possible positions). An example configuration of the system is shown in Figure 2d. To calculate the number of possible positions, we make use of a familiar equation: the number of distinguishable ways of arranging M identical objects in N positions = N! / {M! (N − M)!}. Using this relation we obtain the numbers of possible positions for placing two crowder molecules in the system as 496 and 406 for the folded and unfolded states respectively. K_{eq} for the folding reaction is now 496/406 = 1.22. Repeating the calculations for 5, 10 and 15 crowder molecules, respectively, we obtain values of K_{eq} = 1.70, 3.22 and 7.29, which in free energy terms amount to −0.31, −0.69 and −1.18 kcal/mol at 298 K. Clearly, adding increasing numbers of crowder molecules shifts the protein folding equilibrium increasingly in favor of the folded state. Following this logic, and noting that the 15 crowding molecule situation is qualitatively similar to the total macromolecular concentrations encountered in the *E. coli* cytoplasm, it seems an inescapable conclusion that proteins will be significantly more stable *in vivo* than *in vitro*. Similar lines of thinking can also be used to estimate the likely effects of a highly crowded environment on protein-protein association and aggregation thermodynamics.[9]

Calculations such as those just described provide a good rough estimate of expected behavior, but it is important to be aware of the simplifications involved. One obvious simplification is that the amino acids of our model protein, and our crowder molecules, are restricted to occupying discrete positions on a 2D grid. This simplification could be set aside if, for example, we used computational methods to enumerate or sample the possible positions of the protein and crowder molecules in 3D space: this is now routinely done in studies aimed at modeling the thermodynamics of protein-related equilibria (e.g. Refs. 10–11). A second obvious simplification concerns our modeling of the protein's folding equilibrium. In reality, there are *far* more unfolded conformations than folded conformations, with relative energies that are much less favorable than those of the folded conformations; again, computational modeling methods can address this issue by generating large numbers of reasonable structural models for the unfolded state.[12] A third simplification is that our crowding molecule is treated as a circular disc (it could have been any other shape that fitted entirely within one of the larger squares). At the time of writing, most computational studies that examine crowding effects on biomolecular systems use similar, simple geometric shapes (spheres, rods etc.) to model the crowder molecules. This is almost certainly adequate for the purposes of obtaining estimates of the magnitude of crowding effects, but as we will see below, it is now possible to perform similar studies using "real" protein structures, that therefore more closely mimic the structure of intracellular environments.

Two final assumptions of the above calculations are crucial to note. The first is that the model presented explicitly discounted the possibility of favorable interactions occurring between the protein and the crowder molecules and assumed that their only form of interaction is an antagonistic, steric interaction. In reality, of course, molecules such as proteins interact with each other via any or all means at their disposal: oppositely-charged residues can form transient ion pairs ("salt bridges"), unpaired polar groups (-OH, -NH$_2$ etc.) can form hydrogen bonds to one another, and exposed hydrophobic sidechains can "stick" to one another. Again, we shall see later that it is possible to begin adding in the effects of such interactions into calculations; we will also see that these additional terms can lead us to make quite different conclusions about the potential effects of a cellular environment on the thermodynamics of protein folding. Finally, it should be noted that the model presented above includes no solvent; in reality, solvent molecules would occupy *all* of the apparently unoccupied squares. By leaving them out of the calculations, we have implicitly assumed that the addition of crowder molecules to the system causes no change to the solvent environment. While this is reasonable as a first approximation, it is by no means certain that it will be true of intracellular environments: there is, in fact, continued debate about the extent to which water inside biological cells behaves like "bulk" water (see, e.g. Ref. 13).

4. The Bacterial Cytoplasm as a Model System

The remainder of this chapter will explore one way in which the simple 2D model described above might be replaced by a structurally detailed 3D model of an intracellular environment implemented on a computer. Our ultimate aim is to build a model that allows us, at a molecular level of resolution, to observe directly how macromolecules *in vivo* negotiate their way through a minefield of competing interactions while performing their intended function; one might eventually hope, in particular, to use such a model to uncover the fundamental ways in which behavior occurring *in vivo* might differ from that observed *in vitro*. We assume, for now, that the factors affecting macromolecular behavior inside bacterial cells will be essentially the same as those that affect behavior in eukaryotic cells. Certainly, there are major differences between the two types of cells: the cytoplasm of a eukaryotic cell, for example, differs considerably in its total macromolecular concentration and in its composition (e.g. its cytoskeletal elements are far more complex), but even with such differences, it is still much more similar to the cytoplasm of a prokaryotic cell than it is to a simple aqueous solution.

Undoubtedly the best characterized of the prokaryotes is *Escherichia coli*, a gram-negative bacterium that forms spherocylindrical cells of approximately 2 μm length and 1 μm diameter. The amount of accumulated knowledge on this organism is so vast that it is impossible to do more than scratch the surface of it in a single chapter; instead, the reader is referred to a number of online databases that serve as excellent portals for accessing much of this information (see for example, http://www.ecocyc.org,[14] http://redpoll.pharmacy.ualberta.ca/CCDB/,[15] and http://www.york.ac.uk/res/thomas/[16]). At a basic level, *E. coli* can be considered to have three primary intracellular aqueous-phase environments: the DNA-dominated nucleoid, which is surrounded by and contiguous with the protein-dominated cytoplasm, and the thin periplasmic layer that is sandwiched between the inner and outer membranes. If we include also the lipid-phase inner- and outer-membranes, both of which are highly enriched in embedded proteins, it is clear that even this (relatively) simple organism provides a wide range of physicochemical environments in which to study macromolecular behavior *in vivo*. Accordingly, the rest of this chapter will attempt to use the cytoplasm of *E. coli* as a vehicle for illustrating some of the concepts surrounding molecular behavior inside cells.

5. A Structural Model of the Bacterial Cytoplasm

At least two crucial pieces of information are required for the construction of a molecular model of an intracellular environment such as the bacterial cytoplasm. First, we must know, or be able to guess, the relative abundances of each (macro)molecule to be included in the model. Second, we must have, or be able to build, 3D structures of each molecule. For proteins, answers to the first of these problems comes from the experimental techniques of *quantitative proteomics*, which attempt to quantify the entire complement of proteins (i.e. the proteome) of a given organism under a given set of conditions. Given the extreme compositional complexity of biological cells, this is an enormous challenge. The most recent efforts to quantify proteins in the cytoplasm of *E. coli*, for example, report quantifications of more than 600 proteins,[17] which is a prodigious feat in itself, but given that there are ~ 2900 different proteins thought to be resident in the *E. coli* cytoplasm,[18,19] it is clear that there is still some way to go. Related experimental techniques aim to quantify each of the small molecule metabolites (e.g. glucose, ATP) in an organism; this is the nascent field of *metabolomics*.[20] The second requirement mentioned above — that of needing 3D structures of the molecules of interest — is addressed by the experimental techniques of *structural biology* (X-ray crystallography, NMR spectroscopy and cryoelectron microscopy). The progress that has been made in this field over the last 50 years is truly staggering: at the time of writing, the number of individual structural entries in the protein databank (http://www.rcsb.org) is in the tens of thousands, of which approximately 11% are *E. coli* proteins (it should be noted, however, that many of these are redundant copies). Despite this progress it will be a long time — if ever — before we have high-resolution structures of the entire complement of proteins in the cytoplasm of *E. coli*. For the foreseeable future, therefore, our model will have to contain a mixture of *bona fide* crystal structures of *E. coli* proteins and so-called "homology models" produced by mapping the *E. coli* protein sequence on to the high-resolution structure of an orthologous protein from a different organism.[21]

An illustration of the type of molecular model that can be constructed[22] using currently available structural and proteomic data is shown in Figure 3. This model contains ~1000 randomly arranged macromolecules, each of which is of one of 50 different types of molecules. Individual molecule types have been selected according to their relative abundance in proteomics experiments performed with cells growing on minimal media: the most abundant molecule types in the model are the gene products of TufA, which is the translational elongation factor EF-Tu, and of MetE, which is a methyltransferase involved in methionine biosynthesis; the two proteins are present in copy numbers of 181 and 213 respectively. The least abundant molecule in the model is the very large glutamine synthetase, present in only one copy. Also present are tRNAs, enzymes of the glycolytic and tricarboxylic acid cycle pathways, and the two subunits (30S and 50S) of the protein-synthesizing machine, the ribosome. The combined concentration of the macromolecules in the model is 275 mg/mL, which is close to experimental estimates for the *E. coli* cytoplasm.

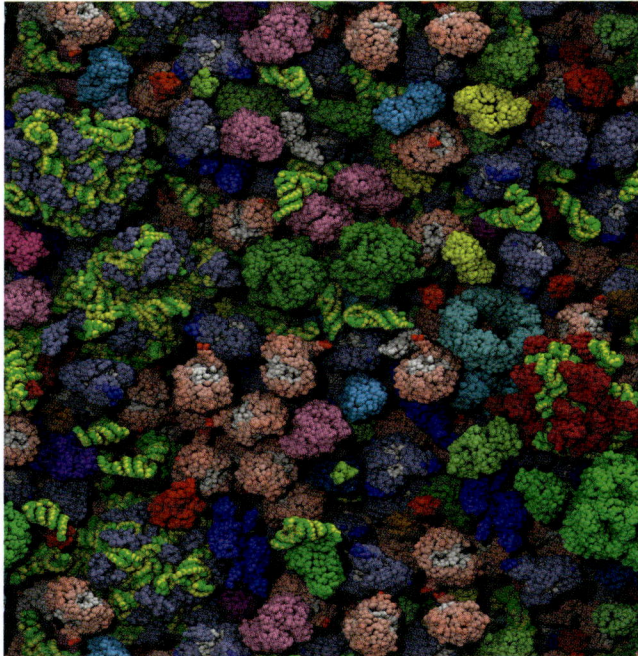

Figure 3. A structural model of the cytoplasm of *Escherichia coli*.[22] RNA components are colored green, with their phosphate groups in yellow; protein components are colored arbitrarily by molecule type. Note the 50S ribosomal subunits (blue + green/yellow) on the left-hand side of the image and the 30S subunit (red + green/yellow) on the right-hand side. This image was prepared with the program VMD.[33] This figure has been adapted with permission from McGuffee & Elcock (Ref. 22).

As a point of reference, it is worth noting that the width of this simulation cell amounts to approximately 1/10th of the diameter of a typical *E. coli* cell. Put like that, it sounds impressive, but in terms of volume it sounds a lot less so: the volume of the simulation cell occupies approximately only 1/1000th of the volume a typical cell. Clearly, therefore, the model shown in Figure 3 represents only the first "baby step" towards modeling entire biological cells. Readers interested in imagining what the interiors of *entire* cells might look like would do well to examine some of the extraordinary and exquisite images made by Dr. David S. Goodsell (see Suggested additional reading material).

6. Some Issues in Simulating Such a System

While Figure 3 is useful in illustrating just how crowded the interior of a typical bacterial cell can be, it is important to note that the true purpose of this particular model is not to produce static images but is instead to provide a dynamic view of the way molecules diffuse, tumble, and interact with each other *in vivo*. There are many computer simulation methods that might be used, in principle, to bring the static model "to life" so it is perhaps worthwhile to very briefly outline some of the factors that might determine one's choice of method; more detailed examinations of different simulation techniques can be found elsewhere (see, e.g. Refs. 23–24). Conceptually, the most straightforward way to simulate the dynamics of macromolecules in the cytoplasm is to use the technique of molecular dynamics (MD) simulation. In its application to aqueous phase systems, such a simulation method usually employs "explicit solvent", which essentially means that the spaces surrounding and within macromolecules are filled — as they are expected to be in real life — with water molecules (and dissolved ions). The essence of MD simulation is then to repeatedly solve Newton's (classical) equations of motion in order to determine the time-dependent trajectory of each individual atom in the system. To do so requires knowledge of the net force acting on *every* atom at each step of the simulation, which, in principle, requires calculating each atom's interaction with every other atom in the system. Computationally, this can be a very expensive undertaking: when explicit water molecules are added to the system shown in Figure 3, for example, we obtain a model that contains approximately 45 million atoms. When one realizes that a typical current MD simulation of a single protein — in an aqueous system containing perhaps a total of 60 000 atoms — takes approximately 1 month to simulate 1 μs on 64 cpu cores (operating in parallel) it can be seen that performing MD simulations of the bacterial cytoplasm model would be highly non-trivial.

Because of the expense associated with the MD approach, use is often made of more approximate, faster simulation methods such as Langevin dynamics (LD) or Brownian dynamics (BD) techniques. The two techniques share much in common: in both, the explicit water molecules, and often also the solution's dissolved ions, are simply removed from consideration. As might be expected, this drastically decreases the number of atoms in the modeled system, and therefore, in principle, results in a huge increase in the (computational) speed of the simulations. But of course, this acceleration comes at a price, and both LD and BD methods must seek to add back in the more crucial features of the missing solvent in order to ensure that the thermodynamic and kinetic behaviors of the solutes are not disastrously compromised. In the case of BD,[25] this involves adding random displacements (abrupt "shoves") to each solute to mimic the Brownian effect that would normally be exerted by the missing solvent molecules. (Note that we cannot simply remove the water molecules and continue to use MD techniques and expect the solutes to behave as they would in solution: in the absence of water, MD will tend to make the solutes move ballistically, which we will recall from earlier, is not the way that solutes really move in aqueous solution).

Water, of course, does far more than just repeatedly crash into solute molecules. Pure water, for example, diminishes (i.e. "screens") long-range electrostatic interactions to approximately 1/80th of the strength that they would have in vacuum, and dissolved salts in solution screen such interactions even more. Clearly, any attempt to model the interactions between charged groups such as the positively and negatively charged amino acids and the phosphate groups of RNAs must take account of this screening effect. Equally importantly, water is also the key ingredient of the hydrophobic effect, which in turn is thought to provide the primary driving force for protein folding and protein–protein association events. Again, any attempt to model folding and association processes as they might occur *in vivo* must make some attempt to mimic the fact that exposed hydrophobic groups have a pronounced tendency to associate with one another in aqueous solution. As is often the case when approximations are involved, there are differing opinions about how best to implicitly model water's effects on electrostatic and hydrophobic interactions, and much current research in computational biophysics is therefore geared toward evaluating and improving upon the wide range of possible approaches.

One final consequence of omitting explicit water molecules is that the hydrodynamic interactions (HI) that act between macromolecules in aqueous solution are lost. Conceptually, HI account for the fact that even when two macromolecules are non-interacting with each other in aqueous solution, the closer they are to one another, the more their motions will tend to become correlated. A simple way of thinking about the origin of this effect is to imagine what would happen if one macromolecule were to be displaced in the direction of the other, for example, by the action of an external force: the displaced macromolecule will displace water molecules that, in turn, will act to displace the other macromolecule. Obviously, if water molecules are excluded from the simulation model then this water-mediated effect will be lost. Computational methods for implicitly modeling hydrodynamic interactions have been intensively investigated in studies of colloidal systems (where solutes might have diameters of ~ 1 μm) (e.g. Ref. 26) and it may be that similar methods will be needed for modeling macromolecular dynamics in intracellular conditions. Such hydrodynamic models are, however, expensive to compute, and it is not necessarily clear at the time of writing how dire the consequences will be for omitting them. In any case, one important point to note is that in contrast to the effect of water on the thermodynamics of electrostatic and hydrophobic interactions, hydrodynamic interactions change only the dynamics, and *not* the thermodynamics, of the system: equilibrium properties, such as free energies of protein-protein associations, should be identical in simulation models that include or exclude HI.

The admittedly rather radical step of omitting all solvent from the model is one way that we can make simulations of the cytoplasm model shown in Figure 3 more feasible with the kinds of computational resources routinely available to researchers. But this still leaves us with a system that contains perhaps 11 million atoms, and that is therefore far from trivial to subject to simulation. A further approximation that one might consider, therefore, is to treat some or all of the modeled macromolecules as *rigid bodies*. This simplification enables dynamical simulations to be accelerated in two ways. First, the forces that *act upon* rigid models of macromolecules can be computed much faster since there is no need to calculate or model any of the interactions that occur *within* the macromolecule itself; this is especially helpful in the case of extremely large members of the model such as the 30S and 50S ribosomal subunits or the GroEL/ES chaperonin complex. Second, depending on the form of the energy function used, the forces that are *exerted by* rigid models of macromolecules can be computed much faster since they can be *precomputed* and stored on a 3D grid that surrounds (and moves with) each rigid molecule: the forces acting on atoms of other molecules that approach the rigid molecule can then simply be read from the corresponding region of the 3D grid.[27,28] Such an approach is obviously only cost-effective if the effort involved in precomputing the 3D grid is expended only once (i.e. before the simulation actually begins); this, in turn, requires that the molecule generating the grid be truly rigid — otherwise, any relative movement of its atoms would necessitate recomputing much, if not all, of the grid. Of course, this rigid body approximation is likely to be more acceptable for some macromolecules than for others: it will definitely *not* be appropriate for modeling mRNAs or intrinsically unstructured proteins, since their degree of conformational flexibility is likely to be very high.

7. A Dynamic Model of the Bacterial Cytoplasm

With all of the above simplifications in hand, it has become possible, at the time of writing, to conduct dynamic simulations of the cytoplasm model shown in Figure 3 on a timescale of approximately 20 µs;[22] movies showing the resulting behavior can be found at the authors' website (http://dadiddly.biochem.uiowa.edu). The overall effect is, as might have been anticipated, rather chaotic: macromolecules undergo relatively short periods of free diffusion interspersed with periods during which they become temporarily stuck — through hydrophobic and/or electrostatic interactions — to other members of the ensemble. Smaller macromolecules can be seen to diffuse comparatively rapidly, while the much larger 30S and 50S ribosomal subunits appear to be effectively stationary over the timescale of the simulations. This lack of motion on the part of the largest members of the model immediately tells us that 20 µs is far from being sufficient for the simulations to completely "sample" the system's possible configurations: in order to achieve proper sampling of the ribosomal subunits, for example, the simulations would probably need to be extended to a timescale of several milliseconds. But that said, for the more mobile and more abundant members of the ensemble, the story is more encouraging: the faster the molecules diffuse, the more rapidly they explore their environment, and the more representatives there are of a particular molecule type, the more complete is our view of this molecule type's potential behavior. Because of this, even simulations of 20 µs are likely to be sufficient for obtaining reasonable estimates of the behavior of the more abundant members of the system.

8. Quantifying Protein Diffusion in the Bacterial Cytoplasm

While the overall behavior of the simulation model appears to fit crudely with our expectations — i.e. with the airport analogy that we so painstakingly constructed earlier — it is obviously important to try to obtain a more quantitative assessment of the extent to which the observed behavior is realistic. This is an issue that arises, or should arise, with all work that is based on computer simulation; the only real way of addressing it is to find some observable that can be measured experimentally and examine whether the value of that observable measured from the simulations matches with it. In the particular case of the bacterial cytoplasm, one good choice of observable is the translational diffusion coefficient of the green fluorescent protein (GFP), since this has already been measured by a number of experimental groups (e.g. Refs. 29–30). Such measurements have been based on the technique of *fluorescence recovery after photobleaching* (FRAP), which briefly summarized involves (a) the use of a laser to photobleach (i.e. destroy the fluorescent properties of) GFP molecules in a selected area of the cell, and (b) the measurement of the time required for the fluorescence in that selected area to recover after the laser pulse. Since the recovery of fluorescence — which occurs on a millisecond timescale — results from GFP molecules that were initially *outside* the laser-illuminated region diffusing *into* it, the rate of recovery is dependent on the translational diffusion of the GFP *in vivo*. The process of extracting an effective diffusion coefficient from the experimentally measured data is somewhat involved, but the key finding of these studies is that GFP's apparent diffusion coefficient *in vivo* is approximately one-tenth of its diffusion coefficient *in vitro*. Clearly, therefore, the crowded cytoplasmic environment strongly affects the abilities of typical macromolecules to freely diffuse.

The translational diffusion coefficient of a molecule in a computer simulation is, at least in principle, one of the easier quantities to measure: one can simply use the Einstein formula, which in 3D reads: $\langle r^2 \rangle = 6D_{trans}\Delta t$, to relate the mean squared distance, $\langle r^2 \rangle$, traveled by the molecule in time Δt, to its translational diffusion coefficient, D_{trans}. In dilute aqueous solution D_{trans} is usually found to be essentially independent of Δt; in highly crowded environments such as the bacterial cytoplasm, however, the apparent diffusion coefficients of macromolecules may appear to depend significantly on Δt. This phenomenon, which is usually termed *anomalous diffusion*, is often a simple manifestation of a macromolecule's motion being frustrated by the macromolecules that form its immediate neighborhood. On a very short timescale, this hindrance may not be apparent because the protein is still essentially free to diffuse within the "cage" formed by its neighbors; on a longer timescale, however, it would soon be apparent that the macromolecule has been prevented from moving as far as one might expect; its apparent translational diffusion coefficient on this longer time scale must therefore be somewhat decreased. In the case of the BD simulations of the cytoplasm model shown in Figure 3, it appears that the crossover between the short-time and long-time regimes occurs on a timescale of a few microseconds; this is similar to the average lifetime of non-specific encounters observed between the members of the ensemble. Importantly, however, it is possible to show that the simulation model — which it should be remembered attempts to account for (favorable) hydrophobic and electrostatic interactions between neighboring macromolecules — can produce a long-time translational diffusion coefficient for GFP that is in good agreement with experimental FRAP measurements.[22]

9. Thermodynamic Stability of Proteins in the Bacterial Cytoplasm

Earlier in this chapter we considered an idealized model showing how the thermodynamics of a protein folding reaction might be affected by the addition of high concentrations of macromolecular crowder molecules. A number of the stated simplifications of that model can be overcome using the cytoplasm model described above: in particular, we now have structurally realistic models of the crowder molecules, we have (hopefully) realistic models of the way that the crowders might be arranged, and we have an energy model that allows for favorable interactions between macromolecules and that seems to be reasonably realistic, at least in so far as it leads to diffusion coefficients of GFP that match well with experiment.[22] In order to use the cytoplasm model to calculate its expected effect on the thermodynamics of a protein folding reaction, the only additional feature that we need is a way of constructing putative models of the protein's unfolded conformations, which is a problem that others have nicely solved (e.g. Ref. 12). The mechanics of the actual calculations of folding thermodynamics need not concern us here but they essentially involve computing the average interaction energies experienced by unfolded and folded-state conformations of the protein of interest when immersed in the cytoplasm; this is achieved by attempting millions of *random* insertions of the conformations into the model. Importantly, the success of these calculations can be assessed for at least two proteins for which there are experimental estimates of the folding thermodynamics *in vivo*[31,32]; here we will focus on only one of these proteins, namely the 136-residue protein *cellular retinoic acid binding protein* (CRABP).

Innovative experimental work carried out on CRABP has indicated that its folding free energy is destabilized by ~ 1.4 kcal/mol *in vivo* relative to *in vitro* conditions.[32] It will be recalled that this experimental result is the exact opposite of what we expected on the basis of our simple 2D crowding model; it is also the exact opposite of what we obtain when we use a strictly steric interaction approach to describe the interaction between CRABP and our more elaborate 3D cytoplasm model.[22] So what is the reason for this qualitative difference between the calculations and experiment? There are several possible explanations. One is that a significant amount of the unfolded protein *in vivo* might be bound by chaperone systems. A second is that our earlier assumption that the solvent is unaffected by the addition of high concentrations of biological macromolecules might be incorrect. A third possibility — not necessarily mutually exclusive with the first — is that there are *favorable* interactions between the unfolded protein conformations and the cytoplasm "crowders" that outweigh the excluded volume effect: it will be recalled that we explicitly excluded this possibility when constructing our simple 2D model in order to focus attention solely on the excluded volume effects. Of the three possibilities, the only one that can be directly addressed with the above cytoplasm model is the third; intriguingly, when we allow for such favorable interactions to occur — with the same energy model used in the BD simulations — we find that the calculations predict that CRABP will be destabilized by ~0.9 – ~1.8 kcal/mol,[22] which is in rather good agreement with experiment.[32]

10. Conclusion

This latter result provides an encouraging indication that computational models such as that outlined here might eventually be routinely capable of making quantitative predictions of the thermodynamics of folding and binding reactions *in vivo*. If that can be achieved then one would have a computational model of an intracellular environment that can meaningfully be said to bridge the biophysical gap that currently exists between *in vitro* and *in vivo* experimentation. In order to achieve this, however, much needs to be done, and hopefully the above description has provided at least an idea of the kinds of issues and challenges that must be faced. Most of the key issues relate to the simplifications currently made in modeling such large systems, and ultimately it is probably fair to say that these will eventually be solved when advances in computing power make the simulation of models such as that shown in Figure 3 possible with explicit-solvent MD methods. Since it will be some time before such power becomes widely available to researchers, however, it is inevitable that some degree of approximation or simplification in the modeling will be involved. We expect that in the coming years significant advances will be made in rapid modeling of hydrodynamic interactions between macromolecules, in modeling of solvation and electrostatic effects *in vivo*, and in developing flexible molecular models that can be applied on the very large scale demanded of attempts to model intracellular environments. Finally, it is important to note that while we have focused on the bacterial cytoplasm here the same challenges — and potential rewards — are also likely to arise in attempts to model other intracellular environments, be they prokaryotic or eukaryotic in origin.

Acknowledgments

The views expressed in this chapter have been shaped over a period of years by the work of many people including, in particular, the following members of the Elcock group: Dr. Sean R. McGuffee, Dr. Tamara Frembgen-Kesner, Shun Zhu and Eli Musselman. This work was supported by the National Institutes of Health (GM087290).

Suggested Additional Reading Material

D. S. Goodsell. *The Machinery of Life*. 2009. Springer Science+Business Media, New York.

References

1. D. S. Goodsell. 1991. Inside a living cell. *Trends Biochem Sci* **16**, 203–206.
2. V. L. Lew, J. E. Raftos, M. Sorette, R. M. Bookchin and N. Mohandas. 1995. Generation of normal human red cell volume, hemoglobin content, and membrane area distributions by "birth" or regulation? *Blood* **86**, 334–341.
3. S. R. Goodman, A. Kurdia, L. Ammann, D. Kakhniashvili and O. Daescu. 2007. The human red blood cell proteome and interactome. *Exp Biol Med* **232**, 1391–1408.
4. J. Cayley, B. A. Lewis, H. J. Guttman and M. T. Record Jr. 1991. Characterization of the cytoplasm of *Escherichia coli* K-12 as a function of external osmolarity. *J Mo Biol* **222**, 281–300.
5. G. H. Wadhams and J. P. Armitage. 2004. Making sense of it all: Bacterial chemotaxis. *Nat. Rev. Mol Cell Biol* **5**, 1024–1037.
6. K. A. Dill and S. Bromberg. 2003. *Molecular Driving Forces: Statistical Thermodynamics in Chemistry & Biology*. Garland Science, New York.
7. H. X. Zhou, G. Rivas and A. P. Minton. 2008. Macromolecular crowding and confinement: Biochemical, biophysical, and potential physiological consequences. *Annu Rev Biophys* **37**, 375–397.
8. A. H. Elcock. 2010. Models of macromolecular crowding effects and the need for quantitative comparisons with experiment. *Curr Opin Struct Biol* **20**, 196–206.
9. S. B. Zimmerman and A. P. Minton. 1993. Macromolecular crowding — biochemical, biophysical, and physiological consequences. *Annu Rev Biophys Biomol Struct* **22**, 27–65.
10. A. H. Elcock. 2003. Atomic-level observation of macromolecular crowding effects: Escape of a protein from the GroEL cage. *Proc Natl Acad Sci U S A* **100**, 2340–2344.

11. M. S. Cheung, D. Klimov and D. Thirumalai. 2005. Molecular crowding enhances native state stability and refolding rates of globular proteins. *Proc Natl Acad Sci U S A* **102**, 4753–4758.

12. A. K. Jha, A. Colubri, K. F. Freed and T. R. Sosnick. 2005. Statistical coil model of the unfolded state: Resolving the reconciliation problem. *Proc Natl Acad Sci U S A* **102**, 13099–13104.

13. M. Jasnin, M. Moulin, M. Haertlein, G. Zaccai and M. Tehei. 2008. Down to atomic-scale intracellular water dynamics. *EMBO Rep* **9**, 543–547.

14. I. M. Keseler, C. Bonavides-Martinez, J. Collado-Vides, S. Gama-Castro, R. P. Gunsalus, D. A. Johnson, M. Krummenacker, L. M. Nolan, S. Paley, I. T. Paulson, M. Peralta-Gil, A. Santos-Zavaleta, A. G. Shearer and P. D. Karp. 2009. EcoCyc: A comprehensive view of *Escherichia coli* biology. *Nucleic Acids Res* **37**, D464–D470.

15. S. Sundararaj, A. Guo, B. Habibi-Nazhad, M. Rouani, P. Stothard, M. Ellison and D. S. Wishart. 2004. The CyberCell Database (CCDB): A comprehensive, self-updating, relational database to coordinate and facilitate *in silico* modeling of *Escherichia coli*. *Nucleic Acids Res* **32**, D293–D295.

16. R. V. Misra, R. S. Horler, W. Reindl, I. I. Goryanin and G. H. Thomas. 2005. EchoBASE: An integrated post-genomic database for *Escherichia coli*. *Nucleic Acids Res* **33**, D329–D333.

17. Y. Ishihama, T. Schmidt, J. Rappsilber, M. Mann, F. U. Hartl, M. J. Kerner and D. Frishman. 2008. Protein abundance profiling of the *Escherichia coli* cytosol. *BMC Genomics* **9**, 102.

18. F. R. Blattner, G. Plunkett III, C. A. Bloch, N. T. Perna, V. Burland, M. Riley, J. Collado-Vides, J. D. Glasner, C. K. Rode, G. F. Mayhew, J. Gregor, N. W. Davis, H. A. Kirkpatrick, M. A. Goeden, D. J. Rose, B. Mau and Y. Shao. 1997. The complete genome sequence of *Escherichia coli* K-12. *Science* **277**, 1453–1462.

19. R. S. P. Horler, A. Butcher, N. Papangelopoulos, P. D. Ashton and G. H. Thomas. 2009. EchoLOCATION: An *in silico* analysis of the subcellular locations of *Escherichia coli* proteins and comparison with experimentally derived locations. *Bioinformatics* **25**, 163–166.

20. B. D. Bennett, E. H. Kimball, M. Gao, R. Osterhout, S. J. van Dien and J. D. Rabinowitz. 2009. Absolute metabolite concentrations and implied enzyme active site occupancy in *Escherichia coli*. *Nature Chem Biol* **5**, 593–599.

21. T. Schwede, J. Kopp, N. Guex and M. C. Peitsch. 2003. SWISS-MODEL: An automated proteinhomology-modeling server. *Nucleic Acids Res.* **31**, 3381–3385.

22. S. R. McGuffee and A. H. Elcock. 2010. Diffusion, crowding & protein stability in a dynamic molecular model of the bacterial cytoplasm. *PLoS Comput. Biol.* **6**:e1000694.

23. T. Schlick. 2002. *Molecular Modeling and Simulation: An Interdisciplinary Guide*. Springer Verlag, New York.

24. H. J. C. Berendsen. 2007. *Simulating the Physical World: Hierarchical Modeling from Quantum Mechanics to Fluid Dynamics*. Cambridge University Press, New York.

25. D. L. Ermak and J. A. McCammon. 1978. Brownian dynamics with hydrodynamic interactions. *J Chem Phys* **69**, 1352–1360.

26. J. F. Brady and G. Bossis. 1988. Stokesian dynamics. *Annu Rev Fluid Mech* **20**, 111–157.

27. R. R. Gabdoulline and R. C. Wade. 1997. Simulation of the diffusional association of barnase and barstar. *Biophys J* **72**, 1917–1929.

28. A. H. Elcock. 2004. Molecular simulations of diffusion and association in multimacromolecular systems. *Methods Enzymol* **383**, 166–198.

29. M. B. Elowitz, M. G. Surette, P. E. Wolf, J. B. Stock and S. Leibler. 1999. Protein mobility in the cytoplasm of *Escherichia coli*. *J Bacteriol* **181**, 197–203.

30. M. C. Konopka, K. A. Sochacki, B. P. Bratton, I. A. Shkel, M. T. Record and J. C. Weisshaar. 2009. Cytoplasmic protein mobility in osmotically stressed *Escherichia coli*. *J Bacteriol* **191**, 231–237.

31. S. Ghaemmaghami and T. G. Oas. 2001. Quantitative protein stability measurement *in vivo*. *Nature Struct Biol* **8**, 879–882.

32. Z. Ignatova, B. Krishnan, J. P. Bombardier, A. M. C. Marcelino, J. Hong and L. M. Gierasch. 2007. From the test tube to the cell: Exploring the folding and aggregation of a β-clam protein. *Biophys J* **88**, 157–163.

33. W. Humphrey, A. Dalke and K. Schulten. 1996. VMD — visual molecular dynamics. *J Mol Graphics* **14**, 33–38.

The Light-Harvesting Apparatus in Purple Photosynthetic Bacteria

2

Introduction to a Quantum Biological Device

Johan Strümpfer, Jen Hsin, Melih Şener, Danielle Chandler and Klaus Schulten

1. Introduction

The chromatophore vesicle of purple bacteria is a quantum biological device consisting of about 200 protein complexes that cooperate to harvest sunlight. It is a biological solar cell at its simplest. A combination of decades of experimental and theoretical efforts provided an atomic level description of the chromatophore and the light-harvesting process that it carries out. The architechture and function of the constituent protein complexes and their integration into the chromatophore is described with a focus on the initial steps of light-harvesting. These steps involve the capture of sunlight in the form of electronic excitation of a chlorophyll and the subsequent migration of the electronic excitation energy across the chromatophore until its energy is stored first as a transmembrane potential and then in the form of chemical energy.

Photosynthesis is the energy source of nearly all life on Earth. Photosynthetic organisms capture sunlight and convert its energy stepwise into more stable forms.[1–6] The first stage of photosynthesis, depicted in Figure 1, involves capture of light energy in the form of an electronic excitation by pigments such as bacteriochlorophylls (BChls) and carotenoids. This form of energy is stable for only a nanosecond (ns), the typical fluorescence lifetime of light-excited molecules,[3] and accordingly must be used within about 100 picoseconds (ps) to avoid significant loss.

As described by Förster and Dexter long ago,[7,8] the excitation can be transferred between carotenoids and BChls, which actually happens in 100 femtoseconds (fs).[9,10] After multiple excitation transfer events between pigment molecules the captured energy arrives at a reaction center (RC). There, it is utilized to transfer electrons across the membrane to hydrate (electron + proton = H) a quinone molecule bound to the RC. This electron transfer in effect forms a transmembrane voltage difference as the electron utilized stems from one side (periplasm in Figure 1) of the membrane, but the proton from the other side (cytoplasm in Figure 1). The transferred electron

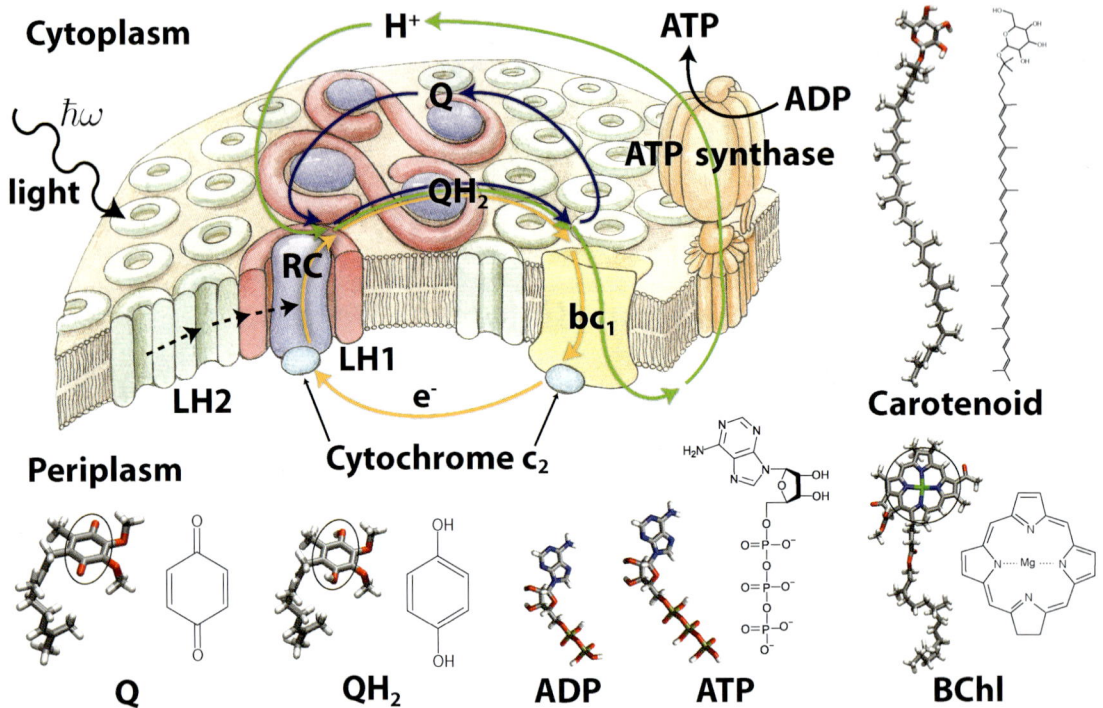

Figure 1. Section of the curved chromatophore membrane illustrating how ATP (adenosine triphosphate) is produced from captured photons. Light is absorbed by molecules of bacterio chlorophyll (BChl) or carotenoid in the circular light-harvesting proteins LH1 and LH2; shown is the absorption of a light quantum $\hbar\omega$ by an LH2. The excitation energy migrates, in the form of electronic excitation spread over several chlorophylls, from LH2 to LH2 to LH1 as shown by dotted arrows, ariving eventually at a protein complex called RC, the photosynthetic reaction center. In the RC, the light energy is used to transfer an electron from the inside (here the bottom or periplasm) of the curved membrane to the out-side (here the top or cytoplasm), thus charging the membrane positive inside and negative outside. The electron (e^-) stemming from the periplasmic side (motion shown as brown arrows) combines with a proton (H^+) stemming from the cytoplasmic side (motion shown as green arrows) to form a hydrogen atom that is added to a molecule of quinone, Q (motion shown as blue arrows). Q binds two hydrogen atoms, thus forming QH_2, and moves to a protein called the bc_1 complex where electron and protons are separated from QH_2, changing it back to Q, and moved jointly to the periplasm (the actual process is more round-about and described in Ref. 11). The electron is then loaded on a protein, cytochrome c_2 (blue), that "helicopters" it back to the RC. The electron hole that was created there previously is then filled by the electron on the cytochrome c_2. At this point, the electrons have completed a circular flow (brown) and do not contribute anymore to the membrane charge, but the protons do! The protons, which have been separated from QH_2, have been moved from the cytoplasm to the periplasm, such that the periplasmic + and cytoplasmic − charge is maintained. This charge is rather stable, even though the protons are eager to flow back from the periplasmic to the cytoplasmic side of the membrane. Indeed, the protons evetually do flow back, passing through a protein complex called ATP synthase, but the energy is then transformed into chemical energy. ATP synthase drives, with the protons' energy, the reaction in which a low energy molecule of ADP (adenosine diphosphate) is transformed into a high energy molecule of ATP by adding a third phosphate (P). Molecules of ATP drive many energy-consuming processes of the bacterial cell, for example cell motion and chemical synthesis.

leaves an electron hole behind on the periplasmic side that needs to be replenished for the process to happen a second time.

The transferred electron is stabilized on the quinone through hydration and cannot tunnel back across the membrane. The quinone, Q, receiving a second hydrogen through the same process and then becoming a hydroquinone, QH_2, detaches from the RC and diffuses to a protein called the cytochrome bc_1 complex. At the bc_1, the hydroquinone is reverted back into a quinone in such a fashion that the voltage difference across the membrane is maintained as a proton gradient across the membrane: electrons and protons move from the cytoplasm to the periplasm; the electrons are shuttled back to the RC to fill the electron holes. The system shown in Figure 1 contains usually a pool of QH_2 to always have electrons ready to fill electron holes at the RC and keep the membrane charged.

In the final stage, a protein complex called ATP synthase utilizes the proton gradient to phosphorylate ADP (adenosine diphosphate) to finally store the absorbed light energy in the very stable form of the bond energy of ATP (adenosine triphosphate). The proteins and molecular cofactors that perform photosynthesis form what is known as a photosynthetic unit.

In purple bacteria, such as *Rhodobacter (Rba.) sphaeroides*, the photosynthetic unit is organized in the form of a nearly spherical membrane vesicle (see Figure 2) of approximately 60 nm diameter.[1,12–14] The vesicle, called a chromatophore, displays significant simplicity compared to its counterpart, the chloroplast, in plants. Also, the constituent proteins that form the chromatophore, which are discussed further in the subsequent sections, are structurally simpler than their plant counterparts. Overall, the chromatophore is evolutionarily more primitive, i.e. easier to explain, than the corresponding, oxygenic plant systems.[15]

In a typical chromatophore vesicle, thousands of BChls distributed over hundreds of proteins form an interaction network sharing electronic excitation prior to the delivery to a reaction center. About 1000 molecules of carotenoid also absorb light and feed its energy into the chlorophyll network[9] as already mentioned above. The role of chlorophyll was already suggested by the work of Emerson, Arnold, and others in the first half of the twentieth century.[16–19] The study of photosynthetic light harvesting has been awarded with several Nobel prizes during

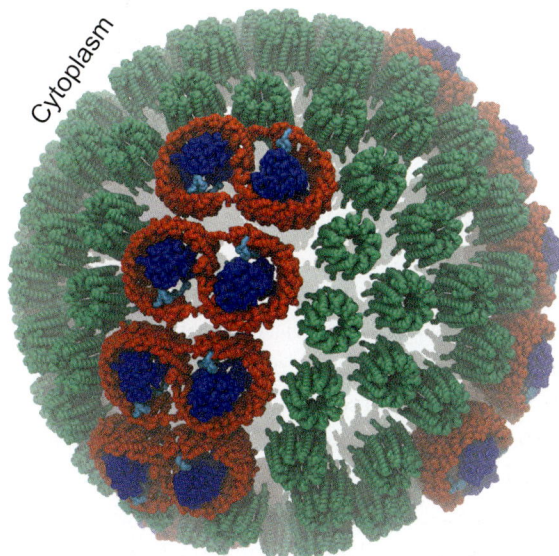

Figure 2. Organization of the chromatophore depicting constituent light-harvesting proteins. Shown is the high-light adapted vesicle model for *Rba. sphaeroides*.[13,14] The three major constituent proteins are colored as follows: LH2: green; LH1: red; RC: blue. The vesicle model depicted here is based on AFM data on intracytoplasmic membranes.[12] Not shown are the lipids making up the membrane that are located between the protein complexes.

the past century, including Wilstätter (1915) and Fischer (1930) (for their study of the chemical properties of Chls), Karrer (1937) and Kuhn (1938) (for their work on the structure and function of carotenoids), Deisenhofer, Huber, and Michel[20] (1988) (for their crystallographic determination of the RC structure), Marcus[21,22] (1992) (for his work on electron transfer reactions), Broyer, Skou and Walker (1997) (for the determination of the structure and function of ATP synthase). Besides Förster and Dexter, the physicists Robert Oppenheimer worked on excitation transfer in light harvesting.[8,17,18] Physicists also learned in great detail that organic molecules like chlorophylls tightly pack together inside the proteins of the chromatophore, share electronic excitation in the form of so-called excitons which are quantum mechanical, i.e. coherent, superpositions of electronic excitations of individual chlorophylls as described further below.[23,24]

The bulk of the chromatophore vesicle consists mainly of protein-pigment complexes, the ring-shaped light-harvesting complexes I and II (LH1 and LH2)[25–29] and RCs,[20,30–36] the latter surrounded by an LH1 ring. These complexes perform the first steps of photosynthesis, often called the light reactions, up to the first transport of an electron across the membrane. The remaining steps, known as the dark reactions, involving the transport of electrons between the RC and bc_1 complexes and the production of ATP, are performed by the bc_1, cytochrome c_2 and ATP synthase protein complexes. Each chromatophore is thought to be equipped with all of the proteins required for photosynthesis[5] and can thus function independently.

The constituent light-harvesting proteins as well as the overall architecture of the chromatophore display a modularity that adds robustness to the system as assembly and repair are simplified. A parts list of the chromatophore is provided in Table 1. For the constituent proteins of the chromatophore, high-resolution atomic structures are known[20,28,37–39] albeit not all for the same species. The supramolecular organization of the chromatophore vesicle, shown in Figure 2, has been determined by combining atomic force microscopy (AFM),[12,36,40–42] cryo-electron microscopy (cryo-EM),[43–46] and linear dichroism (LD)[41] measurements.

Photosynthesis involves the interaction of light with living matter as well as the excitation and flow of electrons. The behavior of electrons is governed by quantum mechanics. Quantum mechanics is thus essential in the description of the early stages of photosynthesis, namely light absorption, migration of electronic excitation energy, and subsequent electron transfer. Since light harvesting is performed at physiological temperature, $T \approx 300$ K, thermal disorder must also be taken into account.

In fact, photosynthetic species have evolved to exploit thermal effects for efficient light-harvesting. For example, the light-harvesting efficiency of a key component of plant photosythesis (photosystem I) is nearly 90% at room temperature, whereas very low, so-called at cryogenic, temperatures it falls below 50%.[47] The correct inclusion of thermal disorder and quantum mechanics in the theoretical models of light-harvesting processes has been the topic of many studies.[7,48–60] The laws of quantum physics constraining light-harvesting processes appear to have been wonderfully exploited by the evolution of photosynthetic systems.[14,61]

The combination of protein structural data with physics-based computational models can accurately portray the light-harvesting function of a photosynthetic protein complex as shown below.[1,58] The greatest challenge lies in the portrayal of hundreds of proteins (see Table 1) that self-assemble and cooperate to form a functional cellular machine, the chromatophore vesicle.

Table 1. Components of a chromatophore.*

LH2 (60–100)	LH1 monomer (10–20)
Protein components:	Protein components:
9 α-subunits	16 α-subunits
9 β-subunits	16 β-subunits
Cofactors:	Cofactors:
27 bacteriochlorophylls	32 bacteriochlorophylls
9 carotenoids	16 carotenoids

RC (10–20)	ATP synthase (1)
Protein components:	Protein components:
1 L-subunit	3 α-subunits
1 M-subunit	3 β-subunits
1 H-subunit	1 γ-subunit
Cofactors:	1 δ-subunit
4 bacteriochlorophylls	1 ε-subunit
2 bacteriopheophytins	1 a-subunit
1 carotenoid	1 b-subunit
2 quinones	10–14 c-subunits

bc1 (5–10)	
Protein components:	Cofactors:
2 Cytochrome b	4 Fe-S centers
2 Cytochrome c_1	6 hemes
2 Rieske-subunits	

* Numbers in parentheses give approximate number of the complexes in the chromatophore.

In the following, first the major components of the chromatophore vesicle are discussed, then the supramolecular architecture of the entire chromatophore and its function are described; the quantum mechanical process of energy harvesting and transfer is introduced at a basic level for a single LH2 protein; finally we demonstrate how a multitude of such proteins form a cooperative network for efficient energy migration across the vesicle.

2. Components of a Chromatophore Vesicle

In this section, the light-harvesting protein complexes of the chromatophore are introduced: the LH2 complex and the LH1-RC core complex. The structure and function of each complex are discussed in the context of the light-harvesting function. We also refer the reader to the parts list collected in Table 1. This structural information is utilized in Section 4 for the construction of a quantum mechanical description of energy transfer.

The light-harvesting complexes of purple bacteria, LH1 and LH2, share certain structural features. They involve a cylindrical, modular arrangement of pairs of transmembrane helices which non-covalently bind BChls and carotenoids, their light-absorbing pigments. In the case of

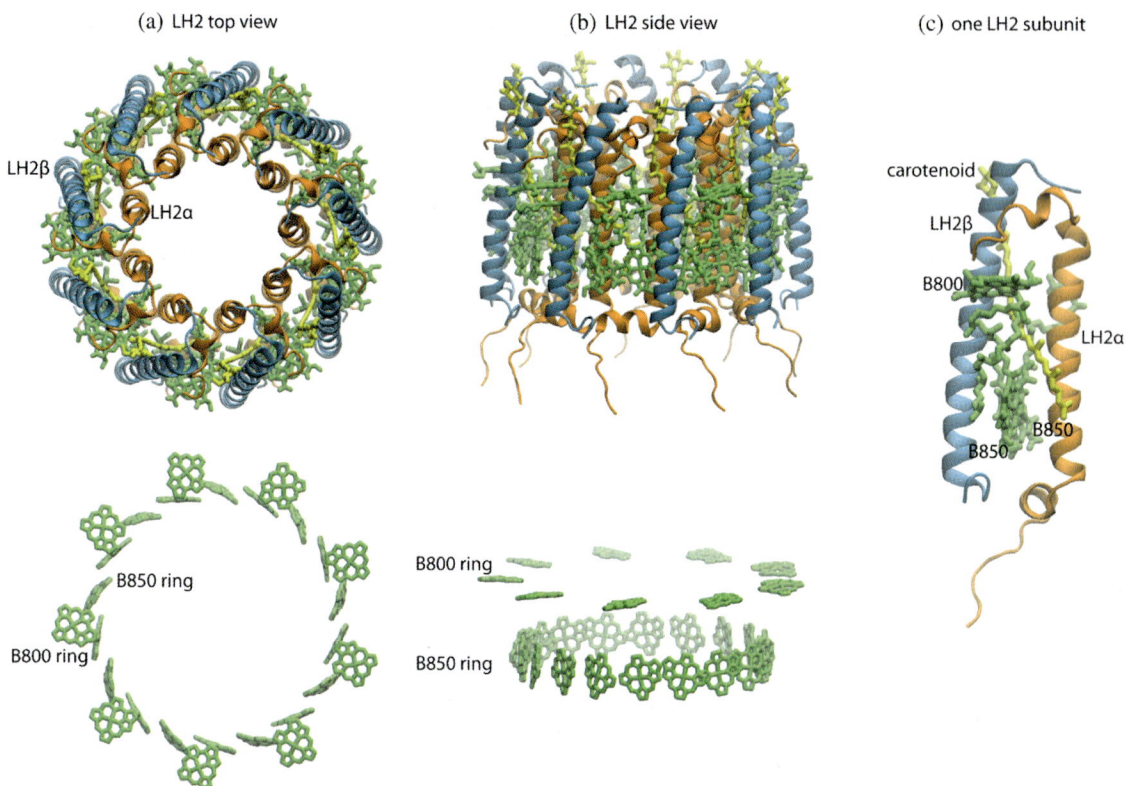

Figure 3. (a) An LH2 complex from *Rba. sphaeroides* as viewed from the top, showing the nine-fold symmetry of the complex and the placement of the pigment molecules. (b) LH2 complex as viewed from the side. (c) LH2 subunit. Each LH2 subunit is composed of two transmembrane helices, called the α-helix (orange-colored) and the β-helix (blue-colored), three bacteriochlorophylls (green-colored) and one carotenoid molecule (yellow-colored). See also Table 1. A review is available.[2]

the bacterium *Rba. sphaeroides*[33,62] a nine-fold symmetry is displayed by the LH2 complex as shown in Figure 3. These proteins are not only responsible for capture and transfer of light energy, but also for the assembly and shaping of the chromatophore.[63–65]

2.1 *The light-harvesting complex LH2*

The light-harvesting process begins with light-harvesting complex II (LH2), also referred to as the peripheral light-harvesting complex antenna. LH2 is a ring-shaped protein complex, formed by eight or nine identical subunits.[28,54] Each subunit contains two separate transmembrane helices which serve as a scaffold for the embedded pigment molecules, of which there are three BChls and one carotenoid per subunit (see Figure 3 and Table 1). Altogether, an LH2 of *Rba. sphaeroides* is made of 54 seprate components. LH2 complexes are found in large numbers (100–200) in the chromatophore and act as a broad light-harvesting antenna that funnels energy to the LH1-RC complexes (see Figure 1). Through the LH2 complexes, the chromatophore not only realizes excellent light absorption, but also makes use of light at different wavelengths, namely the far-red light absorbed by the chlorophylls and yellow-green light absorbed by the carotenoids. While the bacteriocholorphylls in LH1 absorb maximally at 875 nm, the BChls in LH2 absorb maximally at 800 and 850 nm; the carotenoid molecule absorbs maximally at 500 nm. Naturally, excitation is transferred most easily energetically downhill, namely from 500 nm pigments[9] to 800 nm pigments to 850 nm pigments onto 875 nm ones, but opposite transfer is also possible, in particular from 875 nm to 850 nm pigments as the energy difference measures only 0.96 kcal/mol, which is 1.62 $k_B T$ ($k_B T$ = thermal energy at temperature T; k_B is the Boltzmann constant). The absorption of 850–800 nm and 500 nm photons is the result of evolutionary pressure, as purple bacteria are found in murky water where only lights around 800 nm and 500 nm penetrates;[66,67] other wavelengths of light being harvested by plants and algae living "above" the bacteria, i.e. living closer to the sun. A review of the action of LH2 is available.[2]

2.2 The RC-LH1 core complex

Light energy absorbed by LH2 is transported within a few picoseconds to nearby light-harvesting complexes 1 (LH1)[13,14,58] as shown in Figure 1. LH1 also absorbs sunlight directly using the same mechanism as LH2, and, actually, has a very similar structure; it is also made of subunits, containing α- and β-proteins, a carotenoid, but only two bacteriochlorophylls (see Table 1). LH1 is also ring-like, but forms a wider ring. This ring surrounds another protein complex, the reaction center (RC) as already stated above. The RC-LH1 complex is also termed the photosynthetic core complex and performs the first processing of the light energy absorbed. Figures 4 and 5 show the structure of the bacterial photosynthetic core complexes, which exist as a LH1-RC monomer or as a (LH1-RC)$_2$ dimer, depending on species.

2.2.1 Organization of RC-LH1

Recall that LH2 is a ring-like protein complex, its overall structure being preserved across different bacterial species, but with different numbers (8 or 9) of subunits. LH1 comes in a curious

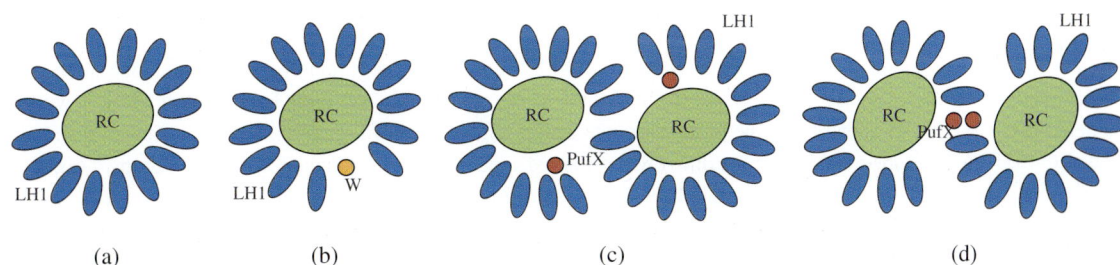

| (a) | (b) | (c) | (d) |

Figure 4. Schematics of different organizations of the bacterial photosynthetic core complexes. (a) A core complex in which the LH1 subunits form a complete ring, as seen in *Rhodospirillum rubrum*.[36] See also Figures 5a and 5b. (b) A core complex in which the LH1 subunits form a ring with a gap, with an extra polypeptide near the gap, as seen in the crystal structure of *Rhodopseudomonas palustris*.[35] (c–d) Two proposed organizations for a dimeric core complex, the dimerization of which requires the extra polypeptide, PufX, shown in red. (c) is drawn according to structural data for a dimeric core complex,[46] which place PufX near LH1 openings. See also Figures 5c and 5d. In (d), PufX is assumed to dimerize and is situated at the symmetry center of the core complex.[71,72] Dimeric core complexes are seen in certain *Rhodobacter* species, the best-known case being *Rba. sphaeroides*.[43]

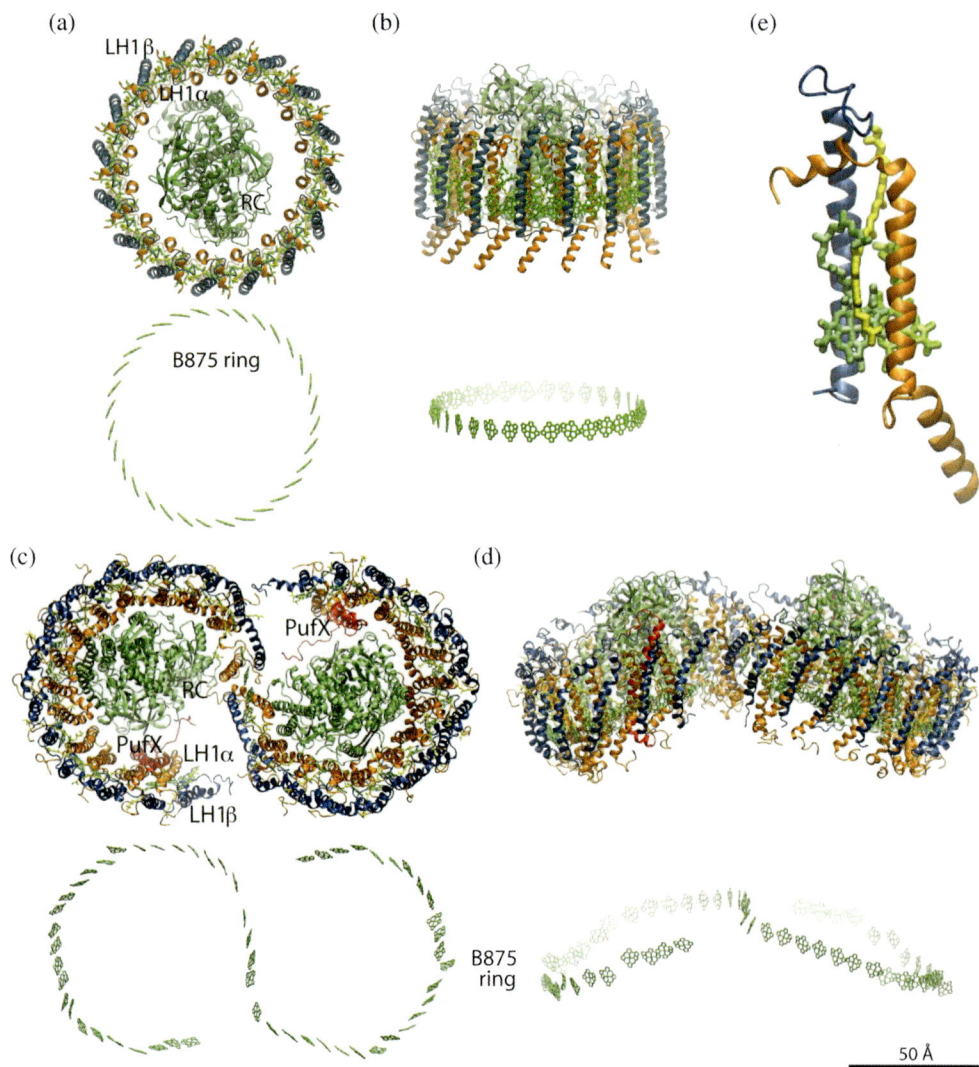

Figure 5. (a) Top view of a modeled monomeric RC-LH1 core complex,[77] closely resembling one found in *Rhodospirillum rubrum*,[36] shown in Figure 4a. (b) Side view of the monomeric core complex model. (c) Top view of a modeled (RC-LH1-PufX)$_2$ dimer;[63,69,70] the protein arrangement resembles that of Figure 4c.[46] (d) Side view of the modeled (RC-LH1-PufX)$_2$ dimer; bending of the dimer was observed experimentally through electron microscopy.[68] (e) LH1 subunit consisting of two transmembrane α-helices (the α- and β-proteins, orange and blue respectively), two BChls (green) and one carotenoid (yellow); the model shown here is from Ref. 69. The top figures (a)–(d) shows the whole system while the bottom ones show only the B875 BChls.

assortment of organizations. In Figure 4, four RC-LH1 arrangements, with the details that are either observed in crystallography[35] or proposed based on electron microscopy[36,46,68–70] and atomic force microscopy[71,72] data, are displayed.

The wide variety of RC-LH1 organization can be classified into two catagories: monomeric RC-LH1 and dimeric (RC-LH1)$_2$. For monomeric RC-LH1, the LH1 subunits surround the RC, forming a ring. The LH1 ring can either be closed (as in the case of Figure 4a) or interrupted by another protein (as in the case of Figure 4b). Dimeric (RC-LH1)$_2$ is primarily seen for the *Rhodobacter* family of bacteria, although some *Rhodobacter* bacteria have monomeric RC-LH1.[73] For (RC-LH1)$_2$, an additional protein has been found, known as PufX. There is no experimentally determined high-resolution structure for a (RC-LH1)$_2$ yet. Two (RC-LH1-PufX)$_2$ structures have been proposed,[45,46,69–76] differing in the location of PufX (Figures 4c and 4d).

2.2.2 *Structures of LH1 and RC*

As already pointed out above, the LH1 complex, much like LH2, is an assembly of identical units; each unit comes with two transmembrane α-helices and the BChls and carotenoid pigments sandwiched inbetween. In Figures 5a and 5b, an RC-LH1 complex is shown;[77] in Figures 5c and 5d, (RC-LH1-PufX)$_2$ is shown;[69,70] in Figure 5e, an LH1 subunit is displayed.[77] As can be seen, each LH1 subunit holds two BChls and one carotenoid. LH1 BChls are often referred to as the B875 BChls, as LH1 has a BChl absorption band at ~875 nm; it also has a carotenoid absorption band of 500 nm. Notably, a high-resolution structure for LH1 is yet to be determined, as the only crystal structure of LH1 (Roszak *et al.*, 2003[35]) does not resolve the terminal ends of LH1. Figure 5 actually shows modeled RC-LH1 structures.[69,70,77] Given that the α- and β-proteins of LH1 and LH2 have a high degree of sequence identity,[63] the models shown in Figure 4 should be accurate in detail, but may be inaccurate in overall shape.

Unlike in the case of LH1, abundant crystollographic studies have been carried out for the bacterial RC. As pointed out above, the Nobel Prize in Chemistry in 1988 was awarded to the scientists who solved the X-ray structure of a bacterial reaction center.[78] This achievement is extraordinarily significant not only in terms of understanding photosynthesis, but also in terms of understanding membrane proteins in general, as the reaction center was the first membrane protein to be crystalized and resolved structurally with atomic resolution. Since then, several crystal structures of the RC for different bacterial species and mutants have been reported.[79–81]

The photosynthetic reaction center (Figure 6), much like LH2 and LH1, is made of protein and cofactor components. The protein components of RC consist of three polypeptide subunits, named L, M and H. Subunits L and M both contain five transmembrane helices, while subunit H is a globular domain that caps the RC towards the cytoplasm (Figure 6a). Scaffolded within subunits L and M are the various cofactors found in RC, including four BChl molecules, two bacteriophaeophytin molecules (which are similar to BChl structurally, but do not have a central Mg atom), a carotenoid, and two quinones (labeled Q_A and Q_B) (Figure 6b).

The photosynthetic RC utilizes light excitation energy absorbed by the LH1 and LH2 complexes to transfer an electron across the membrane and produce a transmembrane charge difference, i.e. a membrane potential. In Figure 6c the mechanism of RC is outlined (see also Figure 1). Excitation energy from the surrounding LH1 is transfered to the "special pair" BChls in RC (Figure 6c); in *Rhodobacter sphaeroides*, these BChls are known as P870 as they absorb photons of wavelength 870 nm. The excited BChl in the RC transfers an eletron to the bacteriopheophytin (BPh) moiety in the RC, which then transfers the electron to Q_A, a quinone. The electron is subsequently delivered to Q_B, which takes on a second electron by repeating the cycle, and is reduced to QH$_2$. At this point, the QH$_2$ molecule detaches from the RC and diffuses through the membrane to the bc$_1$ complex. Another quinone molecule in the membrane enters the RC to replace Q_B, and the whole

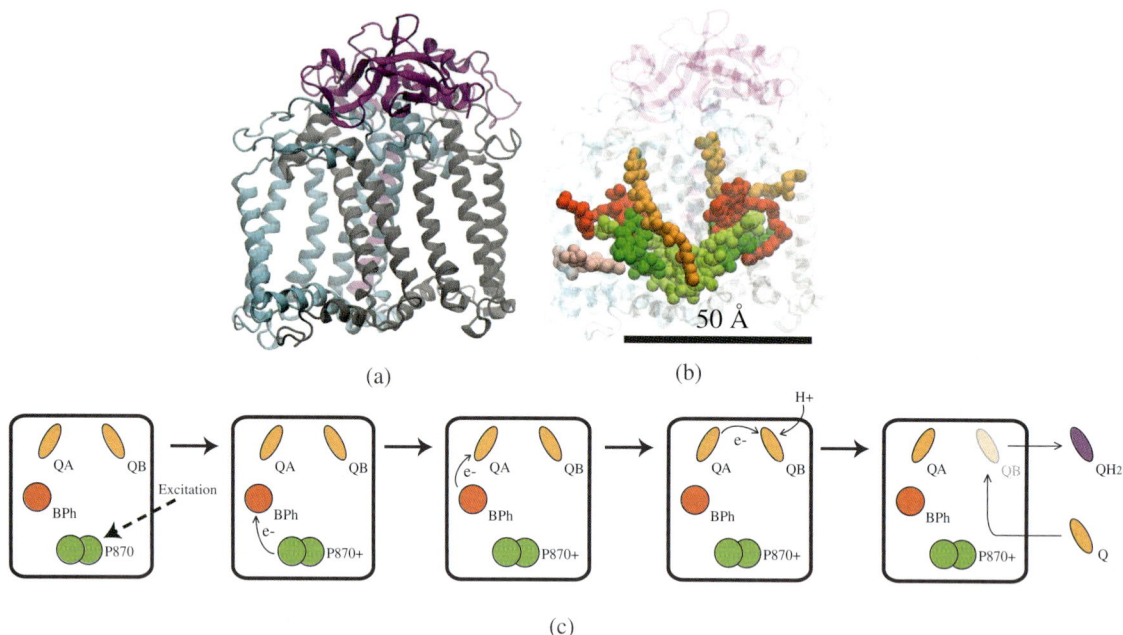

Figure 6. Structure and mechanism of the photosynthetic reaction center (RC). (a) The protein components of RC include an H subunit (purple), an M subunit (light blue), and an L subunit (gray). (b) Cofactors in the RC. The four BChls are colored in different shades of green for distinction; the two bacteriopheophytin (BPh) are shown in red; carotenoid is shown in pink; the quinone molecules are shown in orange. (c) Simplified diagram of energy processing in RC, described in the text; shown are only some of the cofactors (omitted are two BChls and one bacteriopheophytin).

process is repeated. The missing electron in one of the special pair BChls is replenished by the protein cytochrome c_2, which shuttles electrons between the bc_1 complex and RC. Therefore, the flow of electrons is cyclic, as shown in Figure 1. At this point the membrane potential is based on an electron concentration difference between cytoplasm and periplasm.

2.3 *Spatial organization of the chromatophore*

The light-harvesting proteins introduced above assemble to form a chromatophore vesicle that, in *Rba. sphaeroides*, constitutes a spherical bulge in the cell membrane. The formation of the bulge is mediated by the innate curvature of the constituent proteins and their interactions.[64,65] The structural model of the chromatophore shown in Figure 2 has been derived from various sources as already mentioned:[13,14] crystallography data provide the structures of individual proteins, whereas data from atomic force miscroscopy provides images of how the proteins in patches of the chromatophore shell are typically arranged;[12,41,75] electron microscopy data provide information about the overall shape of chromatophores;[43,46,68,82] from spectroscopy data, the relative numbers of light-harvesting complexes are determined.[14]

A chromatophore vesicle (Figure 2 and Table 1) features typically 200 light-harvesting complexes, which together contain 4000 BChls.[14,83,84] Additionally, a chromatophore vesicle contains one ATP synthase and 5–10 bc_1 complexes.[85] These ratios of components depend strongly on growth conditions, in particular light intensity. Atomic force microscopy images[12] reveal that the (RC-LH1-PufX)$_2$ dimers form linear arrays that are surrounded by LH2s.

3. Energy Harvesting and Transfer

Light harvesting begins with the absorption of a photon by the pigments (bacteriochlorophyll and carotenoid) of LH2 or of the LH1-RC complexes. When a photon is absorbed, electrons in the absorbing pigment are promoted to an excited state. The excited electrons can transfer their energy through Coulomb interaction (so-called Förster transfer) or, if close enough, through quantum tunneling (Dexter transfer)[7,8] to a nearby pigment.[2,3,6] The excitation energy transfers between pigments in the vesicle until it reaches one of the reaction centers where it is utilized for electron transfer (see Figure 1).[2,3,6]

LH2 absorbs most of the incoming photons due to its relative abundance in the vesicle.[14,85] The incoming photon's wavelength determines which pigment absorbs it. The B850 BChls absorb 850 nm photons, the B800 BChls absorb 800 nm ones and the carotenoids absorb ~ 500 nm ones. The B800 BChls and carotenoids are well separated and, thus, they interact very weakly with each other. As a result, when these pigments absorb photons, they do so individually, such that there is very little excitation sharing between the pigments.[2,3,6,54] However, the carotenoids and B800 BChls transfer their excitation energy quickly (in a few hundred fs) to the nearby B850 BChls.[2–4]

An excited B850 BChl interacts strongly with B850 BChls in a single LH2 and as a result, the B850 BChls share the excitation. The shared excitations are called excitons and are described in Section 3.1.

3.1 *Excitons*

When a pigment absorbs a photon, one of its electrons is excited such that its total energy ϵ is the sum of its ground state energy ϵ_0 and the absorbed photon energy $\hbar\omega$, where $\hbar$ is the reduced Planck's constant and ω is the photon frequency. The process of absorption is depicted in Figure 7a. An excited pigment in isolation decays within ~1 ns back to the ground state and re-emits light, a process called fluorescence and shown in Figure 7b. The equations for the rates of photon absorption and emission are given in Appendix A; the rates depend on a molecular property called the transition dipole moment **d**. As shown in the appendix, the emission rates are typically of order 1/ns.

Excitation can transfer between two pigments A and B in two ways, via Coulomb interaction or via quantum tunneling.[2,3] The transfer is denoted as $A^*B \rightarrow AB^*$, where * indicates which pigment is excited. The states A^*B and AB^* are associated with energies E_{A^*B} and E_{AB^*}, respectively.

If A and B are separated by >1 nm, excitation transfer occurs via so-called Förster resonant energy transfer (FRET).[7] In this case, the excited electron in pigment A returns to the ground state, while a ground state electron in pigment B becomes excited, as shown in Figure 8a. The interaction energy for Förster transfer V^F is proportional to $1/r_{AB}^3$, where r_{AB} is the center-center separation between pigments A and B. FRET in photosynthesis is reviewed in Ref. 84.

On the other hand, if A and B are close (<1 nm), excitation transfer occurs via the so-called Dexter transfer mechanism where an electron exchange occurs.[8] Here, an excited electron transfers

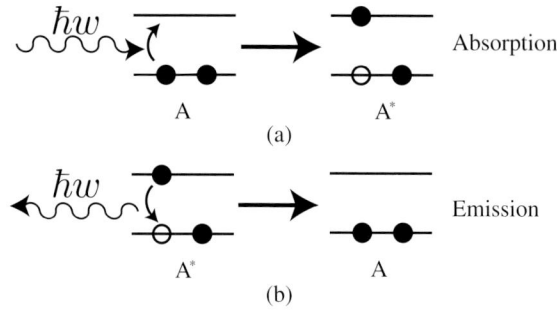

Figure 7. (a) Excitation of a pigment by photon absorption. (b) Excited state decay by photon emission (fluorescence).

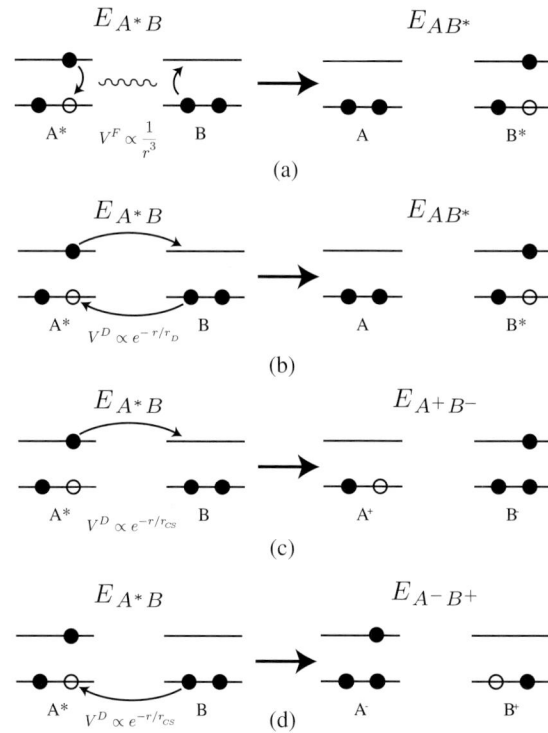

Figure 8. (a) Förster mechanism where excitation transfer occurs via the Coulomb interaction. (b) Dexter mechanism where excitation transfer occurs via an electron exchange. (c–d) Charge separation via the transfer of an electron.

from A to B and, simultaneously, a ground-state electron transfers from B to A, as shown in Figure 8b. The interaction energy V^D between states A^*B and AB^* for Dexter transfer is proportional to $\exp(-r_{edge}/r_D)$, where r_{edge} is the edge-edge distance between A and B and thus decays much faster with distance than Förster transfer. The reason for the fast spatial decay is that Dexter transfer arises from overlap of the orbitals of the electrons undergoing exchange.[9]

There is an alternative possibility for an excited state A^*B to decay, namely electron transfer. In this case the excited electron transfers, as shown in Figure 8c, from the donor A to acceptor B, without B returning a ground state electron. Pigment A thus loses an electron and B gains an electron.

This process is written $A^*B \rightarrow A^+B^-$. It is also possible that A^*B decays through an electron being transfered from B to A, as shown in Figure 8d, i.e. through $A^*B \rightarrow A^-B^+$. The states A^+B^- and A^-B^+ have associated energies $E_{A^+B^-}$ and $E_{A^-B^+}$; the interaction energy for charge separation V^{CS} is proportional to $\exp(-r_{edge}/r_{CS})$, where r_{CS} is an empirical constant typically of order 1 Å.

According to quantum physics (Fermi's Golden Rule[86]) the transfer rate between any two quantum states with energies E_1 and E_2 is given by[87]

$$k_{1 \rightarrow 2} = \frac{2\pi}{\hbar} V_{12}^2 \delta(E_1 - E_2),$$ (1)

where V_{12} is the interaction energy between the two states. In the case of excitation or electron transfer in a biological context, the δ-function in Eq. (1) is replaced by a spectral density S_{12},[88] i.e. the rate is

$$k_{1 \rightarrow 2} = \frac{2\pi}{\hbar} V_{12}^2 S_{12}.$$ (2)

S_{12} accounts for the overlap of the emission spectrum of state 1 and the absorption spectrum of state 2, or a similar quantity in case of electron transfer. In the case of FRET the transfer rate is[7]

$$k_{1 \rightarrow 2}^F = \frac{2\pi}{\hbar} \frac{\kappa^2}{r_{12}^6} S_{12},$$ (3)

where κ^2 depends on the relative orientation of the transition dipole moments of the two pigments, explained in more detail below, and r_{12} is the center-center distance between pigments 1 and 2. The transfer rate in Eq. (3) is derived in Appendix A (see Eq. (A20)).

The dynamics of the above processes depend sensitively on the relative energy of the different states and the interaction energies between them. To illustrate this dependence, we consider the process $A^*B \rightarrow AB^*$. The state energies E_{A^*B}, E_{AB^*} and the interaction energy V_d are collected according to quantum mechanics[88] into a 2×2 Hamiltonian matrix,

$$H = \begin{pmatrix} E_{A^*B} & V \\ V & E_{AB^*} \end{pmatrix}.$$ (4)

The system of two pigments has two stationary states which are given by the eigenvectors of the Hamiltonian matrix. These states are referred to as excitons as they mix coherently the two excited states A^*B and AB^*. In case of $E_{A^*B} = E_{AB^*}$, the exciton states are

$$\mathbf{v}_1 = \frac{1}{\sqrt{2}} \begin{pmatrix} 1 \\ 1 \end{pmatrix}, \quad \mathbf{v}_2 = \frac{1}{\sqrt{2}} \begin{pmatrix} 1 \\ -1 \end{pmatrix},$$

and each pigment will share the excitation equally in each of the exciton states. Indeed the eigenvector, eigenvalue property holds

$$H\mathbf{v}_{1,2} = (E_{A^*B} \pm V)\mathbf{v}_{1,2}$$ (5)

For $E_{A*B} \neq E_{AB*}$, the excitation is asymmetrically distributed. In cases where there is zero interaction energy between the two pigments, the Hamiltonian matrix has only diagonal elements, and the stationary states are

$$\mathbf{v}_1 = \begin{pmatrix} 1 \\ 0 \end{pmatrix}, \quad \mathbf{v}_2 = \begin{pmatrix} 0 \\ 1 \end{pmatrix},$$

where for each state $\mathbf{v}_1$ and $\mathbf{v}_2$ there is no excitation sharing between the pigments. The excitation remains then localized to whichever pigment is initially excited.

For LH2 from *Rba. sphaeroides*, there are 18 B850 BChls (i.e. $N = 18$), each with an excited state that has an energy of $E = 12390$ cm^{-1} (commonly referred to as the site energy).[89] The interaction energy of the nearest-neighbor BChls is an order of magnitude larger than that between next nearest-neighbor BChls. It is thus often assumed that only the interactions between nearest-neighbor BChls need to be taken into account. The interactions between nearest neighbor BChls in the B850 ring are very similar;[89] for simplicity, all nearest-neighbor interactions are approximated by a single value $V = 280$ cm^{-1}. The Hamiltonian matrix for the B850 ring of BChls is then

$$\hat{H}_{B850} = \begin{pmatrix} E & V & 0 & 0 & \cdots & 0 & V \\ V & E & V & 0 & \cdots & 0 & 0 \\ 0 & V & E & V & \cdots & \vdots & \vdots \\ 0 & 0 & V & E & \ddots & 0 & 0 \\ \vdots & \vdots & \vdots & \ddots & \ddots & V & 0 \\ 0 & 0 & \cdots & 0 & V & E & V \\ V & 0 & \cdots & 0 & 0 & V & E \end{pmatrix}. \tag{6}$$

Each of the 18 diagonal entries in the matrix, $\hat{H}_{ii}$, gives the site energy for each pigment. The off-diagonal entries, $\hat{H}_{ij}$, $i \neq j$, describe the interaction energy between pigments i and j. The eigenvectors $\mathbf{v}_\alpha$ corresponding to this Hamiltonian matrix are

$$\mathbf{v}_\alpha = \frac{1}{\sqrt{N}} \sum_{n=1}^{N} \left(\cos \frac{2\pi\alpha n}{N} + \sin \frac{2\pi\alpha n}{N} \right) \hat{\mathbf{e}}_n, \tag{7}$$

where $\hat{\mathbf{e}}_n$ (corresponding to the n-th BChl being excited) are the vectors given by

$$\hat{\mathbf{e}}_1 = \begin{pmatrix} 1 \\ 0 \\ 0 \\ \vdots \\ 0 \end{pmatrix}, \quad \hat{\mathbf{e}}_2 = \begin{pmatrix} 0 \\ 1 \\ 0 \\ \vdots \\ 0 \end{pmatrix}, \quad \cdots \tag{8}$$

It can be shown that the excitonic states defined in Eq. (7) are the eigenstates of $\hat{H}_{B850}$:

$$
\begin{aligned}
H_{B850}\mathbf{v}_\alpha &= \frac{1}{\sqrt{N}}\sum_{n=1}^{N}\Bigg[E\left(\cos\frac{2\pi\alpha n}{N}+\sin\frac{2\pi\alpha n}{N}\right)+V\left(\cos\frac{2\pi\alpha(n-1)}{N}+\sin\frac{2\pi\alpha(n-1)}{N}\right)\\
&\qquad\qquad +V\left(\cos\frac{2\pi\alpha(n+1)}{N}+\sin\frac{2\pi\alpha(n+1)}{N}\right)\Bigg]\,\hat{\mathbf{e}}_n\\
&= \frac{1}{\sqrt{N}}\sum_{n=1}^{N}\Bigg[E\left(\cos\frac{2\pi\alpha n}{N}+\sin\frac{2\pi\alpha n}{N}\right)+2V\cos\frac{2\pi\alpha}{N}\left(\cos\frac{2\pi\alpha n}{N}+\sin\frac{2\pi\alpha n}{N}\right)\Bigg]\,\hat{\mathbf{e}}_n\\
&= \left[E+2V\cos\frac{2\pi\alpha}{N}\right]\mathbf{v}_\alpha,
\end{aligned}
\tag{9}
$$

were the identities

$$
\cos(a\pm b)=\cos a\cos b\mp\sin a\sin b,\quad \sin(a\pm b)=\sin a\cos b\pm\cos a\sin b,
\tag{10}
$$

have been used. The vectors defined in Eq. (7) thus satisfy $\hat{H}_{B850}\mathbf{v}_\alpha=\epsilon_\alpha\mathbf{v}_\alpha$, where ϵ_α are the excitonic energy levels given by

$$
\epsilon_\alpha=E+2V\cos\frac{2\pi\alpha}{N},
\tag{11}
$$

for $\alpha=1,2,\dots,N$. The energy levels ϵ_α and corresponding excitonic states $\mathbf{v}_\alpha$ are shown in Figures 9 and 10.

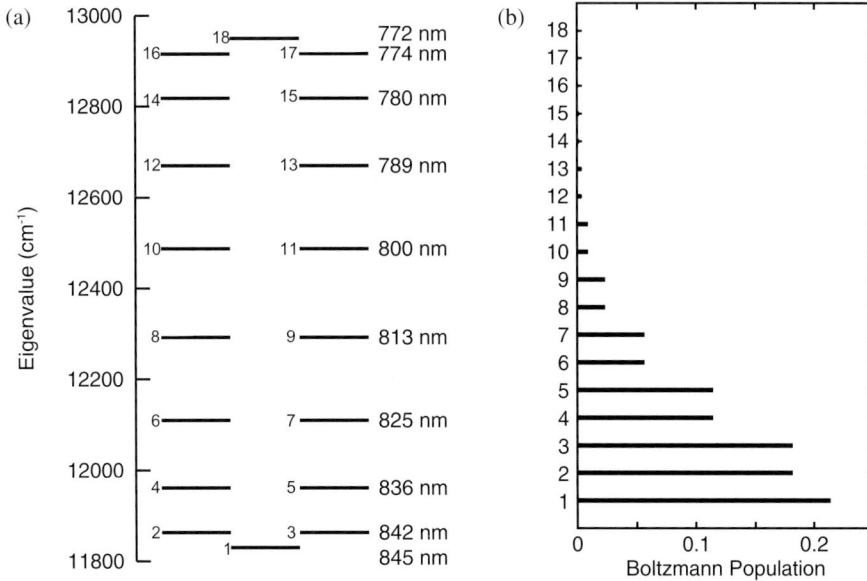

Figure 9. (a) Exciton spectrum ϵ_α and (b) exciton populations p_α of the BChls located in the B850 ring of LH2. The numbering of the energy levels corresponds to that of the exciton states in Figure 10, but not to the index α.

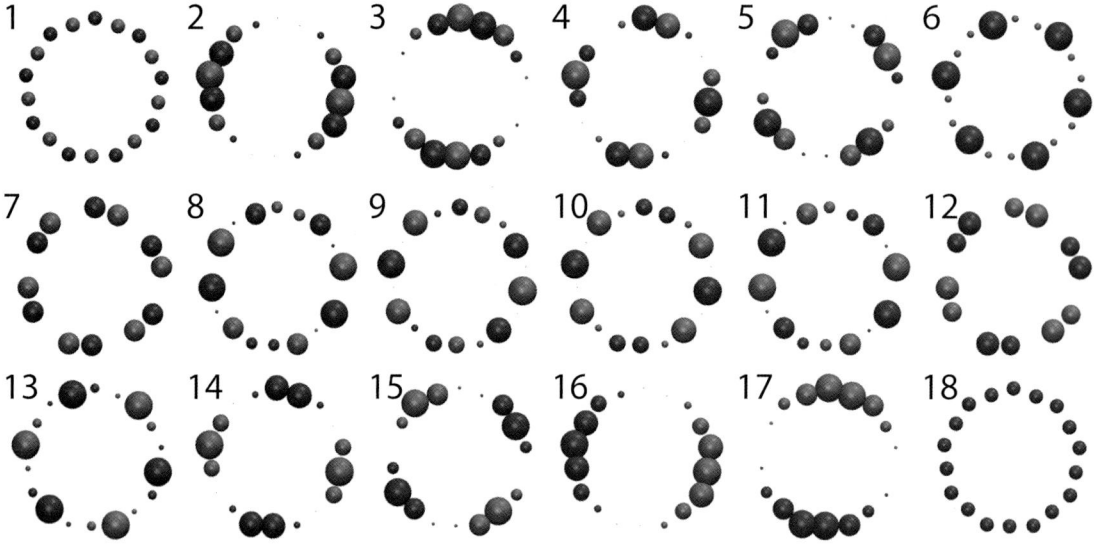

Figure 10. Exciton states $\mathbf{v}_\alpha$ for the B850 ring of LH2 from the Hamiltonian defined in Eq. (6). The size of the spheres indicate the amount of excitation on each BChl for each exciton state. The numbering is in ascending order of energy and corresponds with that in Figure 9, but not to the index α.

When a single B850 BChl is excited, the excitation will spread quickly amongst all the B850 BChls. This is a relaxation process that populates the excitonic states $\mathbf{v}_\alpha$ with equilibrium populations p_α. The BChls are in contact with a fluctuating thermal environment (consisting of proteins, lipids, water and ions) with a temperature $T = 300$ K, and in accordance with statistical mechanics, this determines how much each excitonic state is populated. It holds

$$p_\alpha = \exp(-\beta\epsilon_\alpha) \Big/ \sum_{\delta=1}^{N} \exp(-\beta\epsilon_\delta), \tag{12}$$

where $\beta = 1/k_B T$. It takes approximately 1 ps for the initial state to relax into the equilibrium distribution,[58] from where the exciton transfers then to neighboring LH2 or LH1 complexes on a 10 ps timescale. In the following section, the exciton states $\mathbf{v}_\alpha$ and energy levels ϵ_α are used to calculate the transfer rate between two groups of pigments.

3.2 *Excitation transfer*

The interaction between B850 BChls belonging to neighboring light-harvesting complexes LH2 gives rise to transfer of excitation energy between the two light-harvesting complexes. Since the pigments of neighboring complexes are well separated (with a closest distance of ≈ 3 nm for the

B850 BChls of neighboring LH2s) the transfer rate between the BChls can be calculated according the Förster formula (3). Here the interaction energy is given by[13,54]

$$V_{12} = C \frac{\hat{\mathbf{d}}_1 \cdot \hat{\mathbf{d}}_2 - 3(\hat{\mathbf{d}}_1 \cdot \hat{\mathbf{r}})(\hat{\mathbf{d}}_2 \cdot \hat{\mathbf{r}})}{|\mathbf{r}|^3}, \qquad (13)$$

where $\mathbf{r}$ is the vector from the Mg atom of BChl 1 to the Mg atom BChl 2, $\hat{\mathbf{r}} = \mathbf{r}/|\mathbf{r}|$, $\hat{\mathbf{d}}_1 = \mathbf{d}_1/|\mathbf{d}_1|$ is the direction of the transition dipole moment of BChl 1, and the coupling constant C has the value 348000 cm^{-1}.

Equation (3) can be used to calculate the rate of excitation transfer between single BChls 1 and 2, but cannot be used to calculate the rate between two groups of pigments, such as two B850 rings. As the exciton states are populated according to a Boltzmann distribution, the total transfer rate from a donor group with N_D pigments and an acceptor group with N_A pigments is

$$k = \sum_{\alpha=1}^{N_D} \sum_{\delta=1}^{N_A} p_\alpha k_{\alpha\delta}^{D \to A}, \qquad (14)$$

where p_α is given by Eq. (12). The transfer rate between exciton state $\mathbf{v}_\alpha^{(D)}$ of the donor group and $\mathbf{v}_\delta^{(A)}$ of the acceptor group can be calculated using Eq. (3), resulting in

$$k_{\alpha\delta}^{D \to A} = \frac{2\pi}{\hbar} V_{\alpha\delta}^2 S_{\alpha\delta}, \qquad (15)$$

where the interaction and spectral density terms between two pigments have been replaced by the interaction and spectral density between exciton states $\mathbf{v}_\alpha^{(D)}$ and $\mathbf{v}_\delta^{(A)}$. For donor exciton states $\mathbf{v}_\alpha^{(D)}$ given by

$$\mathbf{v}_\alpha^{(D)} = \sum_{n=1}^{N_D} c_{n\alpha}^D \, \hat{\mathbf{e}}_n^{(D)}, \qquad (16)$$

and acceptor exciton states $\mathbf{v}_\delta^A$ given by

$$\mathbf{v}_\delta^{(A)} = \sum_{m=1}^{N_A} c_{m\delta}^A \, \hat{\mathbf{e}}_m^{(A)}, \qquad (17)$$

the interaction energy is

$$V_{\alpha\delta} = \sum_{n=1}^{N_D} \sum_{m=1}^{N_A} c_{n\alpha}^D c_{m\delta}^A V_{nm}, \qquad (18)$$

where V_{nm} is stated in Eq. (13), with the replacement $1 \to n$ and $2 \to m$. Using Eq. (14), the transfer rate between the B850 rings of two neighboring LH2s is calculated to be 1/7.93 ps^{-1}.

Strümpfer et al.

A similar calculation can be performed for the transfer rate between LH1-RC complexes[13,14,59] and between LH2 and LH1-RC complexes.[13,14] These rates can be employed to construct a model for exciton migration across the whole chromatophore. Equation (14) can be used to determine the exciton transfer rate between light-harvesting complexes. This, however, requires knowledge of many parameters, such as the individual BChl energy levels, inter-BChl coupling strengths, fluorescence and absorption spectra of the light-harvesting complexes and the position and orientations of the light-harvesting complexes in the chromatophore. The parameters that have the strongest influence on excitation transfer between complexes are their relative positions and orientations.[14,84] Parameters are usually determined experimentally or from quantum chemical calculations.[89–94] Once the parameters for all the complexes are known, the transfer rate between each pair of light-harvesting complexes can be calculated. The migration of excitons between light-harvesting complexes can be modeled as random transfer events with transfer rates given by Eq. (15). This has been done for the whole chromatophore,[1,13,14,95] with each significant transfer rate depicted as an LH2-LH2, LH2-LH1 or LH1-LH1 connection, as shown in Figure 11.

3.3 *Chromatophore-wide exciton transfer*

An excitation arising from an absorbed photon needs to arrive at a RC in order to initate the further steps in photosynthesis. This occurs by transferring excitation between the pigments of various molecules until it arrives at the special pair BChls of the RC (see Figure 1). Although excitation transfer occurs between all the pigment molecules in the chromatophore, the

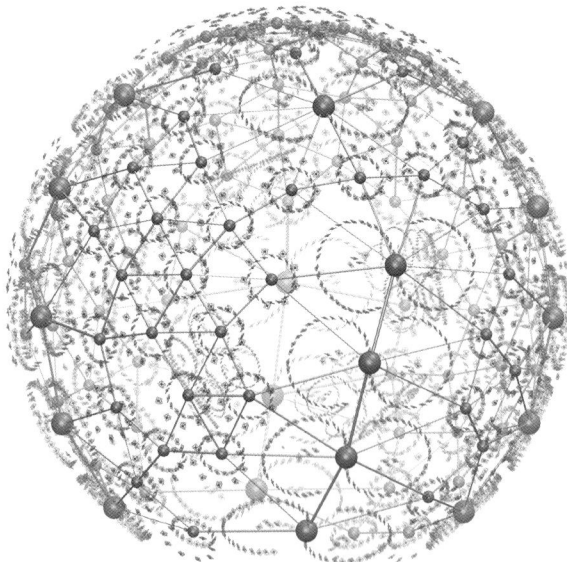

Figure 11. Exciton transfer network spanning an entire *Rb. sphaeroides* chromatophore. Shown are the locations of the BChls in dimeric (LH1-RC)$_2$ core complexes and LH2 complexes. The spheres show the centers of the LH2 and LH1 complexes and the widths of the lines connecting the spheres represent the transfer rates between nearby complexes (thicker lines imply higher transfer rates). The pathways corresponding to very slow transfer rates are not shown. Adapted from Ref. 13.

pigments within a complex interact strongly with each other so as to coherently share excitation between them to form an exciton (see Section 4.2). The transfer of excitation is thus in the form of exciton transfer between the different light-harvesting complexes found in the chromatophore.

With the inclusion of a decay rate to the ground state of 1/ns, where 1 ns is approximately the lifetime of excitation in each light-harvesting complex (see Appendix), and of the electron transfer rate from the special pair to bacteriopheophytin in the RC, which is approximately 1/(3 ps), quantities such as the total excitation lifetime per photon absorbed and the probability that a photon absorbed by the chromatophore charges the membrane can be calculated.[13,14,59] Such calculations revealed that the chromatophores of purple bacteria have very high light-harvesting efficiencies (>90%).

4. Discussion

This review deals with photosynthetic light harvesting, a process that involves the quantum dynamics of electrons and fuels with solar energy nearly all life on Earth. No known process is more deeply grounded in modern physics and more relevant to human life. The greatest fascination in dealing with photosynthetic light harvesting, though, stems from the beautiful integration of many physical scales that turns fundamental quantum physics into a practical quantum biological device. The relevant scales are the scale of the electron (10^0Å), the scale of the molecules (10^1Å), namely chlorophylls and carotenoids, that act with their electrons in absorbing light and funneling its energy to where it is turned into an electrical potential, the scale of the proteins (10^2Å) that act as scaffolds for these molecules, the scale of the chromatophore (10^3Å) that combines the proteins into a functional unit, and finally that of the cell (10^5Å) that organizes the construction of chromatophores, their repair, and utilizes their energy output.

While this multiscale accomplishment, genuine to all living systems in many different ways, poses a challenge to research for possibly as long as people will live, it is deeply rewarding to physicists to see that the laws of quantum physics are being put to such great use as seen in photosynthetic light harvesting. A physicist unfamiliar with the subject might have expected that nature learnt to optimize light absorption by designing molecules with strong transition dipole moments and orienting the molecules to catch light of all polarization. But this physicist learned that living systems tinkering for billions of years also found ways to funnel the electronic excitation energy over 1000Å distances rapidly enough to avoid loss through fluorescence. The path taken involves partially coherent propagation in the form of excitons[54,96] and partially essentially incoherent,[58] random transfers as described by classical kinetics.[13,47,97] Interesting and challenging is that living systems operating at physiological temperature learnt to take advantage of thermal effects that many research physicists tend to avoid by studying systems often at liquid nitrogen or liquid helium temperature, at which temperatures life ceases to exist.[55]

The reader who wishes to dig deeper into the world of photosynthetic light harvesting is encouraged to consult the reviews referenced at the beginning of this chapter. A broad overview of photosynthesis, not only of purple bacteria but all photosynthetic life forms, is given in the book by Blankenship.[5] An introductory review of the smallest light-harvesting protein, LH2, discusses excitons and the use of circular symmetry to avoid photon emission;[2] this is likely the best introduction for a young investigator. A related review discusses not only LH2, but also the role of other molecular components of the photosynthetic chromatophore in purple bacteria; it tells also the story of the discovery of biological fluorescent energy transfer.[3] A review written from a more biological perspective discusses also the genetics of the light-harvesting apparatus.[6] Three reviews have been written recently, one describing the observation of optical properties,[4] one describing the overall structural and functional properties of the chromatophore[1] and one, Ref. 84, describing how photosynthetic systems in bacteria and plants make excellent use of the quantum physics of excitation transfer as described presently only briefly.[1] Lastly, a reader may want to study a research article which describes the constraints placed by optical physics on the biological evolution of efficient light-harvesting systems.[14]

Acknowledgments

This work was supported by grants from National Science Foundation (MCB-0744057, PHY0822613) and National Institute of Health (P41-RR05969).

References

1. M. K. Şener and K. Schulten. 2008. From atomic-level structure to supramolecular organization in the photosynthetic unit of purple bacteria. in *The Purple Phototrophic Bacteria*, C. N. Hunter, F. Daldal, M. C. Thurnauer and J. T. Beatty, eds., Vol. 28 of *Advances in Photosynthesis and Respiration*, 275–294, Springer.

2. X. Hu and K. Schulten. 1997. How nature harvests sunlight. *Physics Today* **50**, 28–34.

3. K. Schulten. 1999. From simplicity to complexity and back: Function, architecture and mechanism of light harvesting systems in photosynthetic bacteria. in *Simplicity and Complexity in Proteins and Nucleic Acids*, H. Frauenfelder, J. Deisenhofer, and P. G. Wolynes, eds. (Berlin), 227–253, Dahlem University Press (pdf available from author upon request).

4. R. J. Cogdell, A. Gall and J. Köhler. 2006. The architecture and function of the light-harvesting apparatus of purple bacteria: From single molecules to *in vivo* membranes. *Quart Rev Biophys* **39**, 227–324.

5. R. E. Blankenship. 2002. *Molecular Mechanisms of Photosynthesis*. Malden, MA: Blackwell Science.

6. X. Hu, T. Ritz, A. Damjanović, F. Autenrieth and K. Schulten. 2002. Photosynthetic apparatus of purple bacteria. *Quart Rev Biophys* **35**, 1–62.

7. T. Förster. 1948. Zwischenmolekulare Energiewanderung und Fluoreszenz. *Ann Phys (Leipzig)* **2**, 55–75.

8. D. Dexter. 1953. A theory of sensitized luminescence in solids. *J Chem Phys* **21**, 836–850.

9. A. Damjanović, T. Ritz and K. Schulten. 1999. Energy transfer between carotenoids and bacteriochlorophylls in a light harvesting protein. *Phys Rev E* **59**, 3293–3311.

10. T. Ritz, A. Damjanović, K. Schulten, J. Zhang and Y. Koyama. 2000. Efficient light harvesting through carotenoids. *Photosyn Res* **66**, 125–144.

11. A. R. Crofts, B. Barquera, G. Bechmann, M. Guergova, R. Salcedo-Hernandez, B. Hacker, S. Hong and R. B. Gennis. 1995. Structure and function in the bc1-complex of *rb. sphaeroides*. in *Photosynthesis: From light to biosphere*, P. Mathis, ed. Kluwer Academic Publ, Dordrecht.

12. S. Bahatyrova, R. N. Frese, C. A. Siebert, J. D. Olsen, K. O. van der Werf, R. van Grondelle, R. A. Niederman, P. A. Bullough, C. Otto and C. N. Hunter. 2004. The native architecture of a photosynthetic membrane, *Nature* **430**, 1058–1062.

13. M. K. Sener, J. D. Olsen, C. N. Hunter and K. Schulten. 2007. Atomic level structural and functional model of a bacterial photosynthetic membrane vesicle. *Proc Natl Acad Sci U S A* **104**, 15723–15728.

14. M. Sener, J. Strumpfer, J. A. Timney, A. Freiberg, C. N. Hunter and K. Schulten. 2010. Photosynthetic vesicle architecture and constraints on efficient energy harvesting. *Biophys J* **99**, 67–75.

15. J. Xiong, W. M. Fischer, K. Inoue, M. Nakahara and C. E. Bauer. 2000. Molecular evidence for the early evolution of photosynthesis. *Science* **289**, 1724–1730.

16. R. Emerson and A. Arnold. 1932. The photochemical reaction in photosynthesis. *J Gen Physiol* **16**, 191–205.

17. T. Förster. 1948. Zwischenmolekulare Energiewanderung und Fluoreszenz. *Ann Phys (Leipzig)*, **2**, 55–75.

18. J. R. Oppenheimer. 1941. Internal conversion in photosynthesis. in *Proceedings of the American Physical Society*, vol. 60 of *Phys Rev* p. 158.

19. W. Arnold and J. R. Oppenheimer. 1950. Internal conversion in the photosynthetic mechanism of blue-green algae. *J Gen Physiol* **33**, 423–435.

20. J. Deisenhofer, O. Epp, K. Mikki, R. Huber and H. Michel. 1985. Structure of the protein subunits in the photosynthetic reaction centre of *Rhodopseudomonas viridis* at 3 A resolution. *Nature* **318**, 618–624.

21. R. A. Marcus. 1956. On the energy of oxidation-reduction reactions involving electron transfer. I. *J Chem Phys* **24**, 966–978.

22. R. A. Marcus. 1956. Electrostatic free energy and other properties of states having nonequilibrium polarization. II, *J Chem Phys* **24**, 979–989.

23. W. L. Liang. 1970. Excitons. *Phys Educ* **5**, 226.

24. J. Frenkel. 1931. On the transformation of light into heat in solids. *Phys Rev* **37**, 17–44.

25. G. McDermott, S. M. Prince, A. A. Freer, A. M. Hawthornthwaite-Lawless, M. Z. Papiz, R. J. Cogdell and N. W. Isaacs. 1995. Crystal structure of an integral membrane light-harvesting complex from photosynthetic bacteria. *Nature* **374**, 517–521.

26. S. Karrasch, P. Bullough and R. Ghosh. 1995. The 8.5 Å projection map of the light-harvesting complex I from *Rhodospirillum rubrum* reveals a ring composed of 16 sub-units. *EMBO J* **14**, 631.

27. X. Hu, D. Xu, K. Hamer, K. Schulten, J. Koepke and H. Michel. 1995. Predicting the structure of the light-harvesting complex II of *Rhodospirillum molischianum*. *Prot Sci* **4**, 1670–1682.

28. J. Koepke, X. Hu, C. Muenke, K. Schulten and H. Michel. 1996. The crystal structure of the light harvesting complex II (B800–850) from *Rhodospirillum molischianum*. *Structure* **4**, 581–597.

29. M. Z. Papiz, S. M. Prince, T. Howard, R. J. Cogdell and N. W. Isaacs. 2003. The structure and thermal motion of the B800–850 LH2 complex from *Rps. acidophila* at 2.0 Å resolution and 100 K: New structural features and functionally relevant motions. *J Mol Biol* **326**, 1523–1538.

30. J. Allen, T. Yeates, H. Komiya and D. Rees. 1987. Structure of the reaction center from *Rhodobacter sphaeroides* R-26: The protein subunits. *Proc Natl Acad Sci U S A* **84**, 6162–6166.

31. K. J. Visscher, H. Bergstrom, V. Sundström, C. Hunter and R. van Grondelle. 1989. Temperature dependence of energy transfer from the long wavelength antenna Bchl-896 to the reaction center in *Rhodospirillum rubrum, Rhodobacter sphaeroides* (w.t. and M21 mutant) from 77 to 177 K, studied by picosecond absorption spectroscopy. *Photosynthesis Research* **22**, 211–217.

32. U. Ermler, G. Fritzsch, S. K. Buchanan and H. Michel. 1994. Structure of the photosynthetic reaction center from *Rhodobacter sphaeroides* at 2.65 Å resolution: Cofactors and protein-cofactor interactions. *Structure* **2**, 925–936.

33. M. J. Conroy, W. Westerhuis, P. S. Parkes-Loach, P. A. Loach, C. N. Hunter and M. P. Williamson. 2000. The solution structure of the *Rhodobacter sphaeroides* LH1 β reveals two helical domains separated by a more flexible region: Structural consequences for the LH1 complex. *J Mol Biol* **298**, 83–94.

34. A. Camara-Artigas, D. Brune and J. P. Allen. 2002. Interactions between lipids and bacterial reaction centers determined by protein crystallography. *Proc Natl Acad Sci U S A* **99**, 11055–11060.

35. A. W. Roszak, T. Howard, J. Southall, A. T. Gardiner, C. J. Law, N. W. Isaacs and R. J. Cogdell. 2003. Crystal structure of the RC-LH1 core complex from *Rhodopseudomonas palustris*. *Science* **302**, 1969–1972.

36. D. Fotiadis, P. Qian, A. Philippsen, P. A. Bullough, A. Engel and C. N. Hunter. 2004. Structural analysis of the reaction center light-harvesting complex I photosynthetic core complex of *Rhodospirillum rubrum* using atomic force microscopy. *J Biol Chem* **279**, 2063–2068.

37. G. Uyeda, A. Camara-Artigas, J. C. Williams and J. P. Allen. 2005. Design of a redox-linked active metal site: Manganese bound to bacterial reaction centers at a site resembling that of photosystem II. *Biochemistry* **44**, 7389–7394.

38. M. Elberry, K. Xiao, L. Esser, D. Xia, L. Yu and C. an Yu. 2006. Generation, characterization, and crystallization of a highly active and stable cytochrome bc_1 complex mutant from *Rhodobacter sphaeroides*. *Biochim Biophys Acta* **1757**, 835–840.

39. H. L. Axelrod, E. C. Abresch, M. Y. Okamura, A. P. Yeh, D. C. Ress and G. Feher. 2002. X-ray structure determination of the cytochrome c2: Reaction center electron transfer complex from *Rhodobacter sphaeroides*. *J Mol Biol* **319**, 501–515.

40. S. Scheuring, J. N. Sturgis, V. Prima, A. Bernadac, D. Levy and J.-L. Rigaud. 2004. Watching the photosynthetic apparatus in native membranes. *Proc Natl Acad Sci U S A* **91**, 11293–11297.

41. R. N. Frese, C. A. Siebert, R. A. Niederman, C. N. Hunter, C. Otto and R. van Grondelle. 2004. The long-range organization of a native photosynthetic membrane. *Proc Natl Acad Sci U S A* **101**, 17994–17999.

42. S. Scheuring, D. Lévy and J.-L. Rigaud. 2005. Watching the components of photosynthetic bacterial membranes and their *in situ* organisation by atomic force microscopy. *Biochim Biophys Acta* **1712**, 109–127.

43. C. Jungas, J. Ranck, J. Rigaud, P. Joliot and A. Verméglio. 1999. Supramolecular organization of the photosynthetic apparatus of *Rhodobacter sphaeroides*. *EMBO J* **18**(3), 534–542.

44. S. J. Jamieson, P. Wang, P. Qian, J. Y. Kirkland, M. J. Conroy, C. N. Hunter and P. A. Bullough. 2002. Projection structure of the photosynthetic reaction centre-antenna complex of *Rhodospirillum rubrum* at 8.5 Å resolution. *EMBO J* **21**, 3927–3935.

45. C. A. Siebert, P. Qian, D. Fotiadis, A. Engel, C. N. Hunter and P. A. Bullough. 2004. Molecular architecture of photosynthetic membranes in *Rhodobacter sphaeroides*: The role of PufX. *EMBO J* **23**, 690–700.

46. P. Qian, C. N. Hunter and P. A. Bullough. 2005. The 8.5 Å projection structure of the core RC-LH1-PufX dimer of *Rhodobacter sphaeroides*. *J Mol Biol* **349**, 948–960.

47. M. K. Şener, D. Lu, T. Ritz, S. Park, P. Fromme and K. Schulten. 2002. Robustness and optimality of light harvesting in cyanobacterial photosystem I. *J Phys Chem B* **106**, 7948–7960.

48. A. G. Redfield. 1965. The theorie of relaxation process. *Adv Magn Reson* **1**, 1.

49. A. O. Cardeira and A. J. Leggett. 1983. Quantum tunnelling in a dissipative system. *J Ann Phys (N.Y.)* **149**, 374–456.

50. H. Sumi and R. A. Marcus. 1986. Dynamical effects in electron transfer reactions. *J Chem Phys* **84**, 4894–4914.

51. A. Warshel, Z. T. Chu and W. W. Parson. 1989. Dispersed polaron simulations of electron transfer in photosynthetic reaction center. *Science* **246**, 112–116.

52. Y. Tanimura and R. Kubo. 1989. Two-time correlation functions of a system coupled to a heat bath with a Gaussian-Markoffian interaction. *J Phys Soc Jpn* **58**(4), 1199–1206.

53. S. Mukamel. 1995. *Principles of Nonlinear Optical Spectroscopy.* Oxford University Press, New York.

54. X. Hu, T. Ritz, A. Damjanović and K. Schulten. 1997. Pigment organization and transfer of electronic excitation in the purple bacteria. *J Phys Chem B* **101**, 3854–3871.

55. A. Damjanović, I. Kosztin, U. Kleinekathoefer and K. Schulten. 2002. Excitons in a photosynthetic light-harvesting system: A combined molecular dynamics, quantum chemistry and polaron model study. *Phys Rev E* **65**, 031919, (24 pages).

56. M. Şener and K. Schulten. 2002. A general random matrix approach to account for the effect of static disorder on the spectral properties of light harvesting systems. *Phys Rev E* **65**, 031916, (12 pages).

57. A. Ishizaki and G. R. Fleming. 2009. Theoretical examination of quantum coherence in a photosynthetic system at physiological temperature. *Proc Natl Acad Sci U S A* **106**(41), 17255.

58. J. Strümpfer and K. Schulten. 2009. Light harvesting complex II B850 excitation dynamics. *J Chem Phys* **131**, 225101, (9 pages).

59. J. Hsin, J. Strumpfer, M. Sener, P. Qian, C. N. Hunter and K. Schulten. 2010. Energy transfer dynamics in an RC-LH1-PufX tubular photosynthetic membrane. *New J Phys* **12**, 085005, (19 pages).

60. J. Strumpfer and K. Schulten. 2011. The effect of correlated bath fluctuations on exciton transfer. *J Chem Phys* **134**, 095102–095111.

61. M. Şener and K. Schulten. 2005. Physical principles of efficient excitation transfer in light harvesting. in *Energy Harvesting Materials* (D. L. Andrews, ed.), pp. 1–26, World Scientific, Singapore.

62. T. Walz, S. J. Jamieson, C. M. Bowers, P. A. Bullough and C. N. Hunter. 1998. Projection structures of three photosynthetic complexes from *Rhodobacter sphaeroides:* LH2 at 6 Å, LH1 and RC-LH1 at 25 Å. *J Mol Biol* **282**, 833–845.

63. D. Chandler, J. Hsin, C. B. Harrison, J. Gumbart and K. Schulten. 2008. Intrinsic curvature properties of photosynthetic proteins in chromatophores. *Biophys J* **95**, 2822–2836.

64. D. E. Chandler, J. Gumbart, J. D. Stack, C. Chipot and K. Schulten. 2009. Membrane curvature induced by aggregates of LH2s and monomeric LH1s. *Biophys J* **97**, 2978–2984.

65. J. Hsin, D. E. Chandler, J. Gumbart, C. B. Harrison, M. Sener, J. Strumpfer and K. Schulten. 2010. Self-assembly of photosynthetic membranes. *ChemPhysChem* **11**, 1154–1159.

66. N. Pfennig. 1967. Photosynthetic bacteria. *Annu Rev Med* **21**, 285–324.

67. N. Pfennig. 1969. *Rhodopseudomonas acidophila,* sp. n., a new species of the budding purple nonsulfur bacteria. *J Bacteriol* **99**(2), 597–602.

68. P. Qian, P. A. Bullough and C. N. Hunter. 2008. Three-dimensional reconstruction of a membrane-bending complex: The RC-LH1-PufX core dimer of *Rhodobacter sphaeroides. J Biol Chem* **283**, 14002–14011.

69. J. Hsin, J. Gumbart, L. G. Trabuco, E. Villa, P. Qian, C. N. Hunter and K. Schulten. 2009. Protein-induced membrane curvature investigated through molecular dynamics flexible fitting. *Biophys J* **97**, 321–329.

70. M. K. Sener, J. Hsin, L. G. Trabuco, E. Villa, P. Qian, C. N. Hunter and K. Schulten. 2009. Structural model and excitonic properties of the dimeric RC-LH1-PufX complex from *Rhodobacter sphaeroides. Chem Phys* **357**, 188–197.

71. S. Scheuring, F. Francia, J. Busselez, B. A. Melandris, J.-L. Rigaud and D. Levy. 2004. Structural role of PufX in the dimerization of the photosynthetic core complex of *Rhodobacter sphaeroides. J Biol Chem* **279**(5), 3620–3626.

72. S. Scheuring, J. Busselez and D. Levy. 2005. Structure of the dimeric PufX-containing core complex of *Rhodobacter blasticus* by *in situ* atomic force microscopy. *J Biol Chem* **280**, 1426–1431.

73. J. Busselez, M. Cottevielle, P. Cuniasse, F. Gubellini, N. Boisset and D. Lévy. 2007. Structural basis for the PufX-mediated dimerization of bacterial photosynthetic core complexes. *Structure* **15**, 1674–1683.

74. R. Frese, J. Olsen, R. Branvall, W. Westerhuis, C. Hunter and R. van Grondelle. 2000. The long-range supraorganization of the bacterial photosynthetic unit: A key role for PufX. *Proc Natl Acad Sci U S A* **97**, 5197–5202.

75. S. Scheuring. 2006. AFM studies of the supramolecular assembly of bacterial photosynthetic core-complexes. *Curr Opin Chem Biol* **10**, 387–393.

76. J. Hsin, C. Chipot and K. Schulten. 2009. A glycophorin A-like framework for the dimerization of photosynthetic core complexes. *J Am Chem Soc* **131**, 17096–17098.

77. X. Hu and K. Schulten. 1998. A model for the light-harvesting complex I (B875) of *Rhodobacter sphaeroides. Biophys J* **75**, 683–694.

78. J. Deisenhofer and H. Michel. 1987. in *The crystal structure of the photosynthetic reaction center from Rhodopseudomonas viridis* (J. Breton and A. Vermeglio, eds.), The photosynthetic bacterial reaction center: Structure and dynamics, pp. 1–3, Plenum Press, London.

79. P. Jordan, P. Fromme, H. T. Witt, O. Klukas, W. Saenger and N. Krauß. 2001. Three-dimensional structure of cyanobacterial photosystem I at 2.5 Å resolution. *Nature* **411**, 909–917.

80. A. Ben-Shem, F. Frolow and N. Nelson. 2003. Crystal structure of plant photosystem I. *Nature* **426**, 630–635.

81. A. Guskov, J. Kern, A. Gabdulkhakov, M. Broser, A. Zouni and W. Saenger. 2009. Cyanobacterial photosystem II at 2.9-resolution and the role of quinones, lipids, channels and chloride. *Nat Struct Mol Biol* **16**(3), 334–342.

82. J. Tucker, C. Siebert, M. Escalante, P. Adams, J. Olsen, C. Otto, D. Stokes and C. Hunter. 2010. Membrane invagination in *Rhodobacter sphaeroides* is initiated at curved regions of the cytoplasmic membrane, then forms both budded and fully detached spherical vesicles. *Mol Microbiol* **76**(4), 833–847.

83. C. N. Hunter, R. van Grondelle, N. G. Holmes, O. T. G. Jones and R. A. Niederman. 1985. Linear dichroism and fluorescence emission of antenna complexes during photosynthetic unit assembly in *Rhodopseudomonas sphaeroides*. *Biochim Biophys Acta* **807**, 44–51.

84. M. Sener, J. Strumpfer, J. Hsin, D. Chandler, S. Scheuring, C. N. Hunter and K. Schulten. 2011. Förster energy transfer theory as reflected in the structures of photosynthetic light harvesting systems. *Chem Phys Chem* **12**, 518–531.

85. T. Geyer and V. Helms. 2006. A spatial model of the chromatophore vesicles of *Rhodobacter sphaeroides* and the position of the cytochrome bc_1 complex. *Biophys J* **91**, 921–926.

86. J. J. Sakurai. 1985. *Modern quantum mechanics*. Addison Wesley Publishing Company, Boston.

87. E. Fermi. 1950. *Nuclear Physics*. University of Chicago Press, Chicago.

88. V. May, O. Kühn, J. Wiley and S. Inc. 2004. *Charge and Energy Transfer Dynamics in Molecular Systems*. Wiley-VCH, Weinheim.

89. A. Freiberg, M. Ratsep, K. Timpmann and G. Trinkunas. 2009. Excitonic polarons in quasi-one-dimensional LH1 and LH2 bacteriochlorophyll a antenna aggregates from photosynthetic bacteria: A wavelength-dependent selective spectroscopy study. *Chem Phys* **357**, 102–112.

90. R. S. Knox and B. Q. Spring. 2003. Dipole strengths in the chlorophylls. *J Photochem Photobiol* **77**, 497–501.

91. M. G. Cory, M. C. Zerner, X. Hu and K. Schulten. 1998. Electronic excitations in aggregates of bacteriochlorophylls. *J Phys Chem B* **102**, 7640–7650.

92. J. Eccles, B. Honig and K. Schulten. 1988. Spectroscopic determinants in the reaction center of *Rhodopseudomonas viridis*. *Biophys J* **53**, 137–144.

93. M. Groot, J. Yu, R. Agarwal, J. R. Norris and G. R. Fleming. 1998. Three-pulse photon echo measurments on the accessory pigments in the reaction center of *Rhodobacter sphaeroides*. *J Phys Chem B* **102**, 5923–5931.

94. T. Ritz, S. Adem and K. Schulten. 2000. A model for photoreceptor-based magnetoreception in birds. *Biophys J* **78**, 707–718.

95. F. Fassioli, A. Olaya-Castro, S. Scheuring, J. Sturgis and N. Johnson. 2009. Energy transfer in light-adapted photosynthetic membranes: From active to saturated photosynthesis. *Biophys J* **97**, 2464–2473.

96. T. Ritz, X. Hu, A. Damjanović and K. Schulten. 1998. Excitons and excitation transfer in the photosynthetic unit of purple bacteria. *J Luminesc* **76–77**, 310–321.

97. M. K. Şener, S. Park, D. Lu, A. Damjanović, T. Ritz, P. Fromme and K. Schulten. 2004. Excitation migration in trimeric cyanobacterial photosystem I. *J Chem Phys* **120**, 11183–11195.

98. R. Loudon. 2001. *The Quantum Theory of Light*, Oxford University Press, Oxford, Third edition.

99. J. D. Jackson. 1999. *Classical Electrodynamics*. John Willey & Sons, New York, Third edition.

Appendix

Light absorption and emission as well as excitation energy transfer are fundamental to the photosynthetic light-harvesting apparatus. Here we introduce a detailed discussion of these quantum mechanical processes. For light absorption and emission, we just provide below the formulas that relate the corresponding rates to the determining molecular property, namely, the transition dipole moment. For excitation energy transfer, we actually derive the respective rate expression, the

Förster resonant energy transfer (FRET) formula. The reader will learn that FRET depends also on the transition dipole moment, namely those of the donor and acceptor molecules involved in the transition.

Rate of light absorption and emission. When a molecule, say D, absorbs a photon of light, one of its electrons changes its state. This is at least the simplest description, sufficient for the present case. Such electron may occupy in the unexcited D the molecular electron orbital $\phi(\mathbf{r})$ and it is moved through photon absorption to a new orbital, not occupied by other electrons, say $\phi^*(\mathbf{r})$, thereby forming the excited D^*. The rate of light absorption is governed by the so-called transition dipole moment defined through the spatial (position $\mathbf{r}$) integral

$$\mathbf{d} = e \int d\mathbf{r}\, \phi^*(\mathbf{r})\mathbf{r}\phi(\mathbf{r}), \tag{A1}$$

where e is the electron charge. Like a molecular dipole moment, $\mathbf{d}$ is of the order electron charge × molecular size. In our description, we assume the simplest type of electron excitation, but even in the realistic case an optical excitation is characterized by just this one molecular vector property, namely the transition dipole moment. $\mathbf{d}$ has a certain orientation relative to the molecular geometry. Molecules can typically undergo more than one optical excitation, in which case each of the different excitations are characterized by their own $\mathbf{d}$. Evolution drove many biological pigments to develop very large $|\mathbf{d}|$ values as these lead to strong absorption.

Light is characterized through a wavelength λ that is related to photon energy through $\hbar\omega = \hbar 2\pi c/\lambda$ where c is the speed of light. It is also characterized typically through polarization $\hat{u}$, a unit vector oriented perpendicular to the light propagation direction. The rate of absorption of such light photon is (for an introduction, see Ref. 98)

$$k_{abs} = \frac{\mathcal{N}_\lambda}{\mathcal{V}} \frac{8\pi^3 e^2 c}{\lambda} |\hat{\mathbf{u}} \cdot \mathbf{d}|^2\, \delta(E_\lambda - \Delta E), \tag{A2}$$

where $\mathcal{N}_\lambda$ is the number of photons in a volume $\mathcal{V}$, e is the electron charge, E_λ is the energy of the incoming photon and ΔE is the electronic excitation energy. $\delta(E_\lambda - \Delta E)$ is the Dirac δ-function for which holds $\int_{-\infty}^{+\infty} dE_\lambda\, \delta(E_\lambda - \Delta E) = 1$.

Accounting for the thermal motion of the molecule requires one to replace $\delta(E_\lambda - \Delta E)$ in Eq. (A2) by a normalized energy density $S_{abs}(E_\lambda)$, for which holds also $\int_{-\infty}^{+\infty} dE_\lambda S_{abs}(E_\lambda) = 1$, and use

$$k_{abs} = \frac{\mathcal{N}_\lambda}{\mathcal{V}} \frac{8\pi^3 e^2 c}{\lambda} |\hat{\mathbf{u}} \cdot \mathbf{d}|^2\, S_{abs}(E_\lambda). \tag{A3}$$

Of great relevance for light harvesting is that $|\mathbf{d}|$ assumes high values for the low-energy electronic excitations in BChls and carotenoids. Also essential is that $S_{abs}(E_\lambda)$ is specifically large in the range of wavelengths that should be absorbed.

Another key issue for effective light harvesting is to prevent the emission of photons, before the excitation energy is used by the light-harvesting apparatus. Emission is described by the same formula as absorption, except that the quantization of light into photons requires one to replace in the formulas above $\mathcal{N}_\lambda$ by $\mathcal{N}_\lambda + 1$. In cases of so-called spontaneous emission with $\mathcal{N}_\lambda = 0$, if one ignores the internal motion of D and accounts for emission of photons with all possible wavelengths and polarizations, one obtains the rate formula (in which mathematical and physical constants have been replaced by their numerical values)

$$k_{emit} = \mathcal{N}_\lambda \left(1.37 \times 10^{19} \frac{1}{s} \right) \lambda^{-3} a_0 d^2 / e^2, \tag{A4}$$

where $a_0 = 0.53$ Å is the Bohr radius. With λ of the order of 4000–8000 Å and $|\mathbf{d}|/e$ typically a few Ångstrom, one can readily estimate that the rate of emission is about 1/ns, i.e. to be effective, the light-harvesting apparatus must utilize the energy of excited chlorophylls considerably faster than 1 ns. Quantum coherence features as they arise in excitons described above, assist the apparatus in regard to optimal absorption and excitation transfer and reduction of emission.[9,54]

Förster resonance energy transfer. In order to describe resonance energy transfer we consider the simplest possible case, namely excitation transfer between two hydrogen atoms, D and A, at positions $\mathbf{R}_D$ and $\mathbf{R}_A$, respectively. The result can be readily generalized to the more complicated case of molecules. Hydrogen D is thought to have absorbed a photon such that its electron is excited from its $1s$ ground state orbital $\phi_D(\mathbf{r}_D)$ to its $2p_z$ orbital, $\phi_D^*(\mathbf{r}_D)$. Hydrogen A is in the ground state with an electron in its $1s$ orbital $\phi_A(\mathbf{r}_A)$. Hydrogen D can transfer excitation energy to the electron on hydrogen A, exciting it from the $1s$ orbital to the $2p_z$ orbital $\phi_A^*(\mathbf{r}_A)$. The position of the atoms and electrons are shown in Figure 12.

The rate for hydrogen D to transfer its excitation energy to hydrogen A is given by

$$k_{D*A \to DA*} = \frac{2\pi}{\hbar} V^2 \int_{-\infty}^{+\infty} dE \, S_D(E) S_A(E), \tag{A5}$$

where V is the interaction energy for the transfer of excitation between hydrogen D and hydrogen A, $S_D(E)$ is the normalized emission spectrum of the electron on hydrogen D in the $2p_z$ orbital and $S_A(E)$ is the normalized absorption spectrum of a ground state electron to reach the $2p_z$ orbital in hydrogen A.

The interaction energy for the transfer of excitation between hydrogen D and hydrogen A, assuming that the atoms are separated by about 1 nm such that electron tunneling can be neglected, is given by the Coulomb interaction integral

$$V = \int d\mathbf{r}_D \int d\mathbf{r}_A \frac{\rho_D(\mathbf{r}_D)\rho_A(\mathbf{r}_A)}{r_{AD}}, \tag{A6}$$

where r_{AD} is the distance from the electron of hydrogen D to the electron of hydrogen A and

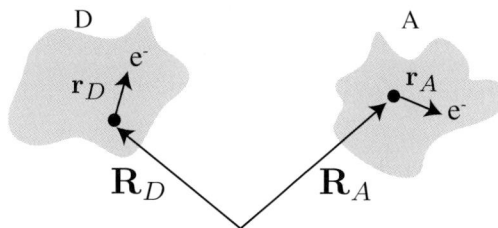

Figure 12. The positions of the donor hydrogen nucleus $\mathbf{R}_D$ and electron $\mathbf{r}_D$ and the acceptor hydrogen nucleus $\mathbf{R}_A$ and electron $\mathbf{r}_A$.

$$\rho_D(\mathbf{r}_D) = e\phi_D^*(\mathbf{r}_D)\phi_D(\mathbf{r}_D) \tag{A7}$$

$$\rho_A(\mathbf{r}_A) = e\phi_A^*(\mathbf{r}_A)\phi_A(\mathbf{r}_A). \tag{A8}$$

Before proceeding, it will be useful to note that due to the orthogonality of orbital pairs $\phi_D(\mathbf{r}_D)$, $\phi_D^*(\mathbf{r}_D)$ and $\phi_A(\mathbf{r}_D)$, $\phi_A^*(\mathbf{r}_D)$ holds

$$c_D = \int \rho_D(\mathbf{r}_D) = 0 \tag{A9}$$

$$c_A = \int \rho_A(\mathbf{r}_A) = 0. \tag{A10}$$

The vector connecting the electrons of hydrogen D and hydrogen A, can be written $\mathbf{r}_{AD} = (\mathbf{R}_D + \mathbf{r}_D) - (\mathbf{R}_A - \mathbf{r}_A)$ or

$$\mathbf{r}_{AD} = (\mathbf{R}_D - \mathbf{R}_A) + (\mathbf{r}_D - \mathbf{r}_A). \tag{A11}$$

Assuming $|\mathbf{R}_D - \mathbf{R}_A| \gg |\mathbf{r}_D - \mathbf{r}_A|$, the factor $1/r_{AD}$ in Eq. (A6) can be written[99]

$$\frac{1}{r_{AD}} = \frac{1}{R} - \frac{\mathbf{r} \cdot \mathbf{R}}{R^3} + \frac{3(\mathbf{r} \cdot \mathbf{R})^2 - r^2 R^2}{2R^5} + O\left(\frac{1}{R^4}\right), \tag{A12}$$

where $\mathbf{R} = \mathbf{R}_D - \mathbf{R}_A$, $\mathbf{r} = \mathbf{r}_D - \mathbf{r}_A$, $R = |\mathbf{R}|$ and $r = |\mathbf{r}|$. Taking only the three leading terms of Eq. (A12) into account, Eq. (A6) can be re-written

$$V = V_1 + V_2 + V_3, \tag{A13}$$

where the three terms are

$$V_1 = \frac{1}{R} \int d\mathbf{r}_D \int d\mathbf{r}_A \rho_D(\mathbf{r}_D)\rho_A(\mathbf{r}_A) \tag{A14}$$

$$V_2 = -\int d\mathbf{r}_D \int d\mathbf{r}_A \rho_D(\mathbf{r}_D)\rho_A(\mathbf{r}_A) \frac{\mathbf{r} \cdot \mathbf{R}}{R^3} \tag{A15}$$

$$V_3 = \int d\mathbf{r}_D \int d\mathbf{r}_A \rho_D(\mathbf{r}_D)\rho_A(\mathbf{r}_A) \frac{3(\mathbf{r} \cdot \mathbf{R})^2 - r^2 R^2}{2R^5}. \tag{A16}$$

The leading term V_1 can be calculated readily

$$V_1 = \frac{1}{R} \int d\mathbf{r}_D \rho_D(\mathbf{r}_D) \int d\mathbf{r}_A \rho_A(\mathbf{r}_A)$$

$$= \frac{1}{R} c_D c_A = 0, \tag{A17}$$

and is found not to contribute to the interaction energy due to $c_D = c_A = 0$. Likewise, V_2 also does not contribute to the interaction energy as is shown by the calculation (again note $c_D = c_A = 0$)

$$V_2 = -\int d\mathbf{r}_D \int d\mathbf{r}_A \rho_D(\mathbf{r}_D) \rho_A(\mathbf{r}_A) \frac{\mathbf{r} \cdot \mathbf{R}}{R^3}$$

$$= -\int d\mathbf{r}_D \int d\mathbf{r}_A \rho_D(\mathbf{r}_D) \rho_A(\mathbf{r}_A) \frac{(\mathbf{r}_D - \mathbf{r}_A) \cdot \mathbf{R}}{R^3}$$

$$= -\int d\mathbf{r}_A \rho_A(\mathbf{r}_A) \int d\mathbf{r}_D \rho_D(\mathbf{r}_D) \frac{\mathbf{r}_D \cdot \mathbf{R}}{R^3} + \int d\mathbf{r}_A \rho_D(\mathbf{r}_D) \int d\mathbf{r}_A \rho_A(\mathbf{r}_A) \frac{\mathbf{r}_A \cdot \mathbf{R}}{R^3}$$

$$= -c_A \frac{(\mathbf{d}_D \cdot \mathbf{R})}{R^3} + c_D \frac{(\mathbf{d}_A \cdot \mathbf{R})}{R^3} = 0, \tag{A18}$$

where the definition of the transition dipole moment in Eq. (A1) has been used.

Lastly, V_3 is calculated (we neglect again terms involving factors c_D and c_A)

$$V_3 = \frac{1}{2R^5} \int d\mathbf{r}_D \int d\mathbf{r}_A \rho_D(\mathbf{r}_D) \rho_A(\mathbf{r}_A) \{3[(\mathbf{r}_D - \mathbf{r}_A) \cdot \mathbf{R}]^2 - |\mathbf{r}_D - \mathbf{r}_A|^2 R^2\}$$

$$= \frac{1}{2R^5} \int d\mathbf{r}_D \int d\mathbf{r}_A \rho_D(\mathbf{r}_D) \rho_A(\mathbf{r}_A) \{3[(\mathbf{r}_D \cdot \mathbf{R})^2 + (\mathbf{r}_A \cdot \mathbf{R})^2 - 2(\mathbf{r}_D \cdot \mathbf{R})(\mathbf{r}_A \cdot \mathbf{R})]$$

$$- (|\mathbf{r}_D|^2 + |\mathbf{r}_A|^2 - 2\mathbf{r}_D \cdot \mathbf{r}_A) R^2\}$$

$$= \frac{1}{2R^5} \int d\mathbf{r}_D \int d\mathbf{r}_A \rho_D(\mathbf{r}_D) \rho_A(\mathbf{r}_A) \{3[-2(\mathbf{r}_D \cdot \mathbf{R})(\mathbf{r}_A \cdot \mathbf{R})] + (2\mathbf{r}_D \cdot \mathbf{r}_A) R^2\}$$

$$= \frac{(\mathbf{d}_D \cdot \mathbf{d}_A) R^2 - 3(\mathbf{d}_D \cdot \mathbf{R})(\mathbf{d}_A \cdot \mathbf{R})}{R^5}, \tag{A19}$$

and is found to be the leading non-zero contribution to the interaction energy V in Eq. (A12) and corresponds to the interaction energy in Eq. (13).

Inserting V_3 into Eq. (A5) yields the well-known formula for Förster resonant excitation transfer (FRET)

$$k_{D^*A \to DA^*} = \frac{2\pi}{\hbar} |\mathbf{d}_D|^2 |\mathbf{d}_A|^2 \frac{[(\hat{\mathbf{d}}_D \cdot \hat{\mathbf{d}}_A) - 3(\hat{\mathbf{d}}_D \cdot \hat{\mathbf{R}})(\hat{\mathbf{d}}_A \cdot \hat{\mathbf{R}})]^2}{R^6} \int_{-\infty}^{+\infty} dE\, S_D(E) S_A(E), \tag{A20}$$

where we have defined $\hat{\mathbf{R}} = \mathbf{R}/R$, $\hat{\mathbf{d}}_D = \mathbf{d}_D/|\mathbf{d}_D|$ and $\hat{\mathbf{d}}_A = \mathbf{d}_A/|\mathbf{d}_A|$. The term

$$\kappa = (\hat{\mathbf{d}}_D \cdot \hat{\mathbf{d}}_A) - 3(\hat{\mathbf{d}}_D \cdot \hat{\mathbf{R}})(\hat{\mathbf{d}}_A \cdot \hat{\mathbf{R}}), \tag{A21}$$

is often averaged over all possible orientations of $\hat{\mathbf{d}}_A$ and $\hat{\mathbf{d}}_A$ to give $\kappa^2 = 2/3$, which simplifies the Förster formula to

$$\kappa_{D^*A \to DA^*} = \frac{4\pi}{3\hbar} \frac{d_D^2 d_A^2}{R^6} \int_{-\infty}^{+\infty} dE\, S_D(E) S_A(E). \tag{A22}$$

Equations (A20) and (A21) correspond to Eq. (3) where the coupling constant C is an empirical value of $2\pi |\mathbf{d}_D|^2 |\mathbf{d}_A|^2/\hbar$. Equations (A20)–(A22) can be readily generalized to molecules, just by employing molecular transition dipole moments and spectra. In the case of the light-harvesting apparatus, all donor and acceptor molecules are spatially fixed such that the specific orientations $\hat{\mathbf{d}}_D$ and $\hat{\mathbf{d}}_A$ of the molecular transition dipole moments matter. Evolution most likely leads to BChl orientations, and by the same token $\hat{\mathbf{d}}_D$ and $\hat{\mathbf{d}}_A$ orientations that optimized excitation transfer rates. For example, in the cases of LH2 and LH1, transition dipole moments of adjacent BChls are always anti-parallel, as this tends to decrease the rates by which the excitons formed emit radiation as opposed to transfer their energy in the apparatus. It is interesting to note that large transition dipole moment values favor not only absorption and emission, but also excitation transfer.

DNA Polymerases

Structure, Function, and Modeling

3

Tamar Schlick

---

1. Introduction

1.1 *Venerable task*

The transmission of our instruction book for life — the DNA genome — from one generation to the next describes both similarities to our parents and differences that make us unique. How this duplication occurs and how this process leads to differences via mutations is a fundamental problem in both basic and applied research. As Watson and Crick observed in their landmark 1953 *Nature* work reporting the structure of the DNA, the double-helical DNA structure lends itself naturally to template-directed replication. Many DNA polymerases are responsible for this venerable task of linking nucleotides in a rapid process that uses each DNA strand as a template for another duplex DNA nearly identical to the original.

Deviations from identity, however, occur when errors are introduced in this copying process — nucleotide deletion, insertion, inversion, or translocation — as well as damage from external sources like radiation or hazard chemicals, including cigarette smoke and environmental pollutants. Such mutations are handled both by proofreading activities of the replicative DNA polymerase themselves as well as specialized cellular repair processes involving a variety of DNA polymerases with expertise in handling specific error types: base excision repair, mismatch repair, chemical adduct removal, double-strand break repair, and many others. Because any error that is not promptly repaired can lead to many human diseases such as skin abnormalities and various cancers, significant research effort has been expended into understanding polymerase replication and repair mechanisms at the atomic level.

Computer modeling plays an important role in these endeavors by helping link static molecular views from X-ray crystallography and NMR with macroscopic kinetic data that measure polymerase efficiency and fidelity. See Refs. 1 and 2 for exceptional works linking crystal forms via a free-energy pathway of dynamic configurational changes. A detailed atomic and molecular-level understanding of how DNA polymerases operate can help suggest possible therapeutic approaches to enhance polymerase fidelity, identify factors that degrade polymerase performance, and assist in the development of pharmacological approaches to diseases such as cancer, heart disease, diabetes, and a variety of neurological conditions that are associated with DNA damage and incorrect DNA repair.

1.2 *Replication and repair fidelity*

In replication, DNA polymerases orchestrate the addition of new nucleotides to a growing chain of DNA by catalyzing a nucleotidyl transfer reaction which increases the primer strand by one base pair. The relative ability of a polymerase to incorporate a *correct* nucleotide rather than an *incorrect* unit from a pool of structurally similar molecules is a measure of its *fidelity*. DNA polymerases catalyze the same nucleotidyl transfer reaction in which an incoming nucleotide unit termed dNTP (2′-deoxyribonucleoside 5′-triphosphate) is inserted at the 3′ end of a growing DNA primer strand (Figure 1). They must select the correct dNTP unit to pair with the templating base (e.g. dGTP opposite template C and other Watson-Crick base pair combinations) from a pool of structurally similar nucleotides (e.g. dATP, dCTP, and dTTP). Intriguingly, the accuracies exhibited by proofreading-deficient DNA polymerases from different families span a wide range of error rates (see Table 1), from 10^{-6} to near unity, namely one error per 10^{6} nucleotides incorporated (for high-fidelity pol α) to nearly one error per nucleotide for low-fidelity pol X. The fidelity of a polymerase is defined as the inverse of the misinsertion frequency (see Table 1 footnote).

For recent reviews, see Refs. 3 and 4.

1.3 *Polymerase structure and function*

DNA polymerases are shaped like a hand with fingers, palm and thumb subdomains[10] (Figure 2). (Spartan pol X lacks the fingers subdomain). The fingers and thumb subdomains are associated

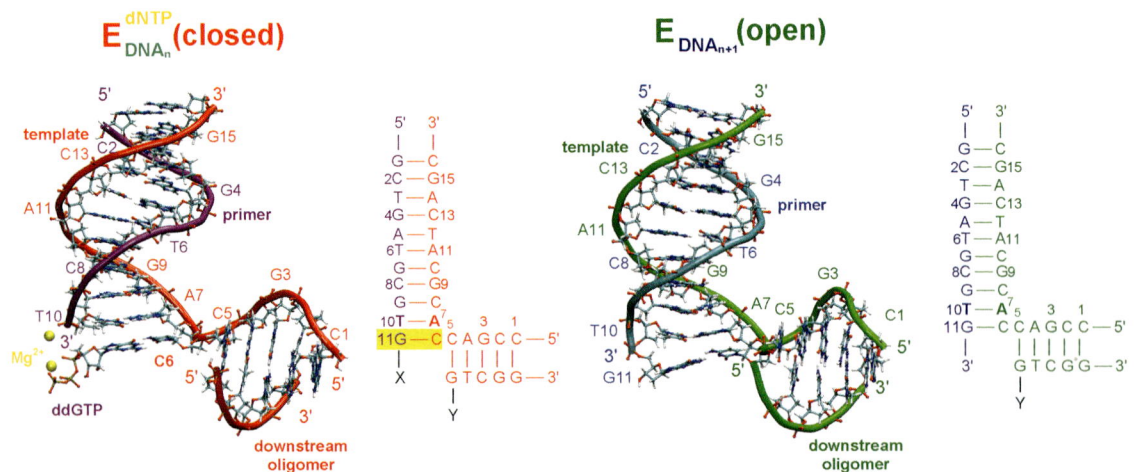

Figure 1. Extension of a growing DNA primer by one nucleotide. Left, correct incoming G opposite template C6. Right, incoming nucleotide covalently bound to primer. See Figure 3 for pathway context of the two DNA shown here within the polymerase complex.

Table 1. Kinetic data for DNA Polymerases.

Enzymes (family)	Fidelity[a]	$k_{pol}(s^{-1})$[b]
Klentaq1 (A)	10^4	20
HIV-1 RT (RT)	10^4	70
Pol β (X)	10^3	10
Pol λ (X)	10^3	4
Pol μ (X)	10^3	0.08
Pol X (X)	~1	0.8
Dpo4 (Y)	10^2	0.3

[a] Fidelity = reciprocal of misinsertion error frequency = $[(k_{pol}/K_d)_c + (k_{pol}/K_d)_i]$ / $(k_{pol}/K_d)_i$, where c and i denote correct and incorrect nucleotide incorporation, and K_d is the apparent equilibrium dissociation constant of dNTP.

[b] k_{pol}: rate of nucleotide incorporation for first-enzyme turnover.

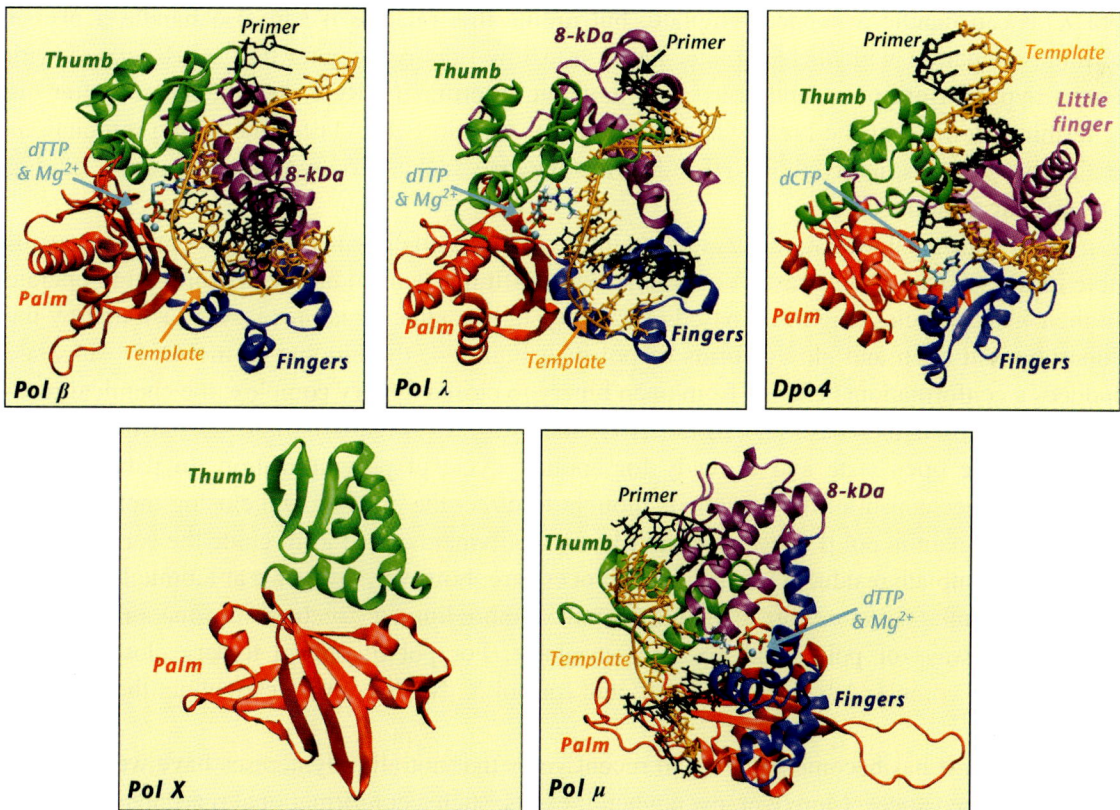

Figure 2. DNA polymerases structures: pol β,[5] pol λ,[6] Dpo4,[7] pol X,[8] and pol μ.[9]

with grasping the DNA and the incoming nucleotide unit, while the palm subdomain is the site of the chemical reaction which increases the DNA chain by one base pair.

Many families of DNA polymerases are recognized (A, B, C, D, X, Y, and reverse transcriptases), including eukaryotic DNA polymerases α, δ, ε in the B family, bacterial polymerase III in the C family, Y-family members η, ι, and κ, and X-family members β, λ, μ, X, and Tdt.

In this article, we focus on several pol X family members (pol β, λ, μ, and X) and one pol Y member (Dpo4) (Figure 2) which we have studied by computer modeling. Each of these family-X enzymes has intriguing characteristics. Pol β is a repair enzyme with moderate fidelity that functions primarily in base excision repair (BER); it is thought to have a tight active site guided by the *induced-fit* mechanism for base-pair recognition (see below). Pol X is a highly error prone polymerase that considers five base pairs intrinsically correct (the four Watson-Crick base pairs plus G:G). Pol λ is a low to moderate fidelity enzyme that plays a back-up role to pol β in BER and possibly has a role in non-homologous end joining (NHEJ), a pathway by which double-stranded DNA breaks are repaired; it has a unique propensity to generate frame-shift errors through single-base deletions rather than base substitutions. Intriguingly, pol λ also has an inherent ability to use substrates with minimal base-pairing homology and can easily tolerate extra-helical nucleotides. Pol μ is also mainly associated with the NHEJ repair process; like pol λ, it can handle a gapped template, but unlike this relative it can also handle a primer terminus unpaired with the template. Dpo4 is a low-fidelity enzyme that is associated with lesion bypass. Both pol λ and pol μ also have an N-terminal BRCT domain (named after the C-terminal domain of the breast cancer suppressor protein BRCA1) for mediating interactions with other proteins involved in NHEJ. All five enzymes lack a proofreading domain for editing DNA replication mistakes.

Many high-resolution X-ray crystallographic structures of enzyme/DNA complexes are available for these DNA polymerase. A common induced-fit mechanism has been proposed to operate in their kinetic cycles (Figure 3) based on key crystallographic structures.[11] As illustrated by the superimposed open and closed forms of pol β in Figure 4, the binding of the dNTP substrate induces a conformational change from open binary to closed ternary complex; the chemical reaction of adding another base pair to the DNA then follows (Figure 1), after-which the enzyme returns to the open state and is ready for another cycle of polymerization. In particular, an induced-fit mechanism of the enzyme is thought to direct the opening and closing motion of the *thumb* subdomain in pol β (*fingers* in right-handed polymerases) so as to guide the correct selection of the template residue. Details of this process are, however, unknown at atomic level.

While similar structures are available for pol λ,[6] more limited structural information about the reaction pathways of pol μ and pol X is available. For pol μ, only a ternary closed pol μ/DNA/ddTTP complex has been resolved,[9] and for pol X, NMR complexes without the substrate are available.[12,13]

However, it has become apparent in recent years that not all polymerases have well defined open and closed states. Moreover, even when it exists, such a conformational rearrangement may involve any combination of protein and DNA motion. The reasons for such preferences remain unclear, but motions are thought to be related to polymerase function and error profile. Alternative frameworks to interpret fidelity, such as conformational sampling in which a larger range of conformational can help select correct substrates, are also actively being explored.

Besides structural information as mentioned above, a wealth of biochemical and kinetic data for these enzymes is also available, including kinetic and error rate data from many sources, such as shown in Table 1.

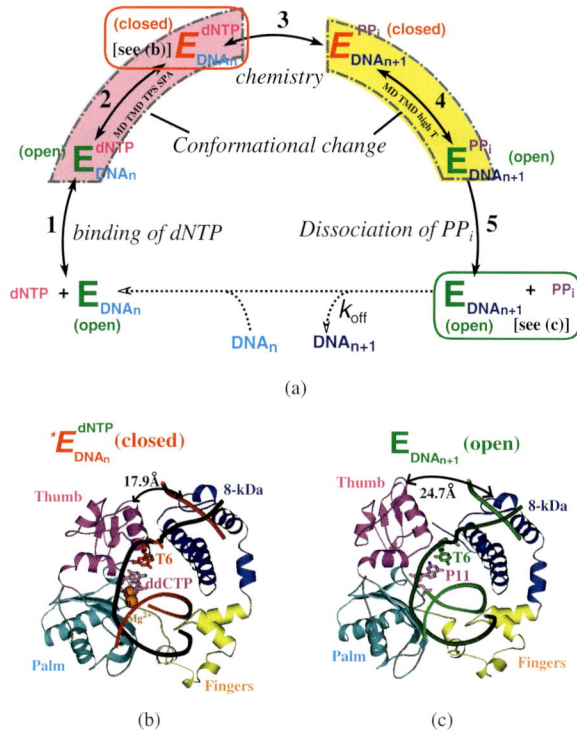

Figure 3. Proposed DNA polymerase pathway for nucleotide insertion (a), with anchoring crystallographic pol β conformations in (b) closed (red) and (c) open (green) states. E: DNA polymerase; dNTP: 2′-deoxyribonucleoside 5′-triphosphate; PP$_i$: pyrophosphate; DNA$_n$ and DNA$_{n+1}$: DNA before and after the nucleotide incorporation to the DNA primer strand. Adapted with permission from "L. Yang, K. Arora, W. A. Beard, S. H. Wilson and T. Schlick. 2004. Critical role of magnesium ions in DNA polymerase β's closing and active site assembly. *J Am Chem Soc* **126**, 8441–8453." Copyright (2004) American Chemical Society.

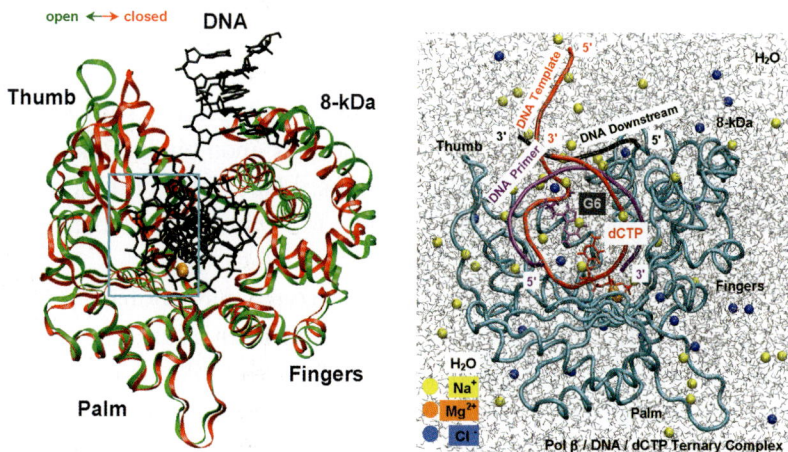

Figure 4. Pol β Structure and Modeling. Overview of the pol β ternary (closed, red) and binary (open, green) states as resolved by crystallography (left) and the solvated, modeled ternary complex (right). Adapted from Journal of Molecular Biology, 317/5, L. Yang, W. A. Beard, S. H. Wilson, S. Broyde and T. Schlick, Polymerase β simulations suggest that Arg258 rotation is a slow step rather than large subdomain motion per se., 651–671, Copyright (2002), with permission from Elsevier.

2. Modeling Studies

2.1 *Need for modeling*

Molecular modeling and simulation are critically needed to interpret and extend experimental findings to provide geometric, thermodynamic, and dynamic explanations that link observed structures to pathways and mechanisms in the goal of interpreting biological function. In particular, it remains a challenge to establish the sequences of motions involved in polymerase cycles and define the sequence of subdomain and residue rearrangements as well as define the rate-limiting step.

Modeling and simulation, while still subject to many approximations and inherent sampling limitations, have nonetheless demonstrated success in many important applications[14–16] from protein folding to RNA structure prediction to membrane dynamics; see Ref. 16 for a recent field perspective and specific examples. With increasing improvements in force fields and algorithms and the growing availability of high-speed computing platforms, modeling and simulation are well on their way to becoming full partner with experiment, if not a field on its own right.[16] Below, a sampling of results from our group's studies on DNA polymerases mechanisms is described.

2.2 *Conformational pathways in pol β, pol λ, pol X, and pol μ*

Dynamics simulations have lended support to the induced-fit mechanism for pol β[17]and pol X[8], in which the correct incoming base triggers the requisite conformational change while an incorrect incoming nucleotide hampers the process (Figure 5). Moreover, studies revealed a cascade of subtle side-chain conformational events following thumb closing but prior to chemistry (Figure 6), including a slow and possibly "gate-keeping" rearrangement of Arg258[1,18–21] that likely directs the catalytic cycle of pol β. This directed sequence enhances the polymerase's selectivity for the correct incoming partner to the template base by allowing an incorrect incoming nucleoside residue to dissociate.[1,22] Significantly, the slow step involving Arg258 has been verified experimentally.[23]

For pol X, which lacks the fingers domain of pol β, we have also shown that the two available NMR structures interconvert to one another at physiological salt.[8] For pol λ, simulations suggest that, besides motions in the thumb subdomain and active-site residues (Ile492, Tyr505, Phe506, Arg517), the DNA undergoes significant correlated motions upon binding the correct nucleotide, when the ions bind too to the primer/template terminus.[24] Dpo4 also utilizes more subtle movements of its little finger and fingers subdomains to bypass small oxidative lesions like 8-oxoguanine.[25]

Recent modeling work on pol μ suggests that the enzyme may not have a well defined open state.[26] The enzyme tends to remain in the closed state even when a substrate is removed from the active site or when the complex is forced to an open-like state. Like its relatives, pol μ exhibits subtle residue rearrangements, with His329, Asp330, Gln440, and Glu443 playing key roles in nucleotide accommodation. The DNA/protein contact environment also appears fragile in this enzyme.

For pol λ, the motions identified above of the primer DNA may be relevant to the inherent tendency for base pair deletions because subtle protein/DNA interactions regulate pol λ stability and hence its error propensity (see below). Thus, though pol λ does not demonstrate large-scale

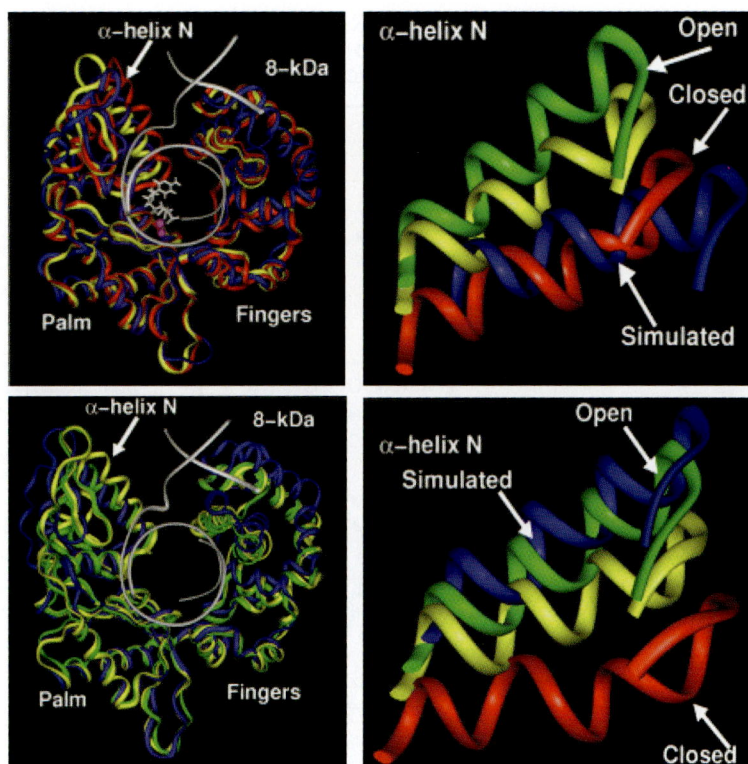

Figure 5. *In silico* evidence for pol β's induced fit mechanism C_α traces of superimposed pol β/DNA complex with dCTP (top left) and without dCTP (bottom left) for the intermediate starting structure (yellow), crystal closed (red), and crystal open (green) and the trajectory final structures (blue).[17] Notable are the residue motions in the thumb subdmain and the 8-kDa domain. The positions of α-helix N in the simulated systems are comparied to the crystal structures in panels on the right (top, with dCTP, and bottom, without dCTP). Adapted from Biophysical Journal, 87/5, K. Arora and T. Schlick, *In silico* evidence for DNA polymerase β's substrate-induced conformational change, 3088–3099, Copyright (2004), with permission from Elsevier.

subdomain movements as pol β, the above correlated motions may also serve as gate-keepers by controlling the evolution of the reaction pathway. For pol μ, the lack of well defined transtions upon binding the substrate may be related to its ability to handle unpaired primer termini. The significant DNA motions in polymerase cycles for both pol λ[24] and Dpo4[25] may suggest a common feature for moderate-to-low fidelity polymerases. See Table 2 for a summary.

2.3 *Pol β's closing pathway*

Our extensive studies on mammalian pol β based on pioneering crystal complexes solved in various forms by the Wilson group[11] have revealed key aspects of the conformational (Figures 6–7) and chemical pathways (Figure 8) of the enzyme and suggested several factors responsible for fidelity discrimination.[27–29] These biological pathways for both correct and incorrect base-pair systems were resolved by developing the enhanced sampling approach *transition path sampling*[30] for biomolecules[1] and an efficient free-energy approach termed *BOLAS*.[31]

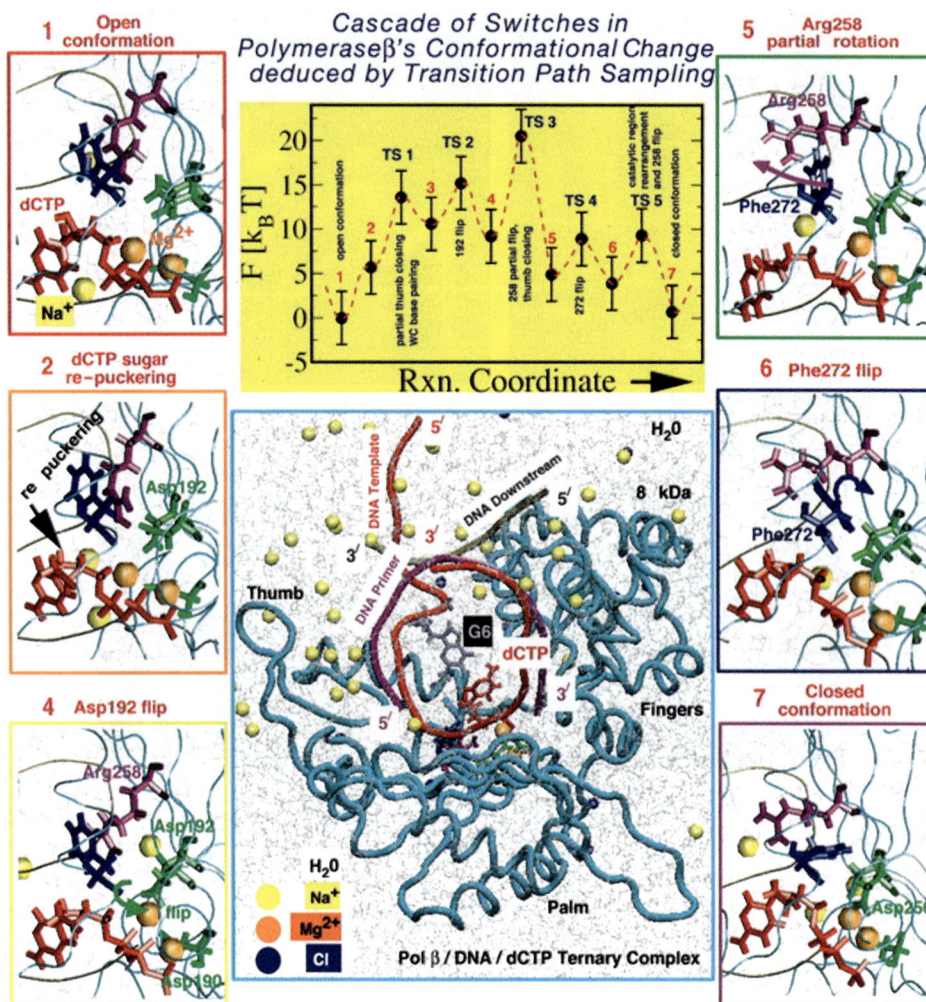

Figure 6. Pol β's closing pathway.[1] Overall captured reaction kinetics profile (from TPS) for the conformational transition of pol β (for G:C) from open (state 1) to closed (state 7) forms showing free energies (in $k_B T$) associated with the different transition state regions.[1] The meta-stable basins (in red) along the reaction coordinate are numbered 1–7. Reprinted from R. Radhakrishnan and T. Schlick. 2004. Orchestration of cooperative events in DNA synthesis and repair mechanism unraveled by transition path sampling of DNA polymerase β's closing, **101**, 5970–5975, Copyright (2004) National Academy of Sciences, U.S.A.

Specifically, transition path sampling simulations for pol β for the correct G:C system[1,22] delineated the specific transition states and energies involved in a complex geometric/energetic landscape (Figure 6). The critical roles of key residues Arg258, Phe272, Asp192, and Tyr271 emerged: a delicate system of checks and balances directs the system to the chemical reaction and likely facilitates enzyme discrimination of the correct from the incorrect unit. Analogous simulations on the mispair G:A (Figure 7) suggest that while free-energy barriers for the matched and mismatched systems are comparable to one another for the conformational pathway, the closed state of the mismatch is much less stable than its open state, unlike the G:C complex, in which the closed and open states have equal stabilities. Thus, different sequences of transition states in the

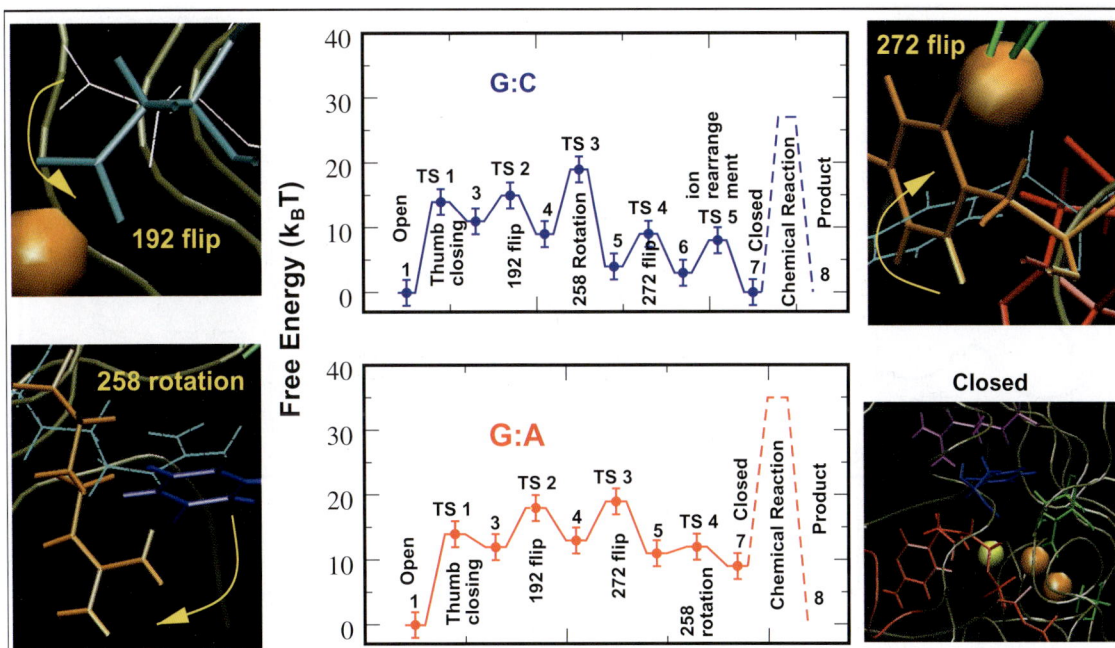

Figure 7. Pol β kinetic profile. Overall captured reaction kinetics profile for pol β's closing transition followed by chemical incorporation of dNTP for G:C and G:A systems.[22] The barriers to chemistry (dashed peaks) are derived from experimentally measured k_{pol} values. The profiles were constructed by employing reaction coordinate characterizing order parameters in conjunction with transition path sampling. The potential of mean force along each reaction coordinate is computed for each conformational event. Adapted with permission from "R. Radhakrishnan, K. Arora, Y. Wang, W. A. Beard, S. H. Wilson and T. Schlick. 2006. Regulation of DNA repair fidelity by molecular checkpoints: "Gates" in DNA polymerase β's substrate selection. *Biochem* **45**, 15142–15156." Copyright (2006) American Chemical Society.

correct versus incorrect basepair complexes dictate different conformational pathways toward an ideal two-metal ion catalysis geometry.

2.4 *Pol β's chemical mechanism for G:C vs. G:A systems*

A novel hybrid molecular/quantum mechanics approach (MM/QM) combining quasi-harmonic free energies with MM/QM dynamics[27] helped delineate a proton-hop mechanism for the chemical reaction of pol β.[27–29] Specifically, a series of transient intermediates is linked via a Grotthuss hopping mechanism of proton transfer between water molecules and the three conserved aspartate residues in the enzyme's active site (Figure 8). In the G:C system, the rate-limiting step is the initial proton hop with a free-energy of activation of at least 17 kcal/mol, which corresponds reasonably to measured k_{pol} values. Fidelity discrimination in pol β can be explained by a significant loss of stability of the closed ternary complex of the enzyme in the G:A system and a much higher activation energy of the initial step of nucleophilic attack, namely de-protonation of the terminal DNA primer O3′H group. Thus, subtle differences in the enzyme active site between matched and mismatched base pairs generate significant differences in catalytic performances.

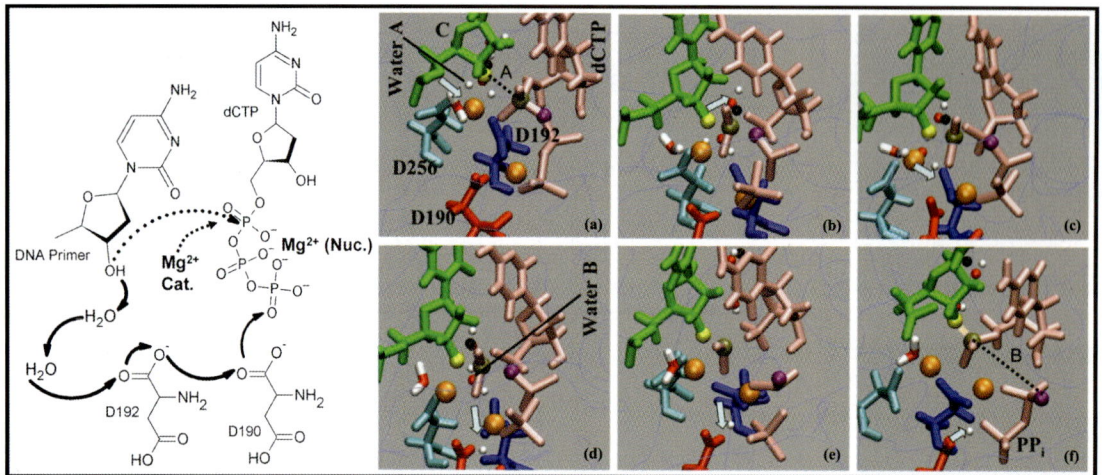

Figure 8. Pol β's chemical synthesis reaction. Left: Schematic drawing of the mechanism of concerted proton hops during phosphoryl transfer in pol β. Solid arrows indicate the migration path of the proton and the dotted arrows represent the nucleophilic attack.[27] Right: Captured reaction intermediates for pol β's phosphoryl transfer in the G:C system:[27] (a) reaction state of the closed nucleotide-bound enzyme state; (b) deprotonation of the O3'H to water; (c,d) proton transfers to Asp192; (e) proton transfers to Asp190; (f) proton reaches the pyrophosphate unit to obtain the final product. The colors represent: cyan (D256), red (D190), blue (D192), pink (dCTP), green (CYT: terminal DNA primer), black (the O3'H-proton), yellow (the O3' oxygen, attacking nucleophile), tan (central phosphorus), purple (leaving O3A oxygen), and orange (Magnesium). The oxygens and hydrogens of water molecules are in red and white, respectively. The arrows denote the location and direction of proton hop. Adapted from Biochemical and Biophysical Research Communications, 350/3, R. Radhakrishnan and T. Schlick, Correct and incorrect nucleotide incorporation pathways in DNA polymerase β's, 521–529, Copyright (2006), with permission from Elsevier.

2.5 *Pol λ's pathway and slippage tendency*

For pol λ, conformational transitions occur via flips of side-chain residues (Ile492, Tyr505, Phe506) and other motions associated with Arg514 and Arg517 in the active site en route to the binary (inactive), though no large-scale subdomain motion was noted.[24] However, significant DNA motion was observed, and this was proposed to be related to the tendency of this polymerase to generate deletion errors.

Dynamics simulations also identified residue Arg517 as crucial to protein/DNA stabilization through mutant simulations (Figure 9). The discrete orientations of the 517 residue can impact protein-coupled DNA stability by forming unfavorable electrostatic interactions, which lower the stability of the ternary complex and move the system toward the binary conformation. Residue 517's critical interactions with the DNA also help interpret pol λ's slippage tendency. Fragile protein/DNA interactions in pol λ might lead to deletion mutations because the DNA "slips". Through side-by-side experiments and additional simulations of other Arg517 mutants in pol λ, we validated the results from simulation and developed the hypothesis that Arg517 plays an important role in modulating deletion error generation.[32,33]

Modeling of pol λ bound to incorrect incoming nucleotides[34] revealed a wide range of DNA motion and protein residue side-chain motions, as well as distinct differences compared to the reference (correct base pair) system (see Figure 10). This led us to suggest key base-checking roles in pol λ of active site residues Arg517, Tyr505 and Phe506. On the basis of

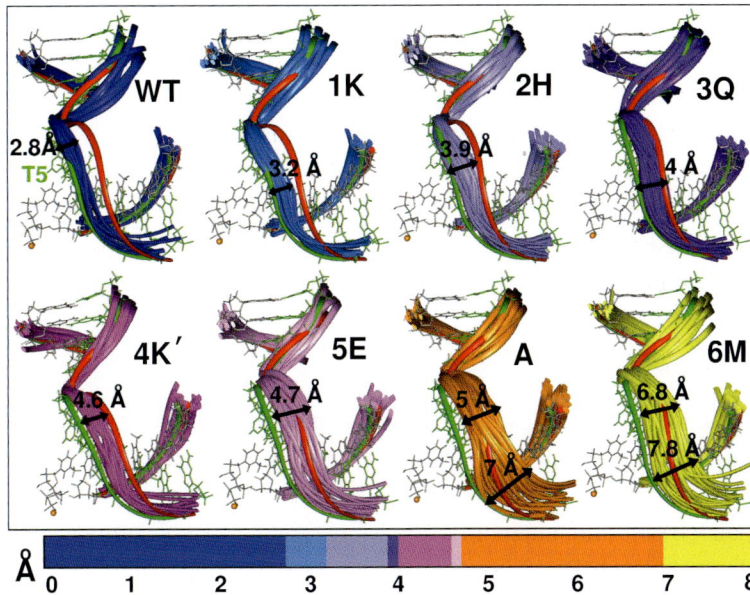

Figure 9. Residue 517 as key regulator of DNA/protein interactions in pol λ. Ranges of DNA motion occurring in each 517 mutant simulation as well as in the wild-type (WT) simulation with the nucleotide-binding ion.[32] Positions of the DNA are shown with reference to the crystal binary (red, PDB entry 1XSL) and ternary (green, PDB entry 1XSN) DNA positions. DNA bases of the template and primer strands are green and grey, respectively. A: Arg517Ala pol λ mutant. Reprinted with permission from "M. C. Foley and T. Schlick. 2008. Simulations of DNA pol λ R517 mutants indicate 517's crucial role in ternary complex stability and suggest DNA slippage origin. *J Amer Chem Soc* **130**, 3967–3977." Copyright (2008) American Chemical Society.

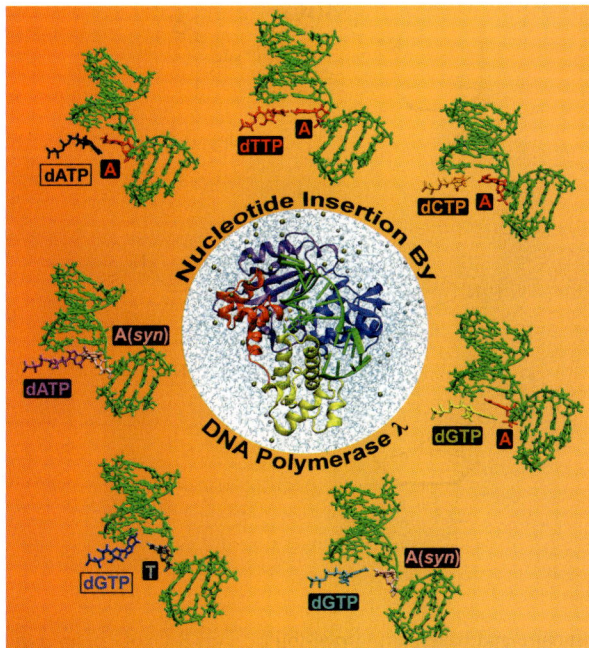

Figure 10. Pol λ's handling of different mismatched base pairs revealed from dynamics simulations.[34]

trends in the electrostatic interactions, we proposed an order of increasing mismatch insertion by pol λ: A:C > A:G > A (*syn*):G > T:G > A (*syn*):A > A:A. This sequence agrees with available kinetic data for incorrect nucleotide insertion opposite template adenine, with the exception of T:G, which may be more sensitive to the insertion context.

Recent modeling of aligned versus misaligned DNA[35] has also suggested that pol λ stabilizes the latter more tightly than aligned DNA, with active-site electrostatics playing a major role in this stabilization. This astonishing fact may be related to this enzyme's ability to handle substrates with minimal base pairing at the cost of generating a high rate of base deletions.

2.6 *Dpo4's handling of 8-oxoG*

Since Dpo4 has a prominent role in lesion bypass, it is natural to study Dpo4 in the context of complexes to small lesions like 8-oxoguanine (8-oxoG) (Figure 11), one of the most common lesions due to oxidative damage such as cigarette smoke and air polutants. These lesions are handled by base excision repair (BER) and nucleotide excision repair (NER) pathways, but the structural details are not well understood. The 8-oxoG lesion has been studied extensively in a variety of contexts both experimentally and by molecular modeling. These lesions can introduce errors because adenines opposite 8-oxoG can lead to mismatched *syn* 8-oxoG:A mispairs (Figure 11) which in turn can lead to C→T transversions.

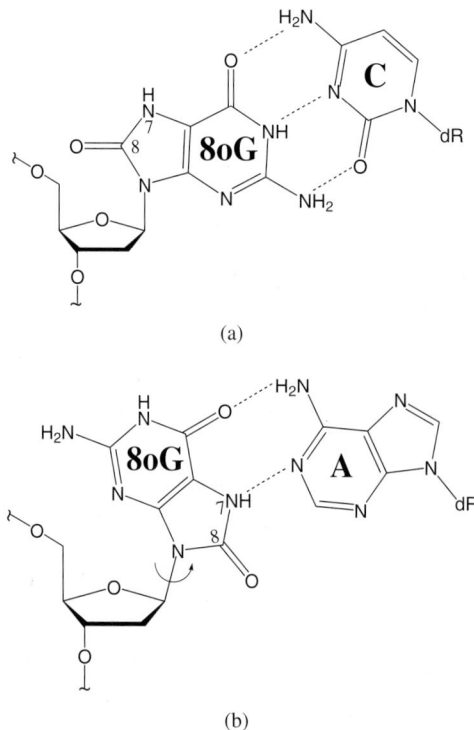

(a)

(b)

Figure 11. 8-oxoG (8oG) structure and base pairing possibilities. (a) 8-OxoG in an *anti* orientation forms a Watson-Crick base pair with dCTP, (b) 8-oxoG in a *syn* orientation forms a Hoogsteen base pair with dATP.

Table 2. Inactive to active complex transitioning.

Enzymes	Motions Involved
Pol β	large-scale thumb motion, subtle protein side-chain motion
Pol λ	large DNA shift, small thumb loop motion, subtle protein side-chain motion
Pol X	large-scale thumb motion, subtle protein side-chain motion
Pol μ	subtle protein side-chain motion
Dpo4	DNA sliding, little finger rotation, subtle protein side-chain motion

Experiments have shown that Dpo4 has a 70-fold preference to insert dCTP rather than dATP opposite 8-oxoG.[36] Since the ternary structure is highly stable before chemistry, both with and without the incoming nucleotide at the active site, we hypothesized that the "induced-fit" mechanism does not operate in Dpo4. A rapid ~12 degree little-finger subdomain rotation occurring in our simulation after chemistry with ions in the active site suggested a low energy barrier for the transition between ternary and binary conformations. In addition, our studies of Dpo4 pathways for the correct insertion of dCTP opposite 8-oxoG using a hybrid quantum-classical mechanics approach revealed that the most favorable reaction path involves initial deprotonation of O3′H via two bridging water molecules to dCTP, overcoming an overall energy barrier of approximately 20 kcal/mol. The proton then migrates to the γ-phosphate oxygen of dCTP as the nucleotide is joined to the primer terminus and PP_i is formed. In contrast, initiating the chemical reaction from the less ideal state of the crystal structure requires a much higher activation energy barrier (29 kcal/mol) due to longer distances for O3′H proton transfer and distorted conformations of the proton acceptors. Compared to the higher fidelity pol β, Dpo4 has a higher chemical reaction barrier, which may result from its more solvent-exposed active site.

2.7 *Pol X mispair incorporation*

MD simulations of pol X bound to different mismatched nascent base pairs (i.e. C:C, A:G, G:G (*anti*), and G:G (*syn*)) were designed to explore a range of incorporation difficulty by pol X: G:G is very frequently inserted by pol X, A:G is moderately inserted, and C:C is relatively infrequently inserted compared to pol X's incorporation of the correct G:C base pair. Intriguingly, our simulations provide an explanation for this G:G preference: in the G:G (*syn*) mismatch simulation, the thumb exhibits a large-scale conformational change from an open to closed state that is similar to what occurs with the correct G:C base pair; in contrast, the A:G and G:G (*anti*) systems only display partial thumb closing and the C:C system maintains the thumb open (Figure 12). Thus, only the G:G (*syn*) base pair fits well within the helix via a Hoogsteen base pair arrangement that is geometrically similar to standard Watson-Crick base pairs. The more open and pliant active site of pol X, compared to pol β, allows pol X to accommodate bulkier mismatches such as G:G, while the more structured and organized pol β active site imposes higher discrimination that results in higher fidelity.

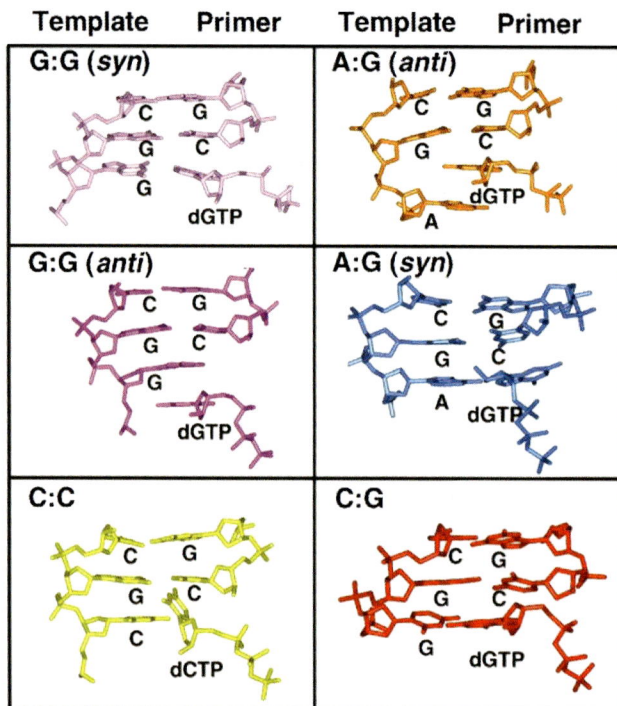

Figure 12. Geometry of template–primer DNA base pairs bound to pol X.[37] Reprinted from Journal of Molecular Biology, 384/5, B. A. S. Benítez, K. Arora, L. Balistreri and T. Schlick, Mismatched base pair simulations for ASFV Pol X/DNA complexes help interpret frequent G:G misincorporation, 1086–1097, Copyright (2008), with permission from Elsevier.

3. Conclusion

Our genomic imprint is both venerable and vulnerable. The rapid duplication of our DNA requires a team of DNA polymerase molecules working in concert, each copying a sliver of a chromosome, with a rescue team of proofreaders and repair agents. The fascinating atomic and molecular aspects of these replication and repair processes are avidly being pursued by both experimentalists and modelers alike. Modeling in this field is well anchored in many available crystal endpoints and kinetic data, and can help shed insights into the operation of these fascinating workhorses. With rapid improvements in modeling and simulation,[16] theoretical techniques are becoming more reliable and hence attractive, but they require expertise in simulation methodology and data analysis to sort through a voluminous amount of information.

In this chapter, a taste of some studies in our group focusing on pol β, γ, μ, X, and Dpo4 have been presented. The modeling tools involve standard dynamics simulations, enhanced sampling protocols, hybrid classical/quantum methods, free energy methods, principal component analysis, and others.

The various simulations have shed insights into the conformational pathways involved in these enzymes — the transition from binary polymerase/DNA to ternary polymerase/DNA/dNTP complexes — and highlighted subtle protein-residue transitions, DNA motions, and protein subdomain motions that are polymerase specific. Studies of these enzymes with mismatches help describe how well or how poorly the active site tolerates mismatched base pair systems, a fact that can be correlated to both the conformational transitions and the fidelity profiles of these enzymes. Indeed, pol λ's architecture is vulnerable to deletion errors, Dpo4's flexible active site handles lesions, pol X's open active site tolerates mispairs easily, and pol μ's lack of specific active-site interactions may allow it to perform repair with unpaired templates.

Mechanistic studies of the chemical reactions helped suggest atomic-level conformational steps in the process as well as associated free energy barriers. Together, such results can be combined to describe the kinetic cycle of polymerases as a monopoly board with conformational, pre-chemistry, and chemistry avenues, as shown in Figure 13. We suggest that while the conformational avenue is dictated by a defined sequence of conformational changes, the "pre-chemistry" avenue[38] involves stochastic sampling of active-site rearrangements to reach the chemical-reaction competent state and can help interpret different fidelity profiles for different polymerases. The experimental confirmation of some modeling predictions, such as the importance of Arg517 for pol λ,[33] the gate-keeping role of Arg258 in pol β,[23] or the G:G *syn* orientation preferred for pol X,[39] lend confidence in the modeling. Undoubtedly, ongoing studies via simulation and experiment will help dissect these separate avenues and the roles of key molecular components in polymerase mechanisms, including alternatives to the induced-fit mechanism. All such findings have important ramifications to understanding DNA synthesis and repair fidelity and, ultimately, the design of drugs to treat human diseases arising from polymerase errors.

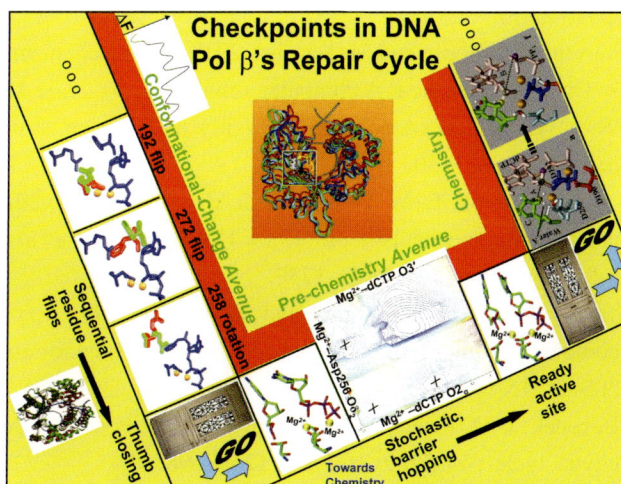

Figure 13. Pol β kinetic pathway. Sequential events and corresponding "gates" controlling pol β's fidelity:[38] the conformational change avenue, which comprises of Arg192 flip, Phe272 flip, and Arg258 rotation accompanying thumb subdomain closing motions upon incoming nucleotide binding; the pre-chemistry avenue, which involves the stochastic reorganization of the protein catalytic region, particularly the coordinating ligands of the two binding metal ions; and the chemistry avenue, where the incoming nucleotide is finally connected onto the primer terminus and the primer is extended by one residues.

Acknowledgment

Support from the National Science Foundation, National Institutes of Health, and Philip Morris USA Inc. and Philip Morris International are gratefully acknowledged. The author thanks Shereef Elmetwaly and Rubisco Li for assistance with the figures.

Suggested Additional Reading Materials

A. A. Golosov, J. J. Warren, L. S. Beese and M. Karplus. 2010. The mechanism of the translocation step in DNA replication by DNA polymerase I: A computer simulation. *Structure* **18**, 83–93.

J. Yamtich and J. B. Sweasy. 2010. DNA polymerase family X: Function, structure, and cellular roles. *Biochim Biophys Acta* **1804**, 1136–1150.

C. E. McKenna, B. A. Kashemirov, L. W. Peterson and M. F. Goodman. 2010. Modifications to the dNTP triphosphate moiety: From mechanistic probes for DNA polymerases to antiviral and anti-cancer drug design. *Biochim Biophys Acta* **1804**, 1223–1230.

R. Radhakrishnan and T. Schlick. 2004. Orchestration of cooperative events in DNA synthesis and repair mechanism unravelled by transition path sampling of DNA polymerase β's closing. *Proc Natl Acad Sci U S A* **101**, 5970–5975.

References

1. R. Radhakrishnan and T. Schlick. 2004. Orchestration of cooperative events in DNA synthesis and repair mechanism unraveled by transition path sampling of DNA polymerase β's closing. *Proc Natl Acad Sci U S A* **101**, 5970–5975.
2. A. A. Golosov, J. J. Warren, L. S. Beese and M. Karplus. 2010. The mechanism of the translocation step in DNA replication by DNA polymerase I: A computer simulation. *Structure* **18**, 83–93.
3. J. Yamtich and J. B. Sweasy. 2010. DNA polymerase family X: Function, structure, and cellular roles. *Biochim Biophys Acta* **1804**, 1136–1150.
4. C. E. McKenna, B. A. Kashemirov, L. W. Peterson and M. F. Goodman. 2010. Modifications to the dNTP triphosphate moiety: From mechanistic probes for DNA polymerases to antiviral and anti-cancer drug design. *Biochim Biophys Acta* **1804**, 1223–1230.
5. M. R. Sawaya, R. Prasad, S. H. Wilson, J. Kraut and H. Pelletier. 1997. Crystal structures of human DNA polymerase β complexed with gapped and nicked DNA: Evidence for an induced fit mechanism. *Biochemistry* **36**, 11205–11215.
6. M. Garcia-Diaz, K. Bebenek, J. M. Krahn, T. A. Kunkel and L. C. Pedersen. 2005. A closed conformation for the pol λ catalytic cycle. *Nat Struct Mol Biol* **12**, 97–98.
7. H. Ling, F. Boudsocq, R. Woodgate and W. Yang. 2001. Crystal structure of a Y-family DNA polymerase in action: A mechanism for error-prone and lesion-bypass replication. *Cell* **107**, 91–102.
8. B. A. S. Benítez, K. Arora and T. Schlick. 2006. Induced-fit mechanism for the interaction of the African swine fever virus DNA polymerase X with its target DNA. *Biophys J* **90**, 42–56.
9. A. F. Moon, M. Garcia-Diaz, K. Bebenek, B. J. Davis, X. Zhong, D. A. Ramsden, T. A. Kunkel and L. C. Pedersen. 2007. Structural insight into the substrate specificity of DNA polymerase μ. *Nat Struct Mol Biol* **14**, 45–53.

10. T. A. Steitz. 1999. DNA polymerases: Structural diversity and common mechanisms. *J Biol Chem* **274**, 17395–17398.

11. W. A. Beard and S. H. Wilson. 2003. Structural insights into the origins of DNA polymerase fidelity. *Structure* **11**, 489–496.

12. M. W. Maciejewski, R. Shin, B. Pan, A. Marintchev, A. Denninger, M. A. Mullen, K. Chen, M. R. Gryk and G. P. Mullen. 2001. Solution structure of a viral DNA repair polymerase. *Nat Struct Biol* **8**, 936–941.

13. A. K. Showalter, I. J. Byeon, M. I. Su and M. D. Tsai. 2001. Solution structure of a viral DNA polymerase x and evidence for a mutagenic function. *Nat Struct Biol* **8**, 942–946.

14. T. Schlick. 2010. *Molecular Modeling: An Interdisciplinary Guide.* Springer-Verlag, New York, Second edition.

15. E. H. Lee, J. Hsin, M. Sotomayor, G. Comellas and K. Schulten. 2009. Discovery through the computational microscope. *Structure* **17**.

16. T. Schlick, R. Collepardo-Guevara, L. A. Halvorsen, S. Jung and X. Xiao. 2011. Biomolecular modeling and simulation: A field coming of age. *Quart Rev Biophys* **44**, 191–228.

17. K. Arora and T. Schlick. 2004. *In silico* evidence for DNA polymerase β's substrate-induced conformational change. *Biophys J* **87**, 3088–3099.

18. L. Yang, W. A. Beard, S. H. Wilson, S. Broyde and T. Schlick. 2002. Polymerase β simulations suggest that Arg258 rotation is a slow step rather than large subdomain motion *per se*. *J Mol Biol* **317**, 651–671.

19. L. Yang, W. A. Beard, S. H. Wilson, B. Roux, S. Broyde and T. Schlick. 2002. Local deformations revealed by dynamics simulations of DNA polymerase β with DNA mismatches at the primer terminus. *J Mol Biol* **321**, 459–478.

20. L. Yang, W. A. Beard, S. H. Wilson, S. Broyde and T. Schlick. 2004. The highly organized but pliant active-site of DNA polymerase β: Compensatory interaction mechanisms in mutant enzymes by dynamics simulations and energy analyses. *Biophys J* **86**, 3392–3408.

21. L. Yang, K. Arora, W. A. Beard, S. H. Wilson and T. Schlick. 2004. The critical role of magnesium ions in DNA polymerase β's closing and active site assembly. *J Am Chem Soc* **126**, 8441–8453.

22. R. Radhakrishnan and T. Schlick. 2005. Fidelity discrimination in DNA polymerase β: differing closing profiles for a mismatched G:A versus matched G:C base pair. *J Amer Chem Soc* **127**, 13245–13252.

23. M. Bakhtina, M. P. Roettger, S. Kumar and M.-D. Tsai. 2007. A unified kinetic mechanism applicable to multiple DNA polymerases. *Biochem* **46**, 5463–5472.

24. M. C. Foley, K. Arora and T. Schlick. 2006. Sequential side-chain residue motions transform the bionary into the ternery state of DNA polymerase A. *Biophys J* **91**, 3182–3195.

25. Y. Wang, K. Arora and T. Schlick. 2006. Subtle but variable conformational rearrangements in the replication cycle of *Sulfolobus solfataricus* P2 DNA polymerase IV may accommodate lesion bypass. *Protein Sci* **15**, 135–151.

26. Y. Li and T. Schlick. 2010. Modeling DNA polymerase μ motions: Subtle transitions before chemistry. *Biophys J* **99**, 3463–3472.

27. R. Radhakrishnan and T. Schlick. 2006. Correct and incorrect nucleotide incorporation pathways in dna polymerase β's. *Biochem Biophys Res Comm* **350**, 521–529.

28. I. L. Alberts, Y. Wang and T. Schlick. 2007. DNA polymerase β catalysis: Are different mechanisms possible? *J Amer Chem Soc* **129**, 11100–11110.

29. M. D. Bojin and T. Schlick. 2007. A quantum mechanical investigation of possible mechanisms for the nucleotidyl transfer reaction catalyzed by DNA polymerase β. *J Phys Chem B* **111**, 11244–11252.

30. C. Dellago and P. G. Bolhuis. 2007. Transition path sampling simulations of biological systems. *Top Curr Chem* **268**, 291–317.

31. R. Radhakrishnan and T. Schlick. 2004. Biomolecular free energy profiles by a shooting/umbrella sampling protocol, "BOLAS". *J Chem Phys* **121**, 2436–2444.

32. M. C. Foley and T. Schlick. 2008. Simulations of dna pol λ R517 mutants indicate 517's crucial role in ternary complex stability and suggest DNA slippage origin. *J Amer Chem Soc* **130**, 3967–3977.

33. K. Bebenek, M. Garcia-Diaz, M. C. Foley, L. C. Pedersen, T. Schlick and T. A. Kunkel. 2008. Substrate-induced DNA strand misalignment during catalytic cycling by DNA polymerase λ. *EMBO Reports* **9**, 459–464.

34. M. Foley and T. Schlick. 2009. The relationship between conformational changes in pol λ's active site upon binding incorrect nucleotides and mismatch incorporation rates. *J Phys Chem B* **113**, 13035–13047.

35. M. C. Foley, V. Padow and T. Schlick. 2010. The extraordinary ability of DNA pol λ to stabilize misaligned DNA. *J Amer Chem Soc* **132**, 13403–13416.

36. O. Rechkoblit, L. Malinina, Y. Cheng, V. Kuryavyi, S. Broyde, N. E. Geacintov and D. J. Patel. 2006. Stepwise translocation of Dpo4 polymerase during error-free bypass of an oxoG lesion. *PLoS Biol* **4**, e11.

37. B. A. S. Benítez, K. Arora, L. Balistreri and T. Schlick. 2008. Mismatched base pair simulations for ASFV Pol X/DNA complexes help interpret frequent G:G misincorporation. *J Mol Biol* **384**, 1086–1097.

38. R. Radhakrishnan, K. Arora, Y. Wang, W. A. Beard, S. H. Wilson and T. Schlick. 2006. Regulation of DNA repair fidelity by molecular checkpoints: "Gates" in DNA polymerase β's substrate selection. *Biochem* **45**, 15142–15156.

39. S. Kumar. 2008. *Biochemical, Mechanistic, and Structural Characterization of DNA Polymerase X from African Swine Fever Virus.* Ph.D. thesis, Department of Chemistry, Ohio State University. http://etd.ohiolink.edu/view.cgi?acc_num=osu1211380265.

Information Processing by Nanomachines

4

Decoding by the Ribosome

Karissa Y. Sanbonmatsu, Scott C. Blanchard and Paul C. Whitford

1. Introduction

The ribosome is the most highly conserved molecular machine in the cell and has the responsibility of implementing the genetic code by converting the information residing in nucleic acid into protein. The ribosome translates the four-letter alphabet of nucleic acid into the twenty-letter alphabet of proteins. Because the ribosome is the only biomolecular complex that performs a non-trivial information-processing operation, it is analogous to a molecular CPU. Many of the ribosome's active sites are identical in every organism. The ribosome is so highly conserved that it is often used to determine phylogeny. Transcription of ribosomal RNA (rRNA) accounts for approximately 60% of the total transcription performed by the cell. In addition, a large portion of cellular metabolism is devoted to the construction of ribosomes.[1]

A large number of ribosome structures have been solved over the past decade.[2-7] The ribosome consists of a small subunit and a large subunit. In bacteria, the small subunit (30S), contains one long RNA (the 16S ribosomal RNA, or rRNA) and approximately 20 proteins (labeled S1, S2 etc.). The large subunit contains one long RNA (23S), one short RNA (5S) and over 30 proteins (labeled L1, L2, . . .). The overall structure of both ribosomal subunits consists of crowded RNA helices, connected end-to-end by RNA junctions (Figure 1). Ribosomal proteins cement the RNA helices together, enhancing stability. Ribosomal proteins typically contain a globular region and a long tail. While the globular regions are often exposed to solvent, the tails extend inside the ribosome to functionally relevant regions.

The small subunit consists of the head, platform, body, spur and decoding center (Figure 2). The mRNA wraps around the neck of the ribosome, located between the head and the platform. The decoding center consists of a handful of universally conserved nucleotides that decodes the information on the mRNA, acting as an effective molecular reading head.

The large subunit contains the central protuberance, the L1 stalk, the L7/L12 stalk, the peptidyl transferase center, the Sarcin-Ricin loop, the GTPase activity center, and the exit tunnel (Figure 3). The peptidyl transferase center is the site of peptide synthesis, producing a nascent polypeptide that moves out of the ribosome into solution through the exit tunnel. The Sarcin-Ricin

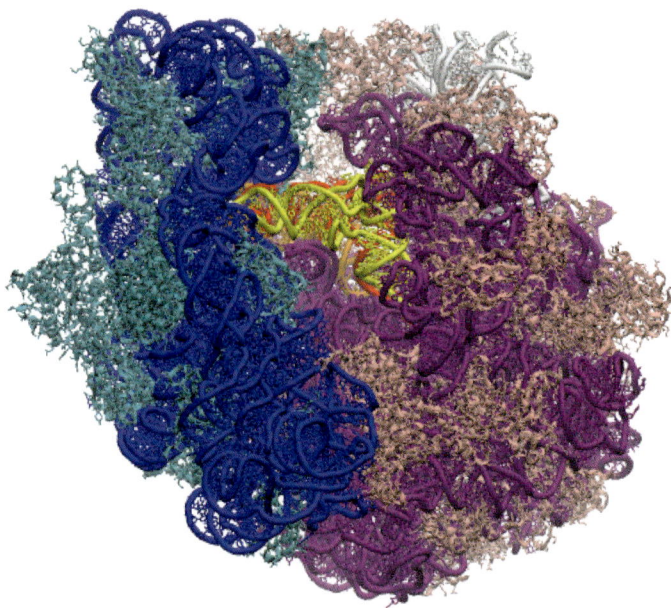

Figure 1. Overview of ribosome structure. Blue: small subunit rRNA; cyan:small subunit ribosomal proteins; magenta: large subunit rRNA; pink: large subunit proteins; white: 5S rRNA; yellow: aminoacyl-tRNA; red: peptidyl tRNA.

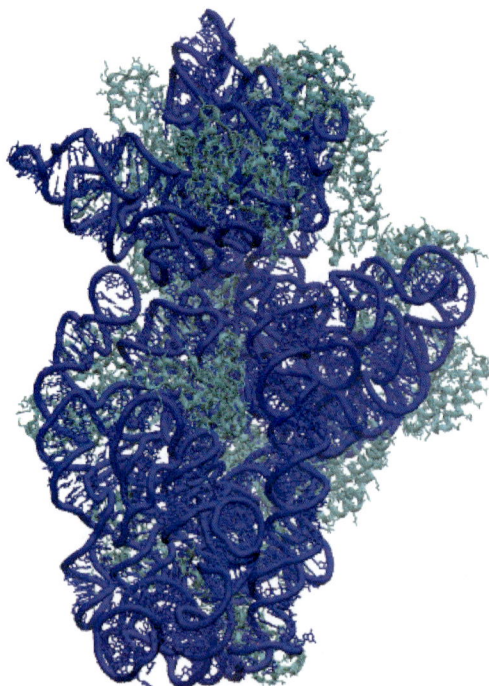

Figure 2. The 30S small ribosomal subunit.

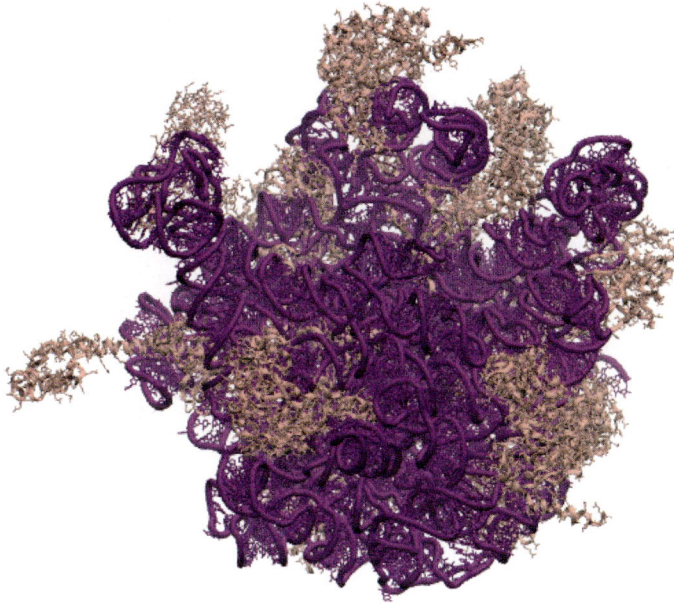

Figure 3. The 50S large ribosomal subunit.

loop and the GTPase activity center are involved in binding factors that aid in tRNA selection and translocation. The functions of the L1 and L7/L12 stalks are not well understood. One possible role may be to recruit factors to the ribosome (L7/L12 stalk) and assist in transfer RNA (tRNA) dissociation from the ribosome (L1 stalk).[8]

Protein synthesis is achieved by a suite of translating molecules or transfer RNAs. Each tRNA carries a specific amino acid to the ribosome for assembly into the new protein. tRNAs are "L"-shaped RNAs, where one arm of the "L" is the anticodon arm, containing the three nucleotide anticodons for decoding the mRNA, and the other arm of the "L" is the acceptor stem, containing the flexible 3′-CCA end that carries the amino acid. The small ribosomal subunit contains three tRNA binding sites: the aminoacyl site (A site), where decoding occurs, the peptidyl site (P site), where the initiator tRNA binds, and the exit site (E site), where the deacylated tRNA detaches from the mRNA (Figure 4). At each site, the anticodon is in close proximity to the mRNA codon, allowing codon-anticodon interactions to occur, usually consisting of three base pairs. The large ribosomal subunit contains three corresponding sites. The aminoacyl-tRNA and peptidyl-tRNA 3′-CCA ends base-pair with nucleotides in the 50S A and P sites, respectively. The 3′ ends are sufficiently close for the peptidyl transferase reaction to occur. The deacylated tRNA 3′-CCA end stacks with bases in the 50S E site, before dissociating from the ribosome. Three tRNAs may simultaneously bind to the ribosome *in vitro*;[9] however it is not clear that this occurs *in vivo*.

2. Overview of Protein Synthesis and Elongation

Protein synthesis has three stages: initiation, where the ribosome is assembled, elongation, where the protein is synthesized, and termination, where the complex is disassembled. The elongation stage has three substeps: tRNA selection, peptide bond formation and translocation.

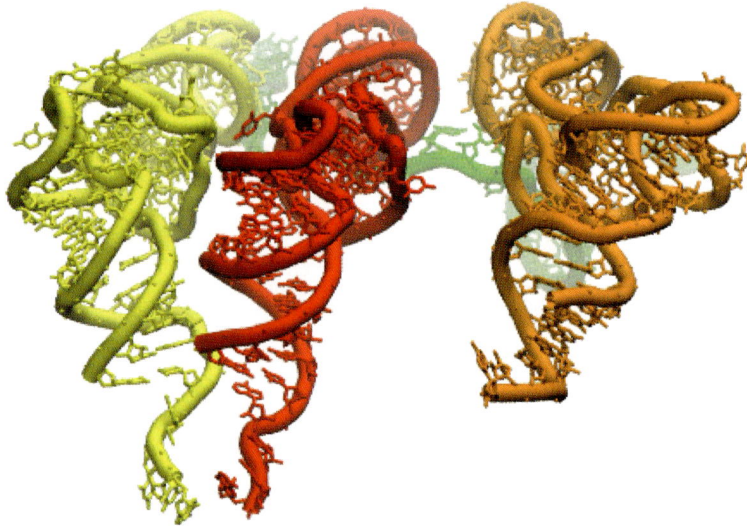

Figure 4. Relative positioning of tRNAs inside the ribosome. Yellow: aminoacyol-tRNA; red: peptidyl-tRNA; orange: exit-site tRNA; green: mRNA.

tRNA selection is the stage where the actual information processing occurs: the ribosome must read the each codon on the mRNA and attach the corresponding amino acid to the nascent polypeptide chain. To accomplish this, the ribosome performs a molecular "look-up table" operation. Here, the molecular look-up table consists of the suite of tRNAs corresponding to the various amino acids. Each tRNA acts as a translating molecule, connecting a codon to a single amino acid. The translation is manifested molecularly in the form of the three-nucleotide anticodon on the RNA portion of the tRNA and the corresponding amino acid attached to the universally conserved 3'-CCA end of the tRNA molecule. The ribosome must recognize correct codon-anticodon "mini-helices" (the codon-anticodon complex forms a stacked three base pair mini-helix), thereby accepting correct amino acids into the new protein. The ribosome must also reject incorrect codon-anticodon complexes.

Once the correct tRNA is accepted by the ribosome, the peptidyl transferase reaction occurs, where the new amino acid is added to the growing protein chain. Next, the ribosome must move exactly three nucleotides along the mRNA to the next codon (translocation). The deacylated tRNA then dissociates after passing through the exit site and the aminoacyl-tRNA is moved into the peptidyl site (P site), leaving the A site vacant. The cycle can then repeat.

The above description applies to a simplified version of protein synthesis — factor free protein synthesis — which has been demonstrated in the lab[10,11] and is thought to have occurred early on in evolution. Factor free protein synthesis is inaccurate, slow and inefficient. In modern day organisms, protein factors help optimize this process, increasing speed and accuracy. In particular, elongation factor Tu (EF-Tu) helps recruit tRNAs to the ribosome. These tRNAs are said to bind to the ribosome in a ternary complex consisting of the aminoacyl-tRNA, EF-Tu and a GTP molecule bound to EF-Tu.[12–14] It was shown that upon binding, the tRNA undergoes a conformational change resulting in a kinked conformation, introducing a second curve in the overall shape of the tRNA, in addition to the elbow of the 'L'.[15–17] Here, the anticodon arm has a severe kink,

(a)

(b)

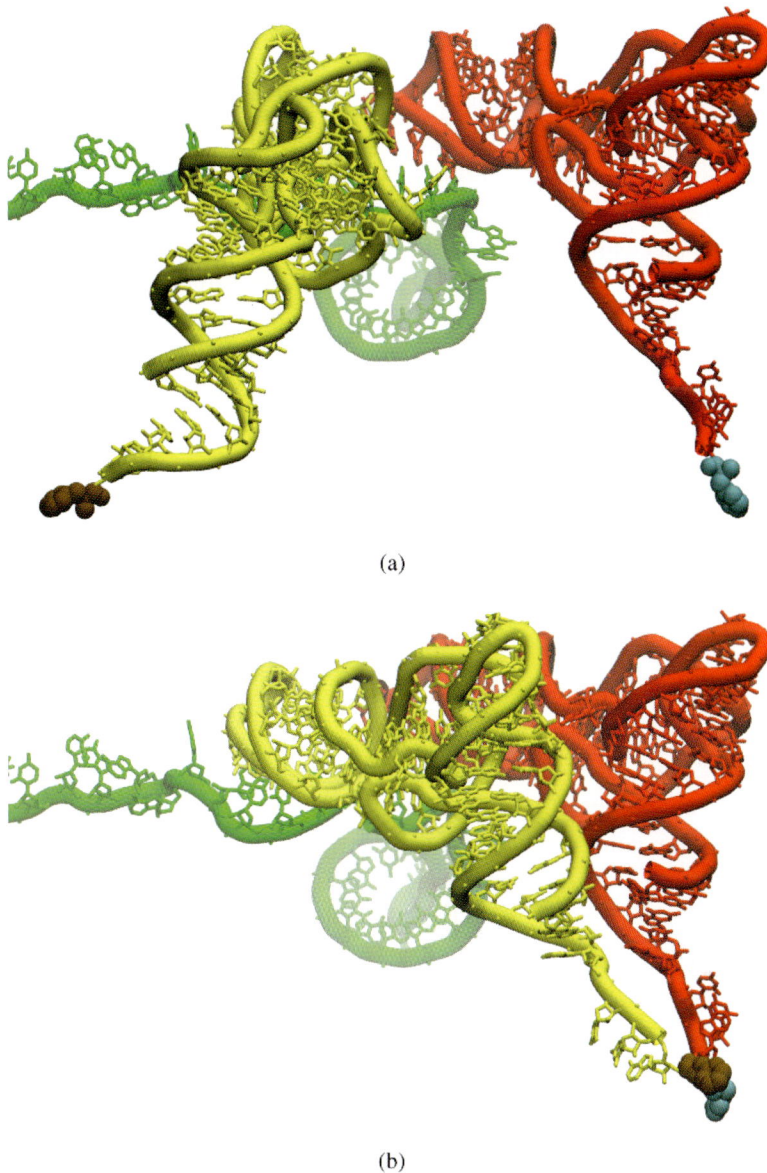

Figure 5. Dynamics of aminoacyl-tRNA (yellow) during accomodations. (a) Before accomodation. (b) After accomodation.

allowing the anticodon to contact the mRNA codon while the acceptor stem is simultaneously tightly bound to EF-Tu (Figure 5). The second factor binding during elongation is elongation factor G (EF-G).[18–22] After EF-Tu dissociates, EF-G binds to the ribosome, trapping it in the so-called "ratchet-like" state, where the small subunit is rotated with respect to the large subunit relative to the initiator configuration.[23,24] Both the factors interact with the Sarcin-Ricin loop and a portion of the L7/L12 stalk.[16,23,25]

Intense activity in eukaryotic ribosome research has been driven by medical applications. In eukaryotic ribosomes, the large subunit (60S) and the small subunit (40S) combine to form the 80S ribosome. Cryogenic electron microscopy (cryo-EM) reconstructions of the 80S have been completed and the higher resolution X-ray structure is well under way. Bacterial and eukaryotic ribosomes are quite similar, with the major exception near the solution side of the exit tunnel of the large subunit that must dock to the endoplasmic reticulum.[26] There is a significant difference between bacterial and eukaryotic translation initiation.[27] While bacterial initiation entails only three initiation factors, eukaryotic initiation requires many more.

Understanding the mechanism of translation by the ribosome is of critical importance for the development of new antibiotics. In the USA, about half of all antibiotics used target the ribosome.[28] Currently, "superbugs" are ubiquitous in hospitals, resistant to many antibiotics. These resistant bacteria include multidrug-resistant bacteria (MDR), extensive multidrug-resistant bacteria (XDR) and methicillin-resistant *Staphylococcus aureus* (MRSA). Ribosome antibiotics attack the various active sites including the decoding center and the peptidyl transferase center.[29–31]

3. Decoding by the Ribosome

Initially, it was proposed that tRNA selection is accomplished by Watson-Crick base pairing. However, differences in stability between Watson-Crick and non–Watson-Crick pairing are not sufficient to account for observed translational error rates. Furthermore, Streptomycin A and ribosomal ambiguity mutations demonstrate that changes in ribosomal sequence affect error rates, suggesting that the ribosome plays a role in discrimination. Thus, to explain observed error rates in translation (ranging from approximately 5×10^{-5} to 3×10^{-3} for *Escherichia coli*, depending on codon usage, codon context, and tRNA levels), the distinctions in stability between correct and incorrect codon-anticodon pairs must be amplified by the ribosome.

During the 70s and 80s, the decoding problem was dominated by the two-step kinetic proofreading model, described by Ninio and Hopfield. This consists of (1) an initial selection step, and (2) a proofreading step, separated by hydrolysis of GTP by the EF-Tu-aminoacyl-tRNA ternary complex. The general framework of the two-step kinetic model was verified via cell-free experiments measuring binding, GTP hydrolysis, and peptide formation for correct and incorrect tRNA. More recently, Rodnina and co-workers have shown that each of these reaction steps is preceded by a rate-limiting conformational change, consistent with an induced-fit mechanism of discrimination. They have also shown that the initial selection step is preceded by a distinct codon-independent binding step. Those tRNAs which survive the GTP hydrolysis test are called near-cognate, while those which do not are called non-cognate. The tRNAs which survive all steps are called cognate. Near-cognate tRNAs typically have one non–Watson-Crick pair in the codon-anticodon minihelix, while cognates either have none or a G:U wobble in the third position of the codon-anticodon interaction.

Recently, the focus of the decoding problem has turned to structural approaches. Chemical probing experiments of Noller and mutagenesis experiments of Puglisi have shown which ribosome bases are likely to participate in decoding (A1492, A1493 and G530). The pioneering work of Ramakrishnan, Noller, and Cate has determined which chemical groups on the ribosome participate in decoding and which conformational changes occur. By diffusing in the anticodon stem loop (ASL) of tRNA (the portion of tRNA which contains the anticodon) into small ribosomal subunit crystals, Ramakrishnan determined the tRNA-mRNA-rRNA hydrogen bonds involved in recognition of correct tRNA by the ribosome. In particular, the Ramakrishnan group has shown that two ribosome bases (small subunit bases A1492 and A1493) which normally reside in an RNA helix (helix 44 of the small subunit), flip out of the helix to interact simultaneously with the tRNA anticodon and the mRNA codon. They have also obtained structures of the small subunit in complex with near-cognate tRNAs. These structures require the presence of the error-inducing antibiotic paromomycin to assist in binding the tRNA. The antibiotic mimics the cognate geometry, allowing near-cognate tRNAs to bind strongly to the small subunit, and may not shed light on how the ribosome rejects these near-cognate tRNAs. Crystals with no antibiotic showed no density for the near-cognate anticodon stem loops. The Cate structure suggests that non-cognate tRNAs do bind in the fully accommodated state to the ribosome. Our previous molecular dynamics simulation study has examined near-cognate tRNA-ribosome interactions in atomic detail, in the absence of antibiotics.

Finally, a recent cryo-electron-microscopy study has examined the conformational changes of tRNA and the elongation factor Tu upon ternary complex binding.

4. Dynamic Regions of the Ribosomal Complex

The feat of moving large ligands (~ 60 Å in length from anticodon to elbow) through the ribosome from site to site with exquisite precision, while simultaneously discriminating between correct and incorrect amino acids, has perplexed researchers for almost half a century. Recently it has become clear that dynamics play an enormous role in this process.[32–34]

For elongation to occur, tRNAs bind to the ribosome, move into the A site, proceed to the P site, continue to the E site, and dissociate from the ribosome.[35] The ribosome produces precise alignment of the tRNA with the mRNA codon of the 30S and the peptidyl transferase center on the 50S.[3,36–38] The 3′-CCA end of the tRNA is surrounded by a dense thicket of ribosomal RNA in the 50S A site and the P site. In the 50S E site, nucleotides of the 3′-CCA end are intercalated between two ribosomal nucleotides. This precise alignment raises the question: How can the tRNA be surrounded by many specific ribosome interactions, yet move to two other completely different states with equally complex interactions? The solution is that both the ribosome and the tRNA are extremely dynamic, as evidenced by single molecule FRET, cryo-EM and X-ray crystallography data.[2,32–34,39–41]

4.1 *Dynamic regions of the ribosome*

The ribosome itself has been shown to have dynamic regions. The motion of a region of a subunit with respect to the same subunit is called intrasubunit motion. The motion of the two subunits with respect to each other is called intersubunit motion.[42] The large-scale intersubunit motion performed by the ribosome is the so-called "ratchet-like" pivoting of the 30S subunit with respect to the 50S subunit (Figure 6).[42] Cryo-EM studies of the 70S ribosome in the presence of EF-G showed a substantial rotation of the 30S relative to its initiator configuration upon EF-G binding.[42] It was thought that this motion causes translocation of the tRNA and mRNA through the ribosome. FRET

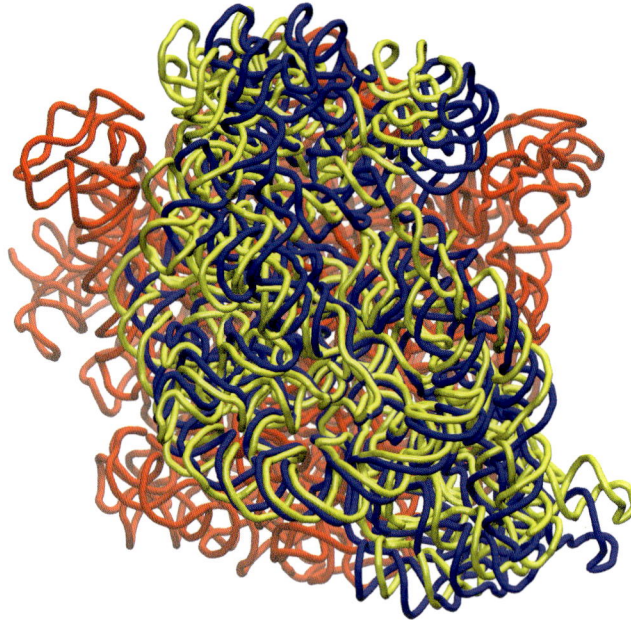

Figure 6. Ratchet-like motion of small ribosomal subunit (blue/yellow) relative to large ribosomal subunit (red).

experiments have shown that ratchet-like fluctuations occur at equilibrium in absence of EF-G.[43] EF-G traps the ribosome in the ratcheted position.

Intrasubunit motions have been observed within the small subunit. We note that the decoding center of the small subunit contains two universally conserved nucleotides: A1492 and A1493. These appear to be dynamic in NMR studies,[44] fluorescence studies[45] and are observed to be disordered in X-ray crystallography systems.[37,46]

The head domain of the small subunit exists in multiple conformations[2,47] and is dynamic.[43,48] Cate observed in alternative crystal forms that the head of the small subunit was rotated around a neck region with respect to its body. Spahn and co-workers have shown that head-swivel is critical for translocation.[47] In particular, cryo-EM studies of EF-G bound ribosomes have shown that two EF-G bound populations exist: (1) the 30S body pivoted with respect to the 50S (ratcheted) and (2) the 30S body pivoted and the 30S head rotated (head swiveled). The study shows that the majority of translocation movement occurs during 30S head swivel, almost orthogonal motion to the "ratchet-like" pivot of the 30S with respect to the 50S (Figure 7). FRET experiments show that the head undergoes enormous fluctuations in solution.[43,48]

The 50S undergoes large intrasubunit conformational changes. Structural biology studies show the L1 stalk and L7/L12 stalk in two distinct configurations: factor bound and factor free.[42,49] During ternary complex binding, the L7/L12 stalk moves towards the central protuberance (Figure 8). During EF-G binding and ratcheting, the L1 stalk moves towards the central protuberance. Single molecule studies have shown the L1 stalk to undergo enormous and rapid fluctuations that are independent of tRNA fluctuations.[50–52] Upon EF-G binding, these fluctuations tend to be more synchronized. Because the L1 stalk interacts with the P/E position and E/E position tRNAs,

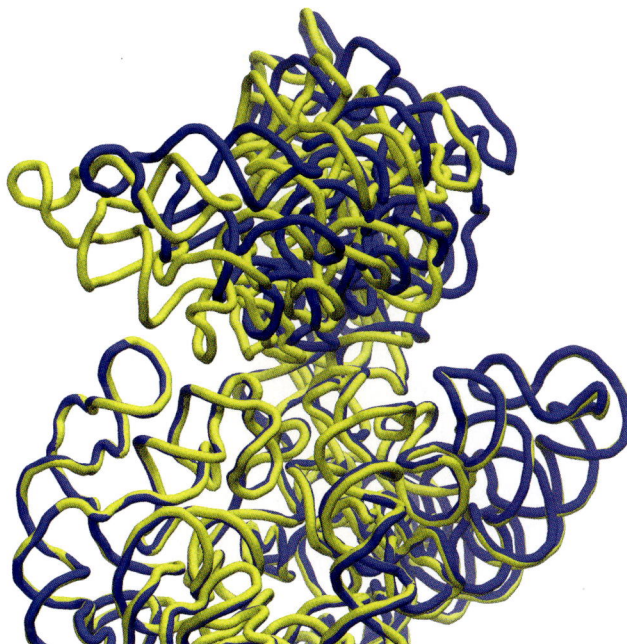

Figure 7. Dynamics of small subunit unit head domain before and after head-swivel motion.

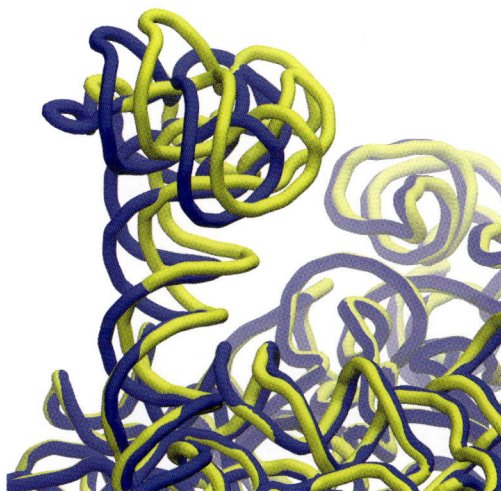

Figure 8. Dynamics of the L1 stalk in all-atom structure-based simulations. Blue and yellow display snapshots of ribosome at different times during simulation.

the L1 stalk may be involved with E site tRNA dissociation. Single-molecule FRET studies also show the L7/L12 stalk to fluctuate rapidly.

The L7/L12 protein complex is dynamic. This consists of an alpha helix (L10) bolstered by several L7/L12 NTD dimers.[8,53] Each of the L7/L12 NTD domains is connected to its CTD by a long disordered flexible linker. The CTD undergoes large fluctuations that may aid in capturing translation factors in a fly-casting mechanism.[54]

4.2 *tRNA as an active player in decoding: Flexible tRNAs*

The aminoacyl-tRNA binds to the ribosome in a highly kinked configuration called the A/T state where the "A" represents the anticodon interacting with the 30S subunit A site and the "T" represents the ternary complex interacting with the 50S subunit at the GTPase activity center.[15,16] Single-molecule studies show rapid tRNA fluctuations between several intermediates occurring before accommodation, where the tRNA moves into the fully bound "A/A" position (anticodon interacting with the 30S A site and 3′-CCA end interacting with the 50S A site). The transitions include transitions between the initial binding configurations, the codon-anticodon recognition configuration, the GTPase activation configuration and the A/T state. The aminoacyl-tRNA samples several different ensembles of configurations prior to accommodation.[55]

Cryo-EM reconstructions of intermediates during translocation demonstrate tRNA flexibility. Following accommodation and peptidyl transferase, the tRNA in the A/A position moves to the A/P position (anticodon in 30S A site and 3′-CCA end in 50S P site), while the tRNA in the P/P position moves to the P/E position (anticodon in the 30S P site and 3′-CCA end in the 50S E site). Cryo-EM studies of the EF-4 complex show the tRNA to be in a curved configuration. This curved state differs from the curved configuration required for the A/T position.[56] The two curved configurations are suggestive of an energy storage-release mechanism that allows the tRNA to propel itself through the ribosome.[57] The enthalpic energy source would of course be combined with the much larger source of gated thermal fluctuations resulting from the thermal bath.[58]

Single-molecule FRET studies have shown the tRNAs to fluctuate between three conformations in the pre-translocation/post-accommodation stage.[32] In the first configuration ("classical"), the deacylated tRNA is in the P/P position and the peptidyl tRNA is in the A/A position. In the second configuration, the deacylated tRNA is in the P/E position and the peptidyl tRNA is in the A/A position.[32] In the third configuration, the deacylated tRNA is in the P/E position and the peptidyl tRNA exists in the A/P position. This is known as "hybrid states" reported by Noller and co-workers.[35] Single molecule studies display the two tRNAs asynchronously fluctuating rapidly between these three configurations.[32]

5. Computational Studies of Decoding by the Ribosome

To obtain the free energy landscape of the ribosome and a full understanding of elongation, the range of ribosome experiments must be integrated into a coherent picture. Molecular dynamics simulation enables the integration of several forms of experiments.

Rapid kinetics experiments have been able to isolate the various substeps of elongation, giving us the framework for how translation procedes.[19] Single-molecule FRET experiments use FRET pairs to obtain time-dependent distance constraints on specific conformational changes.[34] X-ray structures provide high-resolution atomic models of the ribosome in various snap shots of the ribosome during ribosome function (classical, EF-G bound, and EF-Tu bound).[2-7] Cryo-EM reconstructions allow us to construct atomic models of the ribosome in a wide variety of functional states.[39,41,47] Molecular dynamics simulations integrate these data into a coherent scenario.

Modeling efforts[59,60] and advances in supercomputing have enabled advances that help to clarify the relationship between static ribosome structures,[47,61] structural fluctuations about these local energetic minima,[62,63] and ribosome function.[58,64,65]

Molecular dynamics simulations and computational studies have explored the decoding center,[66–69] isolated tRNA[70] and the inner core of the 30S subunit,[71,72] kink turns,[73,74] the ribosome tunnel,[75] and drug binding.[76–78] Adaptive-mesh refinement Poisson-Boltzmann studies based have examined the electrostatic potential of the large and small subunits.[79] Normal mode studies have investigated the motions of the ribosome within the approximations of the Gaussian network model.[62,79,80] The real-space refinement technique has been combined with cryo-EM structures to study the beginning and end states of a conformational change that occurs during translocation.[81] Explicit solvent molecular dynamics simulations have helped elucidate functional mechanisms of biomolecular systems[82–86] and have been validated with experiments.[87] High performance computing platforms, together with the GROMACS program,[88] have enabled us to simulate the ribosome in explicit solvent (3.2×10^6 atoms) for an aggregate sampling of >2 microseconds (~ 2100 ns), producing the largest simulation of a biomolecular complex to date,[58] more than 20-fold larger than previous efforts, which simulated the ribosome for 85 ns[89] and a small portion (1.25×10^5 atoms) for 2.1 microseconds.[90]

5.1 *Early simulations of decoding and tRNA selection*

In 2001–2002, we performed a limited explicit solvent study of the fidelity on the ribosome, simulating the tRNA-mRNA-ribosome decoding center complex for ultrafast time scales (~ 10 ns).[91] We were able to observe stable cognate (correct) codon-anticodon interactions and unstable near-cognate (incorrect) codon-anticodon interactions. We concluded that the decoding bases (G530, A1492 and A1493) measure the geometry of the codon-anticodon minihelix minor groove and test the stability of the codon-anticodon interaction. This first functional study of the ribosome using molecular dynamics established the foundation for more extensive studies.

5.2 *REMD simulations of the decoding center produce exhaustive sampling*

Recent studies show that at least 1 microsecond of replica sampling is required for exhaustive sampling of the decoding helix.[77] Microsecond replica sampling likely amounts to over 25 microseconds of standard molecular dynamics.[92] Without exhaustive sampling, the free energy cannot be calculated accurately. In this study, the thousands of nucleotide flipping events for A1492 and A1493 were observed. Here, "flipping" refers to the movement of a nucleotide from inside helix 44 to solution. The simulations show that these nucleotides flip into and out of helix 44 of the small subunit when no ligands are bound to the small subunit. The barrier in free energy for base flipping of these nucleotides was observed to be on the order of a few kcal/Mol, suggesting that the nucleotides continuously flip until ligands bind, trapping them in the flipped out configuration.[77,93]

Aminoglycoside antibiotics bind to the decoding center (on small subunit helix 44).[29,44,94] These antibiotics bind inside the decoding region of helix 44, preventing A1492 and A1493 from residing in their flipped in conformation. We have shown that rather than an induced-fit scenario, where the drug induces these nucleotides to flip out of their helix, a stochastic gating or conformational selection scenario is more likely. Here, the nucleotides continuously flip. Aminoglycoside binding traps the nucleotides in their flipped-out state. Upon correct tRNA binding, an intricate hydrogen bond network exists between G530, A1492, A1493, the tRNA anticodon, and the mRNA codon. Since G530, A1492 and A1493 sense the Watson-Crick geometry of the codon-anticodon interaction, these nucleotides act as a molecular reading head. Aminoglycosides stabilize this reading head in the accepting configuration with A1492 and A1493 flipped out of their helix, allowing them to interact productively with the tRNA leading to acceptance of the tRNA into the ribosome ("accommodation"). Aminoglycoside interactions lead to the incorporation of incorrect amino acids into the nascent protein. As a result, widespread errors in proteins manufactured are produced by the aminoglycoside-bound ribosome. These errors are known as misreading error defects.

5.3 *High performance computing systems are required to simulate the intact ribosome*

Even with fast supercomputers, such as the petaFlop RoadRunner in the Los Alamos National Laboratory, a full simulation of the ribosome in explicit solvent for physiological time scales (~ 100 ms–1 s) is not possible. Assuming currently used algorithms, we estimate this would require a ~ 10^3–10^4 fold speed-up or a 1–10 exaFlop machine, which may be available between 2019–2024 according to DOE SCIDAC projections. We use a multifaceted simulation approach to tackle ribosome dynamics. For short time scale stability of hydrogen bonds, we perform standard explicit solvent molecular dynamics simulations of localized regions of the ribosome.[91] To study localized conformational changes, exhaustive sampling techniques such as replica exchange explicit solvent molecular dynamics are used.[77,93,95] To study the structural stability and to obtain a baseline for validation of reduced force field techniques, we perform multiple extensive explicit solvent simulations of the entire ribosome. Finally, to study large-scale conformational changes we use a variety of reduced techniques. To obtain a first look, we use targeted molecular dynamics simulations in explicit solvent, uncovering steric interactions occurring during the large-scale conformational change. The results have been validated experimentally.[96,97] We also use a reduced-potential method, called structure-based simulation.[58,98–100] The unrestrained simulations display spontaneous large-scale conformational changes with an estimated total sampling of ~ 200 ms. We obtain reasonable agreement with single-molecule FRET, cryo-EM, and X-ray crystallography B-factors.

5.4 *Simulations of large-scale conformational changes of tRNAs*

We first turned to accommodation, the simplest large-scale conformational change involved in tRNA movement through the ribosome — accommodation. Here, the aminoacyl-tRNA moves from the A/T state to the fully bound A/A state. This 70Å movement requires the kinked anticodon arm to relax to its native conformation. In 2002–2003, we performed explicit solvent targeted molecular dynamics simulations, with a total sampling of ~ 20 ns.[65] The simulations revealed several important new features of accommodation. Our study showed explicitly that flexibility of the 3′-CCA end is required for accommodation.[65] The simulation revealed that many steric barriers are present in the accommodation pathway. The 3′-CCA end must be extremely flexible to be transported from the A/T state into the A/A state. The 3′-CCA end must navigate the intricate and universally conserved accommodation corridor consisting of large subunit rRNA helices H89, H90 and H92. The accommodation corridor was identified by our simulations[65] and could not be identified from X-ray crystallography or cryo-EM, which only characterize the initial and final states of accommodation.

Recently, we have simulated spontaneous accommodation events. Structure-based simulations were used.[58,98–100] A potential defined by the X-ray structure of the ribosome with the aminoacyl-tRNA occupying the A/A state was used. The tRNA was started in the A/T position. The tRNA and ribosome were then allowed to fluctuate freely under the A/A based potential. The estimated total sampling was on the order of hundreds of milliseconds. The tRNA was observed to be highly dynamic, undergoing reversible excursions from the A/T state towards the A/A state and back. These same reversible excursions were also observed single-molecule FRET studies of accommodation.[58] In our simulations, it was clear that H89 must fluctuate to make way for the accommodating tRNA. The study shows that the 3′-CCA end is disordered, suggesting that the 3′-CCA end itself presents an entropic barrier to accommodation.

To validate our structure-based potential, extensive multiple explicit solvent simulations were performed with total sampling of ~ 2100 ns. Simulations were performed with the tRNA in the A/A state and also in the A/T state. The configurational space sampled in the A/A state explicit solvent simulation was similar to the A/A basin sampled by the structure-based simulation. Similar results were achieved for the A/T state. Fluctuations of the explicit solvent and structure-based simulations were similar. Both simulations also showed agreement with X-ray B-factors,[6] with the exception of the 50S stalks, which have been shown to undergo large amplitude fluctuations.[50,51]

Finally, we used the structure-based method to investigate L1 stalk fluctuations, revealing large amplitude fluctuations both in displacement and twist, consistent with the high fluctuations observed by single-molecule FRET.

6. Towards Energy Landscapes of Decoding

Our simulations investigate principles important for ribosome dynamics: functional ensembles, multiple pathways, stochastic gating, large amplitude reversible excursions, and entropic barriers. Structure-based simulations display a wide variety of accommodation pathways. The pathway distribution is determined by a combination of the intrinsic properties of the tRNA and the ribosome. The structure-based and explicit solvent simulations show large ensembles of configurations representing the A/A and A/T positions that are consistent with single-molecule FRET experiments. Our study reveals that reversible excursions of the aminoacyl-tRNA combined with stochastic gating of helix 89 allow for accommodation. For accommodation to occur, the entropic barrier introduced by the 3′-CCA end must be over-powered by the enthalpic minima of the peptidyl transferase center, the enthalpic minima of the anticodon arm, or the entropic "kick" produced by the presence of EF-Tu, which prevents aminoacyl-tRNA from escaping the ribosome. While extremely flexible components introduce entropic barriers, this flexibily softens the system relative to a molecular machine consisting of more rigid components. Softening may act to fine-tune the free energy landscape.

Similar principles may operate during translocation. Here, the energy landscape seen by the tRNA is thought to be dynamic. Ratchet-like movement of the 30S body, rotation of the 30S head and the opening and closing of various gates will significantly alter the barriers in a dynamic fashion. FRET experiments suggest that, in the absence of translation factors, components of the ribosome fluctuate relatively independently at different time scales. It is in the rare moments that these fluctuations occur productively and simultaneously that translocation may occur.[50,51] EF-G may enhance synchronization.

In factor-free translation, the thermal bath in the environment surrounding the ribosome contains enough energy to produce the large-scale fluctuations that move the tRNAs through the ribosome. In the case of factor-free translation, the only reaction is the peptidyl transferase reaction. Thermal fluctuations overcome enthalphic and entropic barriers involved in moving the aminoacyl-tRNA from solution to the accommodated A/A position. Following peptide bond formation, thermal fluctuations move the deacylated tRNA through the hybrid state to the exit site. The structure of the ribosome carefully controls this movement. Thermal fluctuations allow the deacylated tRNA to exit the ribosome. Elongation factors streamline this process. EF-Tu carefully positions the tRNA in a partially bound position, allowing for high-fidelity accuracy testing without requiring accommodation into the A/A position. EF-G may enhance translocation by decreasing fluctuations in the reverse direction of translocation.

Suggested Additional Reading Materials

K. Y. Sanbonmatsu, S. Joseph and C. S. Tung. 2005. Simulating movement of tRNA into the ribosome during decoding. *Proc Natl Acad Sci U S A* **102**, 15854–15859.

P. C. Whitford, P. Geggier, R. B. Altman, S. C. Blanchard, J. N. Onuchic and K. Y. Sanbonmatsu. 2010. Accommodation of aminoacyl-tRNA into the ribosome involves reversible excursions along multiple pathways. *RNA* **16**, 1196–1204.

References

1. J. R. Warner. 1999. The economics of ribosome biosynthesis in yeast. *Trends Biochem Sci* **24**, 437–440.
2. B. S. Schuwirth, M. A. Borovinskaya, C. W. Hau, W. Zhang, A. Vila-Sanjurjo, J. M. Holton and J. H. Cate. 2005. Structures of the bacterial ribosome at 3.5 A resolution. *Science* **310**, 827–834.
3. G. Yusupova, L. Jenner, B. Rees, D. Moras and M. Yusupov. 2006. Structural basis for messenger RNA movement on the ribosome. *Nature* **444**, 391–394.
4. M. Selmer, C. M. Dunham, F. V. T. Murphy, A. Weixlbaumer, S. Petry, A. C. Kelley, J. R. Weir and V. Ramakrishnan. 2006. Structure of the 70S ribosome complexed with mRNA and tRNA. *Science* **313**, 1935–1942.
5. F. Schluenzen, A. Tocilj, R. Zarivach, J. Harms, M. Gluehmann, D. Janell, A. Bashan, H. Bartels, I. Agmon, F. Franceschi and A. Yonath. 2000. Structure of functionally activated small ribosomal subunit at 3.3 angstrom resolution. *Cell* **102**, 615–623.
6. A. Korostelev, S. Trakhanov, M. Laurberg and H. F. Noller. 2006. Crystal structure of a 70S ribosome-tRNA complex reveals functional interactions and rearrangements. *Cell* **126**, 1065–1077.

7. N. Ban, P. Nissen, J. Hansen, P. B. Moore and T. A. Steitz. 2000. The complete atomic structure of the large ribosomal subunit at 2.4 A resolution. *Science* **289**, 905–920.

8. M. Diaconu, U. Kothe, F. Schlunzen, N. Fischer, J. M. Harms, A. G. Tonevitsky, H. Stark, M. V. Rodnina and M. C. Wahl. 2005. Structural basis for the function of the ribosomal L7/12 stalk in factor binding and GTPase activation. *Cell* **121**, 991–1004.

9. L. B. Jenner, N. Demeshkina, G. Yusupova and M. Yusupov. 2010. Structural aspects of messenger RNA reading frame maintenance by the ribosome. *Nat Struct Mol Biol* **17**, 555–560.

10. L. P. Gavrilova, O. E. Kostiashkina, V. E. Koteliansky, N. M. Rutkevitch and A. S. Spirin. 1976. Factor-free ("non-enzymic") and factor-dependent systems of translation of polyuridylic acid by *Escherichia coli* ribosomes. *J Mol Biol* **101**, 537–552.

11. A. S. Spirin, O. E. Kostiashkina and J. Jonak. 1976. Contribution of the elongation factors to resistance of ribosomes against inhibitors: Comparison of the inhibitor effects on the factor-free translation systems. *J Mol Biol* **101**, 553–562.

12. M. V. Rodnina and W. Wintermeyer. 2001. Ribosome fidelity: tRNA discrimination, proofreading and induced fit. *Trends Biochem Sci* **26**, 124–130.

13. M. V. Rodnina and W. Wintermeyer. 2001. Fidelity of aminoacyl-tRNA selection on the ribosome: Kinetic and structural mechanisms. *Annu Rev Biochem* **70**, 415–435.

14. K. B. Gromadski and M. V. Rodnina. 2004. Kinetic determinants of high-fidelity tRNA discrimination on the ribosome. *Mol Cell* **13**, 191–200.

15. M. Valle, J. Sengupta, N. K. Swami, R. A. Grassucci, N. Burkhardt, K. H. Nierhaus, R. Agrawal and J. Frank. 2002. Cryo-EM reveals an active role for aminoacyl-tRNA in the accommodation process. *EMBO J* **21**, 3557–3567.

16. M. Valle, A. Zavialov, W. Li, S. M. Stagg, J. Sengupta, R. C. Nielsen, P. Nissen, S. C. Harvey, M. Ehrenberg and J. Frank. 2003. Incorporation of aminoacyl-tRNA into the ribosome as seen by cryo-electron microscopy. *Nat Struct Mol Biol* **10**, 899–906.

17. T. M. Schmeing, R. M. Voorhees, A. C. Kelley, Y. G. Gao, F. V. T. Murphy, J. R. Weir and V. Ramakrishnan. 2009. The crystal structure of the ribosome bound to EF-Tu and aminoacyl-tRNA. *Science* **326**, 688–694.

18. A. Savelsbergh, V. I. Katunin, D. Mohr, F. Peske, M. V. Rodnina and W. Wintermeyer. 2003. An elongation factor G-induced ribosome rearrangement precedes tRNA-mRNA translocation. *Mol Cell* **11**, 1517–1523.

19. W. Wintermeyer, F. Peske, M. Beringer, K. B. Gromadski, A. Savelsbergh and M. V. Rodnina. 2004. Mechanisms of elongation on the ribosome: Dynamics of a macromolecular machine. *Biochem Soc Trans* **32**, 733–737.

20. B. Wilden, A. Savelsbergh, M. V. Rodnina and W. Wintermeyer. 2006. Role and timing of GTP binding and hydrolysis during EF-G-dependent tRNA translocation on the ribosome. *Proc Natl Acad Sci U S A* **103**, 13670–13675.

21. A. Savelsbergh, D. Mohr, U. Kothe, W. Wintermeyer and M. V. Rodnina. 2005. Control of phosphate release from elongation factor G by ribosomal protein L7/12. *EMBO J* **24**, 4316–4323.

22. A. Savelsbergh, M. V. Rodnina and W. Wintermeyer. 2009. Distinct functions of elongation factor G in ribosome recycling and translocation. *RNA* **15**, 772–780.

23. J. Frank and R. K. Agrawal. 2000. A ratchet-like inter-subunit reorganization of the ribosome during translocation. *Nature* **406**, 318–322.

24. Y. G. Gao, M. Selmer, C. M. Dunham, A. Weixlbaumer, A. C. Kelley and V. Ramakrishnan. 2009. The structure of the ribosome with elongation factor G trapped in the posttranslational state. *Science* **326**, 694–699.

25. M. Valle, A. Zavialov, J. Sengupta, U. Rawat, M. Ehrenberg and J. Frank. 2003. Locking and unlocking of ribosomal motions. *Cell* **114**, 123–134.

26. C. M. Spahn, M. G. Gomez-Lorenzo, R. A. Grassucci, R. Jorgensen, G. R. Andersen, R. Beckmann, P. A. Penczek, J. P. Ballesta and J. Frank. 2004. Domain movements of elongation factor eEF2 and the eukaryotic 80S ribosome facilitate tRNA translocation. *EMBO J* **23**, 1008–1019.

27. N. Sonenberg and T. E. Dever. 2003. Eukaryotic translation initiation factors and regulators. *Curr Opin Struct Biol* **13**, 56–63.

28. A. Mankin. 2006. Antibiotic blocks mRNA path on the ribosome. *Nat Struct Mol Biol* **13**, 858–860.

29. A. P. Carter, W. M. Clemons, D. E. Brodersen, R. J. Morgan-Warren, B. T. Wimberly and V. Ramakrishnan. 2000. Functional insights from the structure of the 30S ribosomal subunit and its interactions with antibiotics. *Nature* **407**, 340–348.

30. F. Schlunzen, R. Zarivach, R. Harms, A. Bashan, A. Tocilj, R. Albrecht, A. Yonath and F. Franceschi. 2001. Structural basis for the interaction of antibiotics with the peptidyl transferase centre in eubacteria. *Nature* **413**, 814–821.

31. M. Pioletti, F. Schlunzen, J. Harms, R. Zarivach, M. Gluhmann, H. Avila, A. Bashan, H. Bartels, T. Auerbach, C. Jacobi, T. Hartsch, A. Yonath and F. Franceschi. 2001. Crystal structures of complexes of the small ribosomal subunit with tetracycline; edeine and IF3. *EMBO J* **20**, 1829–1839.

32. J. B. Munro, R. B. Altman, N. O'Connor and S. C. Blanchard. 2007. Identification of two distinct hybrid state intermediates on the ribosome. *Mol Cell* **25**, 505–517.

33. S. C. Blanchard, H. D. Kim, R. L. Gonzalez Jr., J. D. Puglisi and S. Chu. 2004. tRNA dynamics on the ribosome during translation. *Proc Natl Acad Sci U S A* **101**, 12893–12898.

34. S. C. Blanchard. 2009. Single-molecule observations of ribosome function. *Curr Opin Struct Biol* **19**, 103–109.

35. D. Moazed and H. F. Noller. 1989. Intermediate states in the movement of transfer RNA in the ribosome. *Nature* **342**, 142–148.

36. N. Demeshkina, L. Jenner, G. Yusupova and M. Yusupov. 2010. Interactions of the ribosome with mRNA and tRNA. *Curr Opin Struct Biol* **20**, 325–332.

37. J. M. Ogle, D. E. Brodersen, W. M. Clemons Jr., M. J. Tarry, A. P. Carter and V. Ramakrishnan. 2001. Recognition of cognate transfer RNA by the 30S ribosomal subunit. *Science* **292**, 897–902.

38. R. M. Voorhees, A. Weixlbaumer, D. Loakes, A. C. Kelley and V. Ramakrishnan. 2009. Insights into substrate stabilization from snapshots of the peptidyl transferase center of the intact 70S ribosome. *Nat Struct Mol Biol* **16**, 528–533.

39. W. Zhang, M. Kimmel, C. M. Spahn and P. A. Penczek. 2008. Heterogeneity of large macromolecular complexes revealed by 3D cryo-EM variance analysis. *Structure* **16**, 1770–1776.

40. W. Zhang, J. A. Dunkle and J. H. Cate. 2009. Structures of the ribosome in intermediate states of ratcheting. *Science* **325**, 1014–1017.

41. N. Fischer, A. L. Konevega, W. Wintermeyer, M. V. Rodnina and H. Stark. 2010. Ribosome dynamics and tRNA movement by time-resolved electron cryomicroscopy. *Nature* **466**, 329–333.

42. J. Frank and R. K. Agrawal. 2000. A ratchet-like inter-subunit reorganization of the ribosome during translocation. *Nature* **406**, 318–322.

43. D. N. Ermolenko, Z. K. Majumdar, R. P. Hickerson, P. C. Spiegel, R. M. Clegg and H. F. Noller. 2007. Observation of intersubunit movement of the ribosome in solution using FRET. *J Mol Biol* **370**, 530–540.

44. D. Fourmy, S. Yoshizawa and J. D. Puglisi. 1998. Paromomycin binding induces a local conformational change in the A-site of 16S rRNA. *J Mol Biol* **277**, 333–345.

45. M. Kaul and D. S. Pilch. 2002. Thermodynamics of aminoglycoside-rRNA recognition: The binding of neomycin-class aminoglycosides of the A site of 16S rRNA. *Biochemistry* **41**, 7695–7706.

46. V. Berk, W. Zhang, R. D. Pai and J. H. Cate. 2006. Structural basis for mRNA and tRNA positioning on the ribosome. *Proc Natl Acad Sci U S A* **103**, 15830–15834.

47. A. H. Ratje, J. Loerke, A. Mikolajka, M. Brünner, P. W. Hildebrand, A. L. Starosta, A. Dönhöfer, S. R. Connell, P. Fucini, T. Mielke, P. C. Whitford, J. N. Onuchic, Y. Yu, K. Y. Sanbonmatsu, R. K. Hartmann, P. A. Penczek, D. N. Wilson and C. M. Spahn. 2010. Head swivel on the ribosome facilitates translocation via intra-subunit tRNA hybrid sites. *Nature* **468**, 713–716.

48. Z. K. Majumdar, R. Hickerson, H. F. Noller and R. M. Clegg. 2005. Measurements of internal distance changes of the 30S ribosome using FRET with multiple donor-acceptor pairs: Quantitative spectroscopic methods. *J Mol Biol* **351**, 1123–1145.

49. M. Valle, A. Zavialov, W. Li, S. M. Stagg, J. Sengupta, R. C. Nielsen, P. Nissen, S. C. Harvey, M. Ehrenberg and J. Frank. 2003. Incorporation of aminoacyl-tRNA into the ribosome as seen by cryo-electron microscopy. *Nat Struct Biol* **10**, 899–906.

50. J. B. Munro, R. B. Altman, C. S. Tung, J. H. Cate, K. Y. Sanbonmatsu and S. C. Blanchard. 2010. Spontaneous formation of the unlocked state of the ribosome is a multistep process. *Proc Natl Acad Sci U S A* **107**, 709–714.

51. J. B. Munro, R. B. Altman, C. S. Tung, K. Y. Sanbonmatsu and S. C. Blanchard. 2010. A fast dynamic mode of the EF-G-bound ribosome. *EMBO J* **29**, 770–781.

52. P. V. Cornish, D. N. Ermolenko, D. W. Staple, L. Hoang, R. P. Hickerson, H. F. Noller and T. Ha. 2009. Following movement of the L1 stalk between three functional states in single ribosomes. *Proc Natl Acad Sci U S A* **106**, 2571–2576.

53. E. V. Bocharov, A. G. Sobol, K. V. Pavlov, D. M. Korzhnev, V. A. Jaravine, A. T. Gudkov and A. S. Arseniev. 2004. From structure and dynamics of protein L7/L12 to molecular switching in ribosome. *J Biol Chem* **279**, 17697–17706.

54. B. A. Shoemaker, J. J. Portman and P. G. Wolynes. 2000. Speeding molecular recognition by using the folding funnel: The fly-casting mechanism. *Proc Natl Acad Sci U S A* **97**, 8868–8873.

55. P. Geggier, R. Dave, M. B. Feldman, D. S. Terry, R. B. Altman, J. B. Munro and S. C. Blanchard. 2010. Conformational sampling of aminoacyl-tRNA during selection on the bacterial ribosome. *J Mol Biol* **399**, 576–595.

56. S. R. Connell, M. Topf, Y. Qin, D. N. Wilson, T. Mielke, P. Fucini, K. H. Nierhaus and C. M. Spahn. 2008. A new tRNA intermediate revealed on the ribosome during EF4-mediated back-translocation. *Nat Struct Mol Biol* **15**, 910–915.

57. J. Frank, J. Sengupta, H. Gao, W. Li, M. Valle, A. Zavialov and M. Ehrenberg. 2005. The role of tRNA as a molecular spring in decoding, accommodation, and peptidyl transfer. *FEBS Lett* **579**, 959–962.

58. P. C. Whitford, P. Geggier, R. B. Altman, S. C. Blanchard, J. N. Onuchic and K. Y. Sanbonmatsu. 2010. Accommodation of aminoacyl-tRNA into the ribosome involves reversible excursions along multiple pathways. *RNA* **16**, 1196–1204.

59. A. Malhotra, R. K. Tan and S. C. Harvey. 1990. Prediction of the three-dimensional structure of *Escherichia coli* 30S ribosomal subunit: A molecular mechanics approach. *Proc Natl Acad Sci U S A* **87**, 1950–1954.

60. A. Malhotra, R. K. Tan and S. C. Harvey. 1994. Modeling large RNAs and ribonucleoprotein particles using molecular mechanics techniques. *Biophys J* **66**, 1777–1795.

61. E. Villa, J. Sengupta, L. G. Trabuco, J. LeBarron, W. T. Baxter, T. R. Shaikh, R. A. Grassucci, P. Nissen, M. Ehrenberg, K. Schulten and J. Frank. 2009. Ribosome-induced changes in elongation factor Tu conformation control GTP hydrolysis. *Proc Natl Acad Sci U S A* **106**, 1063–1068.

62. F. Tama, M. Valle, J. Frank and C. L. Brooks III. 2003. Dynamic reorganization of the functionally active ribosome explored by normal mode analysis and cryo-electron microscopy. *Proc Natl Acad Sci U S A* **100**, 9319–9323.

63. Y. Wang, A. J. Rader, I. Bahar and R. L. Jernigan. 2004. Global ribosome motions revealed with elastic network model. *J Struct Biol* **147**, 302–314.

64. K. Y. Sanbonmatsu and S. Joseph. 2003. Understanding discrimination by the ribosome: stability testing and groove measurement of codon-anticodon pairs. *J Mol Biol* **328**, 33–47.

65. K. Y. Sanbonmatsu, S. Joseph and C. S. Tung. 2005. Simulating movement of tRNA into the ribosome during decoding. *Proc Natl Acad Sci U S A* **102**, 15854–15859.

66. K. Sanbonmatsu and S. Joseph. 2003. Understanding discrimination by the ribosome: Stability testing and groove measurement of codon-anticodon pairs. *J Mol Biol* **328**, 33–47.

67. V. I. Lim and J. F. Curran. 2001. Analysis of codon: Anticodon interactions within the ribosome provides new insights into codon reading and the genetic code structure. *RNA* **7**, 942–957.

68. M. S. VanLoock, R. K. Agrawal, I. S. Gabashvili, L. Qi, J. Frank and S. C. Harvey. 2000. Movement of the decoding region of the 16S ribosomal RNA accompanies tRNA translocation. *J Mol Biol* **304**, 507–515.

69. M. S. VanLoock, T. R. Easterwood and S. C. Harvey. 1999. Major groove binding of the tRNA/mRNA complex to the 16S ribosomal RNA decoding site. *J Mol Biol* **285**, 2069–2078.

70. P. Auffinger, S. Louise May and E. Westhof. 1999. Molecular dynamics simulations of solvated yeast tRNA(Asp). *Biophys J* **76**, 50–64.

71. W. Li, B. Ma and B. Shapiro. 2003. Binding interactions between the core central domain of 16S rRNA and the ribosomal protein S15 determined by molecular dynamics simulations. *Nucleic Acids Res* **31**, 629–638.

72. J. A. Mears, J. J. Cannone, S. M. Stagg, R. R. Gutell, R. K. Agrawal and S. C. Harvey. 2002. Modeling a minimal ribosome based on comparative sequence analysis. *J Mol Biol* **321**, 215–234.

73. F. Razga, J. Koca, J. Sponer and N. B. Leontis. 2005. Hinge-like motions in RNA kink-turns: The role of the second a-minor motif and nominally unpaired bases. *Biophys J* **88**, 3466–3485.

74. F. Razga, N. Spackova, K. Reblova, J. Koca, N. B. Leontis and J. Sponer. 2004. Ribosomal RNA kink-turn motif — a flexible molecular hinge. *J Biomol Struct Dyn* **22**, 183–194.

75. H. Ishida and S. Hayward. 2008. Path of nascent polypeptide in exit tunnel revealed by molecular dynamics simulation of ribosome. *Biophys J* **95**, 5962–5973.

76. X. Ge and B. Roux. 2010. Absolute binding free energy calculations of sparsomycin analogs to the bacterial ribosome. *J Phys Chem B* **114**, 9525–9539.

77. A. C. Vaiana and K. Y. Sanbonmatsu. 2009. Stochastic gating and drug-ribosome interactions. *J Mol Biol* **386**, 648–661.

78. M. Dlugosz and J. Trylska. 2009. Aminoglycoside association pathways with the 30S ribosomal subunit. *J Phys Chem B* **113**, 7322–7330.

79. J. Trylska, R. Konecny, F. Tama, C. L. Brooks III and J. A. McCammon. 2004. Ribosome motions modulate electrostatic properties. *Biopolymers* **74**, 423–431.

80. P. Chacon, F. Tama and W. Wriggers. 2003. Mega-Dalton biomolecular motion captured from electron microscopy reconstructions. *J Mol Biol* **326**, 485–492.

81. H. Gao, J. Sengupta, M. Valle, A. Korostelev, N. Eswar, S. M. Stagg, P. Van Roey, R. K. Agrawal, S. C. Harvey, A. Sali, M. S. Chapman and J. Frank. 2003. Study of the structural dynamics of the *E. coli* 70S ribosome using real-space refinement. *Cell* **113**, 789–801.

82. M. Karplus and J. A. McCammon. 2002. Molecular dynamics simulations of biomolecules. *Nat Struct Biol* **9**, 646–652.

83. S. Berneche and B. Roux. 2001. Energetics of ion conduction through the K+ channel. *Nature* **414**, 73–77.

84. M. A. Young, S. Gonfloni, G. Superti-Furga, B. Roux and J. Kuriyan. 2001. Dynamic coupling between the SH2 and SH3 domains of c-Src and Hck underlies their inactivation by C-terminal tyrosine phosphorylation. *Cell* **105**, 115–126.

85. R. A. Bockmann and H. Grubmuller. 2002. Nanoseconds molecular dynamics simulation of primary mechanical energy transfer steps in F1-ATP synthase. *Nat Struct Biol* **9**, 198–202.

86. J. A. McCammon, B. R. Gelin and M. Karplus. 1977. Dynamics of folded proteins. *Nature* **267**, 585–590.

87. S. Sporlein, H. Carstens, H. Satzger, C. Renner, R. Behrendt, L. Moroder, P. Tavan, W. Zinth and J. Wachtveitl. 2002. Ultrafast spectroscopy reveals subnanosecond peptide conformational dynamics and validates molecular dynamics simulation. *Proc Natl Acad Sci U S A* **99**, 7998–8002.

88. D. Van Der Spoel, E. Lindahl, B. Hess, G. Groenhof, A. E. Mark and H. J. Berendsen. 2005. GROMACS: Fast, flexible, and free. *J Comput Chem* **26**, 1701–1718.

89. L. G. Trabuco, E. Schreiner, J. Eargle, P. Cornish, T. Ha, Z. Luthey-Schulten and K. Schulten. 2010. The role of L1 stalk-tRNA interaction in the ribosome elongation cycle. *J Mol Biol* **402**, 741–760.

90. L. G. Trabuco, C. B. Harrison, E. Schreiner and K. Schulten. 2010. Recognition of the regulatory nascent chain TnaC by the ribosome. *Structure* **18**, 627–637.

91. K. Y. Sanbonmatsu and S. Joseph. 2003. Understanding discrimination by the ribosome: Stability testing and groove measurement of codon-anticodon pairs. *J Mol Biol* **328**, 33–47.

92. K. Y. Sanbonmatsu and A. E. Garcia. 2002. Structure of Met-enkephalin in explicit aqueous solution using replica exchange molecular dynamics. *Proteins* **46**, 225–234.

93. K. Y. Sanbonmatsu. 2006. Energy landscape of the ribosomal decoding center. *Biochimie* **88**, 1053–1059.

94. Q. Vicens and E. Westhof. 2003. Crystal structure of geneticin bound to a bacterial 16S ribosomal RNA A site oligonucleotide. *J Mol Biol* **326**, 1175–1188.

95. A. E. Garcia and K. Y. Sanbonmatsu. 2001. Exploring the energy landscape of a beta hairpin in explicit solvent. *Proteins* **42**, 345–354.

96. A. Meskauskas and J. D. Dinman. 2007. Ribosomal protein L3: Gatekeeper to the A site. *Mol Cell* **25**, 877–888.

97. J. L. Baxter-Roshek, A. N. Petrov and J. D. Dinman. 2007. Optimization of ribosome structure and function by rRNA base modification. *PLoS One* **2**, e174.

98. P. C. Whitford, J. K. Noel, S. Gosavi, A. Schug, K. Y. Sanbonmatsu and J. N. Onuchic. 2009. An all-atom structure-based potential for proteins: bridging minimal models with all-atom empirical forcefields. *Proteins* **75**, 430–441.

99. P. C. Whitford, A. Schug, J. Saunders, S. P. Hennelly, J. N. Onuchic and K. Y. Sanbonmatsu. 2009. Nonlocal helix formation is key to understanding S-adenosylmethionine-1 riboswitch function. *Biophys J* **96**, L7–9.

100. J. K. Noel, P. C. Whitford, K. Y. Sanbonmatsu and J. N. Onuchic. 2010. SMOG@ctbp: Simplified deployment of structure-based models in GROMACS. *Nucleic Acids Res* **38**(Suppl.), W657–661.

Chaperonins

5

The Machines Which Fold Proteins

Del Lucent, Martin C Stumpe and Vijay S Pande

---

1. Introduction

In so much as the ribosome is a machine that makes the parts for other machines, molecular chaperones are machines that help put together those parts to assemble new machines. Molecular chaperones are a class of proteins that exist within all cellular life. The cell contains a variety of these molecules, but they all share a common goal of maintaining the stability of the cellular proteome. Various tasks are required to meet this goal including shuttling unfolded proteins around the cell, dis-aggregating proteins, unfolding misfolded proteins, acting as proline isomerases, and most importantly acting as a catalyst for proteins to assemble themselves into their native state ("protein folding").

This ability to help other proteins fold ("foldase activity") is of particular importance to cellular viability since many proteins are large and require the assistance of chaperones to reach their native state amidst the crowded macromolecular milieu of the cytosol.[1,2] Additionally, cells are frequently subject to varying conditions of temperature, pH, and ionic concentration that stress the stability of the cellular proteome. The upregulation of chaperones serves as a mechanism to allow survival of the cell in an adverse environment.[3] For this reason, chaperones were historically referred to as heat shock proteins. Furthermore, while not all proteins actually require chaperones for folding, the existence of chaperones allows for much more flexibility in sequence space: evolution can get from one sequence to a different sequence by a path on which a given protein might not be able to fold on its own, but with the help of chaperones. This opens up much more room for functional mutations, which often have a negative impact on folding and stability.

Among the various molecular chaperones are a special class known as chaperonins. Chaperonins are large multimeric complexes known for their ability to engulf misfolded proteins and allow them to reach their native states more effectively. One may ask if such a large protein-folding machine is truly needed given that there are a multitude of other molecular chaperones that share similar duties with the chaperonins. Proteomic analysis reveals that chaperonins are in fact needed, as they are necessary catalysts for the folding of a number of essential proteins.[2,4,5] For this reason, without functioning chaperonins a cell will inevitably die. This is a contrast to other molecular chaperones that can effectively compensate for each other if one is

87

inactivated.[4,6] Similarly, chaperones usually do not service only a specific kind of target protein each, but instead can facilitate the folding of multiple different proteins. Chaperonin selectivity, on the other hand, can vary from completely general to highly specific (as we shall discuss below).

2. GroEL

The canonical example of a chaperonin (and the molecule on which we will focus much of our discussion) is the prokaryotic chaperonin GroEL. GroEL is best characterized as two large rings stacked back to back. Each ring is composed of seven identical subunits arranged radially around a large central cavity (Figure 1). This cavity constitutes an active site for protein folding. Additionally, each of these rings has the ability to open and close with the aid of ATP and a capping protein known as GroES. The closed chaperonin chamber can trap a protein inside and assist in its folding. Each subunit has a mass of approximately 60 kilodaltons and contains three domains: an equatorial ATPase domain, a hinge domain, and an apical substrate-binding domain. Each subunit in a given ring undergoes a large-scale conformational change upon the binding of an ATP molecule. This includes shifting the substrate-binding domain so as to allow the chamber to close (with the aid of GroES). This closed chamber is larger in volume than the open chamber and its surface can be roughly characterized as hydrophilic in nature, as opposed to the open conformation, which is mostly hydrophobic (Figure 4a).

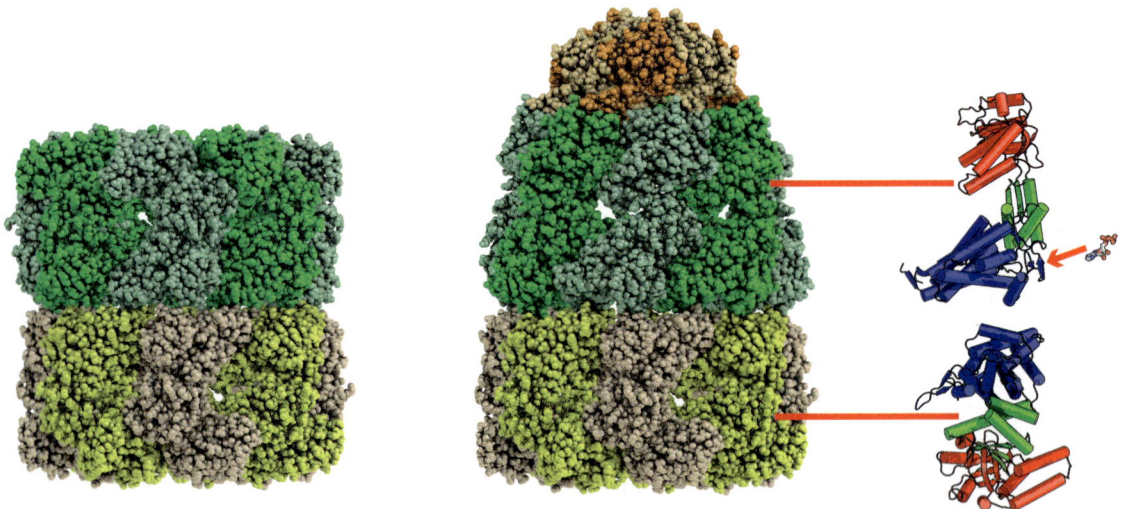

Figure 1. Displayed above is the structure of the open (left side) and closed (right side) conformations of the GroEL complex. Subunits of the cis ring are rendered in alternating light and dark shades of green, while subunits of the trans cavity are rendered in alternating yellow and tan. The capping protein GroES is rendered as alternating shades of light and dark orange. Although GroEL/GroES is a homo-oligomer, this coloration allows easy recognition of subunit interfaces. On the right, a subunit from the open and closed conformations are rendered as cartoons, as well as the approximate location of the ATP-binding pocket. For these subunits, the equatorial domain is colored blue, the linker domain green, and the apical domain red.

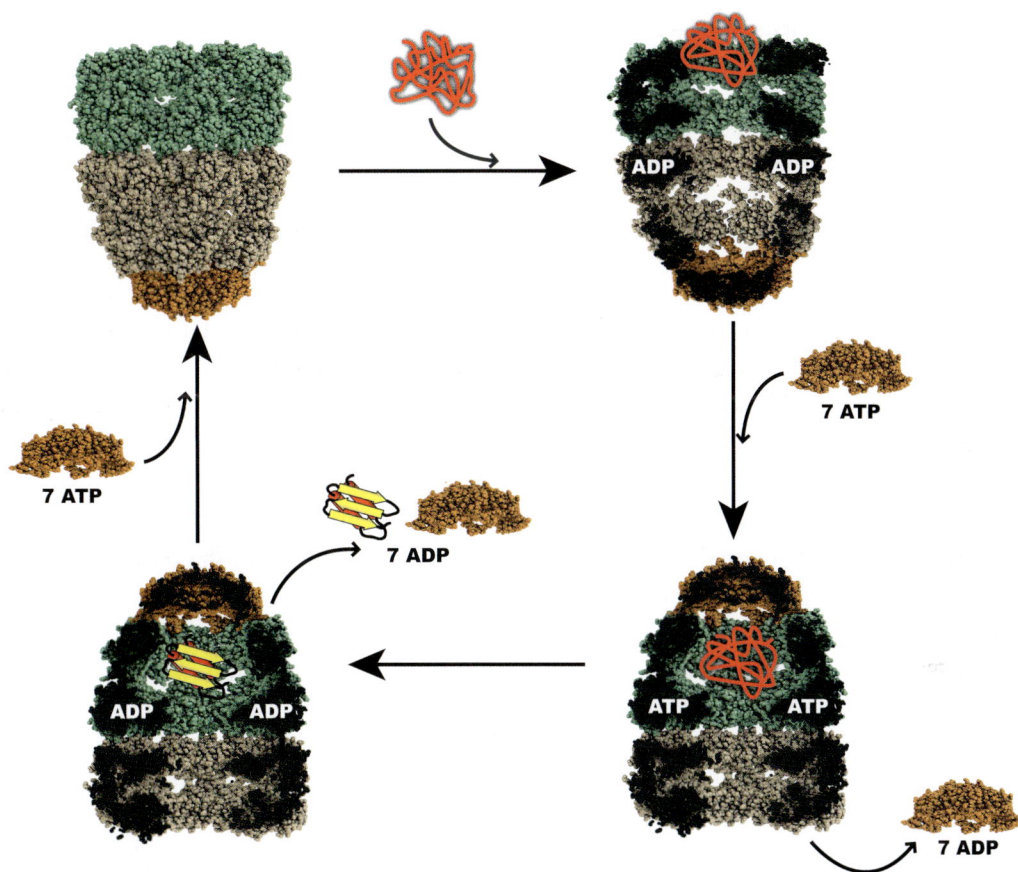

Figure 2. Shown here is the reaction cycle of the chaperonin GroEL. The cis ring is rendered in green while the trans ring is rendered in tan. GroES is rendered in orange. The un/misfolded protein is shown schematically as a random coil (in red) while the folded protein is shown schematically as red helices and yellow sheets.

GroEL has an interesting, albeit slightly complex reaction cycle (Figure 2). The hallmark of this cycle is a negative cooperativity between the rings. This has led to a characterization of GroEL as a sort of two-stroke engine, wherein one ring is open while the other is closed. Furthermore, the closing of one ring facilitates the opening of the other.[7] The reaction cycle begins with an unfolded or misfolded protein binding to the open ring (we will call the cis ring). In the case of GroEL, this is usually achieved through lipophilic interaction of exposed hydrophobic moieties of the substrate protein with the apical domains of the cis ring.[2] This binding also aids in the unfolding of substrate protein from a potentially kinetically trapped state. The binding of ATP to each subunit triggers the closing of the cis ring (and opening of the trans ring). This closing action translocates the protein into the cis cavity and, with the aid of GroES, seals the cavity. The hydrolysis of ATP in each subunit acts as a rate-limiting step for ATP binding in the trans cavity and subsequent opening of the cis cavity. Once the cis cavity opens, GroES, seven ADP molecules and the substrate protein are released into the cytosol. The substrate protein may have reached its native conformation by this point, or may still be unfolded or misfolded. In the case of the latter, the chaperonin can re-sequester the substrate protein and allow additional attempts at folding.

3. How Do Chaperonins Work?

In spite of this complex reaction cycle, which is relatively well characterized biochemically, the precise mechanism by which this process facilitates folding is a topic of much debate. There have been a number of conflicting experiments and simulations performed in recent years regarding how a chaperonin can facilitate protein folding. The emerging paradigm for chaperonin function, however, is what one could call a protein folding "Swiss Army knife", where different substrate proteins take advantage of different chaperonin mediated mechanisms to reach their native state. Thus, the "conflicting evidence" really represents the actual diversity of tools the chaperonin uses to perform its task. This seems to be a reasonable way to characterize chaperonin-mediated protein folding, considering the incredible number of cellular proteins that are chaperonin substrates.

3.1. *The Anfinsen cage hypothesis*

The simplest possible mechanism by which a chaperonin can affect protein folding is referred to as the Anfinsen cage hypothesis.[8] Christian Anfinsen postulated that all of the information needed for a protein to correctly reach its functional conformation is contained in its one-dimensional protein sequence.[9] If a protein cannot fold properly in the cell, this may simply be the result of interference from other macromolecules in the crowded cellular milieu. Thus, a chaperonin simply serves as an isolation chamber for protein folding, where inside, the protein is allowed to fold in essentially infinite dilution. This is in fact true for a number of proteins, which would fold productively *in vitro* (i.e. in a buffered solvent with low protein concentration) but aggregate *in vivo*. The presence of chaperonins prevents this aggregation and allows productive folding.

There are, however, a number of proteins that cannot fold without the aid of a chaperonin, even at extremely low concentrations.[10] Furthermore, there are many proteins that fold more easily when allowed to bind a chaperonin, but are not actually enclosed in the chaperonin cavity.[11] The Anfinsen cage model is insufficient in explaining these observations. A new model was needed to account for these observations, and from this necessity came what is known as iterative annealing.

3.2. *Iterative annealing*

The premise of iterative annealing is that the chaperonin can act as an unfoldase (it can unfold misfolded proteins). This has been demonstrated experimentally for a number of substrate proteins. But how can acting as an unfoldase lead to productive folding and increased cellular concentration of functional native proteins? This paradox can be easily explained by considering a balance of rates. If a protein misfolds faster than it folds, the equilibrium concentration of protein is shifted to the misfolded state. This state is usually kinetically trapped, in that reaching the native state from this misfolded state is incredibly slow. If a chaperonin can bind and unfold these trapped proteins, it can essentially give them a new chance to fold productively. If the rate of chaperonin-mediated unfolding is greater than the rate of reaching the native state from the misfolded state, then the equilibrium can be shifted in favor of folded protein.[12] For this to work, the chaperonin would need to have an optimal affinity for the substrate protein. If the chaperonin binds to the misfolded protein too tightly, the protein would spend most of its time stuck to the chaperonin and not be given sufficient chance to refold.[13,14]

But how can binding by a chaperonin unfold a protein? There are two general hypotheses for this. The first is that the chaperonin's conformational change is used to perform mechanical work on the protein, stretching it out from its misfolded state. The second hypothesis states that binding hydrophobic moieties of the misfolded state induces unfolding by stabilizing the unfolded state (or destabilizing the misfolded state). There is some experimental evidence for both of these models and it stands to reason that both can be viable ways to induce unfolding by the chaperonin for different substrate proteins.[3]

3.3. *The landscape modulation hypothesis: Confinement*

It would seem that the combination of iterative annealing and infinite dilution (Anfinsen cage) could account for a large amount of the biochemical evidence for the chaperonin's ability to fold proteins. Interestingly, current research has revealed that there must be even more tools in the chaperonin's protein folding repertoire. It is known that for some substrate proteins, the chaperonin can accelerate protein folding in a single round of chaperonin activity.[15–17] This acceleration indicates that for some proteins, the closed chaperonin cavity is not a passive cage, but rather an active force guiding the protein towards its native state (i.e. it acts as a foldase). We refer to this model for chaperonin function as landscape modulation (the free energy landscape the protein explores is modulated or modified by encapsulation inside the chaperonin).[14,18]

How may encapsulation facilitate foldase activity? What components of the chaperonin cavity are likely to affect folding? To answer these questions, it is reasonable to start from a simple model. The simplest way to envision the chaperonin cavity is as an inert confining volume. If we were to simplify the process of protein folding, the energy landscape would consist of two macro-states: the folded state and the unfolded state. Under this model, it is reasonable to assume that conformational entropy is more dominant in the unfolded state than the folded state (the density of micro-states of the unfolded ensemble is much larger than that of the folded state). By enclosing the unfolded substrate protein in a smaller volume, some expanded micro-states (conformations with a large radius of gyration) are effectively removed from the ensemble. This reduces the entropy of the unfolded state and concordantly increases its free energy. The native state, on the other hand, should be dominated by a very small number of compact conformations and thus should not be thermodynamically perturbed by confinement like the unfolded state. Thus, by destabilizing the unfolded state, the relative stability of the native state is increased. Furthermore, in the case where the transition state is not significantly perturbed by confinement (relative to the unfolded state) a net increase in folding rate should occur (Figure 3). This model has been extensively characterized theoretically[14,18] and has been recently explored experimentally via mutagenesis of the chaperonin subunit C-terminus so as to increase or decrease the volume of the cavity.[17] One major shortcoming of this model however is that it does not explicitly consider the fact that solvent molecules are trapped inside the chaperonin cavity along with the substrate protein. Recent protein folding simulations have shown that including solvent molecules may change the results one would see from polymeric confinement.[19] Nonetheless the idea for confinement-induced stabilization due to polymer entropy effects is still a reasonable model for chaperonin mediated protein folding.

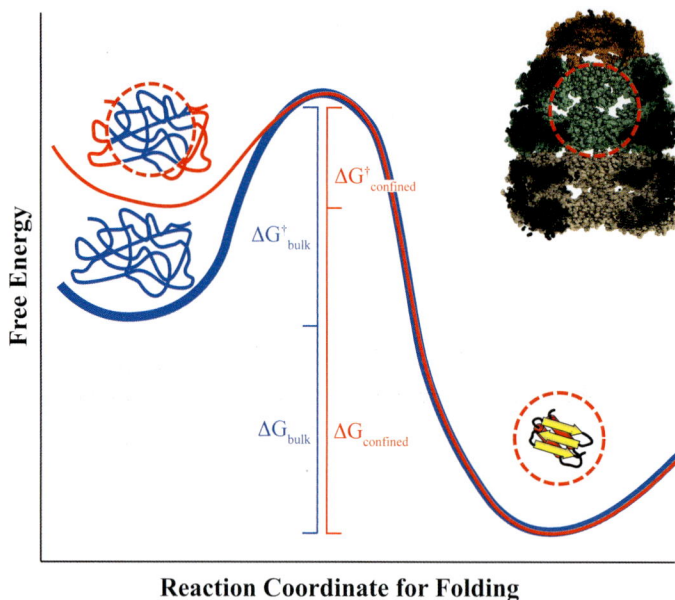

Figure 3. This diagram shows the theoretical basis for confinement induced folding mediated by polymer entropy. The blue line represents the free energy for the protein folding transition. The red line represents the perturbation to this landscape as a result of polymeric confinement. The confining volume is shown as a dashed red circle. The relative effects of confinement on the folding free energy (ΔG) and the transition state ($\Delta G^{\dagger}$) are also shown.

3.4. *The landscape modulation hypothesis: Chaperonin-enhanced hydrophobic effect*

We know that the chaperonin is not an inert confining volume, but rather a hydrophilic confining surface (Figure 4a). What effect would this surface have on protein folding? We know that if the confined protein can be considered a hydrophobic globule of sorts, it stands to reason that a hydrophilic cavity will prevent the globule from sticking to the sides of the cavity. But again, if we consider the thermodynamics of the confined solvent (water molecules that are inside the chaperonin cavity along with the substrate protein), this picture changes. Water is known to organize itself differently when close to a hydrophilic versus a hydrophobic surface.[20–23] Near a hydrophilic surface (the chaperonin cavity surface), the water will want to orient its dipole in the direction of the electrostatic field created by the surface and/or form potential hydrogen bonds with this surface. On the other hand, water near a hydrophobic surface (the surface of the unfolded protein) would want to do whatever it could to preserve its own network of hydrogen bonds, since it cannot form hydrogen bonds with the hydrophobic surface. As we bring these two surfaces together (Figure 4b) the two competing effects essentially put frustrating thermodynamic demands on the confined water. This mediates a net repulsive force between the two surfaces. In the context of a chaperonin and substrate protein, we would see this force manifest as an increased hydrophobic effect for the confined protein.[23–25]

This theoretical characterization of the chaperonin-enhanced hydrophobic effect has been recently verified experimentally for a mutant of maltose-binding protein confined to GroEL. In

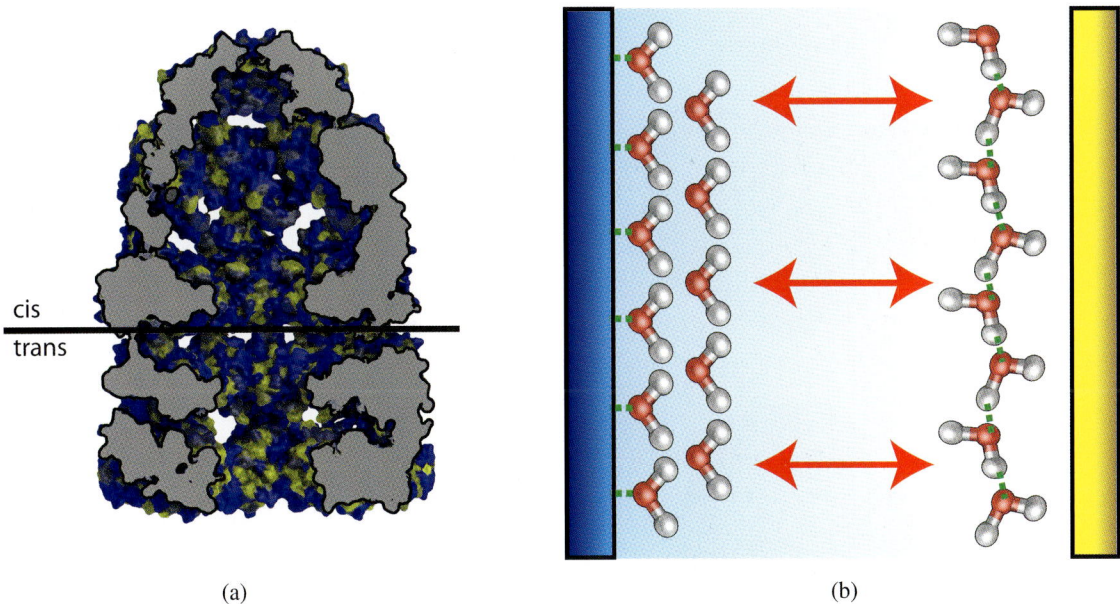

(a)   (b)

Figure 4. Panel a shows the structure of GroEL cut in half. The molecular surface is rendered and colored according to Kyte-Doolittle hydropathic index (with blue being hydrophilic and yellow hydrophobic). Panel b shows a simple model demonstrating the idea of solvent mediated foldase behavior. Water molecules between two surfaces (one hydrophobic and one hydrophilic) are orienting themselves differently. Red arrows signify the solvent mediated repulsion between the two surfaces arising as a result of the frustration that water endures when confined to such a microenvironment. The color gradient behind the water molecules shows the relative solvent density between the surfaces.

these experiments, charged residues lining the chaperonin cavity were mutated (from negative to positive and negative to neutral-polar).[17] The resulting changes in foldase activity were highly correlated with the hydrophilicity of the mutant GroEL cavity surface.[26] Among the suite of proteins that GroEL folds, there is an enrichment of TIM barrel folds,[4] a protein motif that is known to be highly dependant on hydrophobic collapse in its folding mechanism.[27] When considering this fact, it stands to reason that although this model has only been verified for a single substrate protein, it may actually be a somewhat general phenomenon.

4. Other Chaperonins

Together, all of these models and results paint a picture of GroEL as a sort of cellular MacGyver, that uses it's protein folding Swiss Army knife to fix misfolded proteins on the fly as the cell goes about its various adventures. Indeed, the number of potential GroEL substrates is quite large (50% of the proteome) although a much smaller number actually require GroEL to correctly reach their native conformations in the absence of environmental stress. Having many different methods at its disposal seem to allow GroEL to act as a protein-folding general practitioner and meet the needs of a great variety of substrate proteins.

But if GroEL can be thought of as a protein-folding general practitioner, what happens when there is significant need for a protein-folding specialist? By contrasting GroEL (the

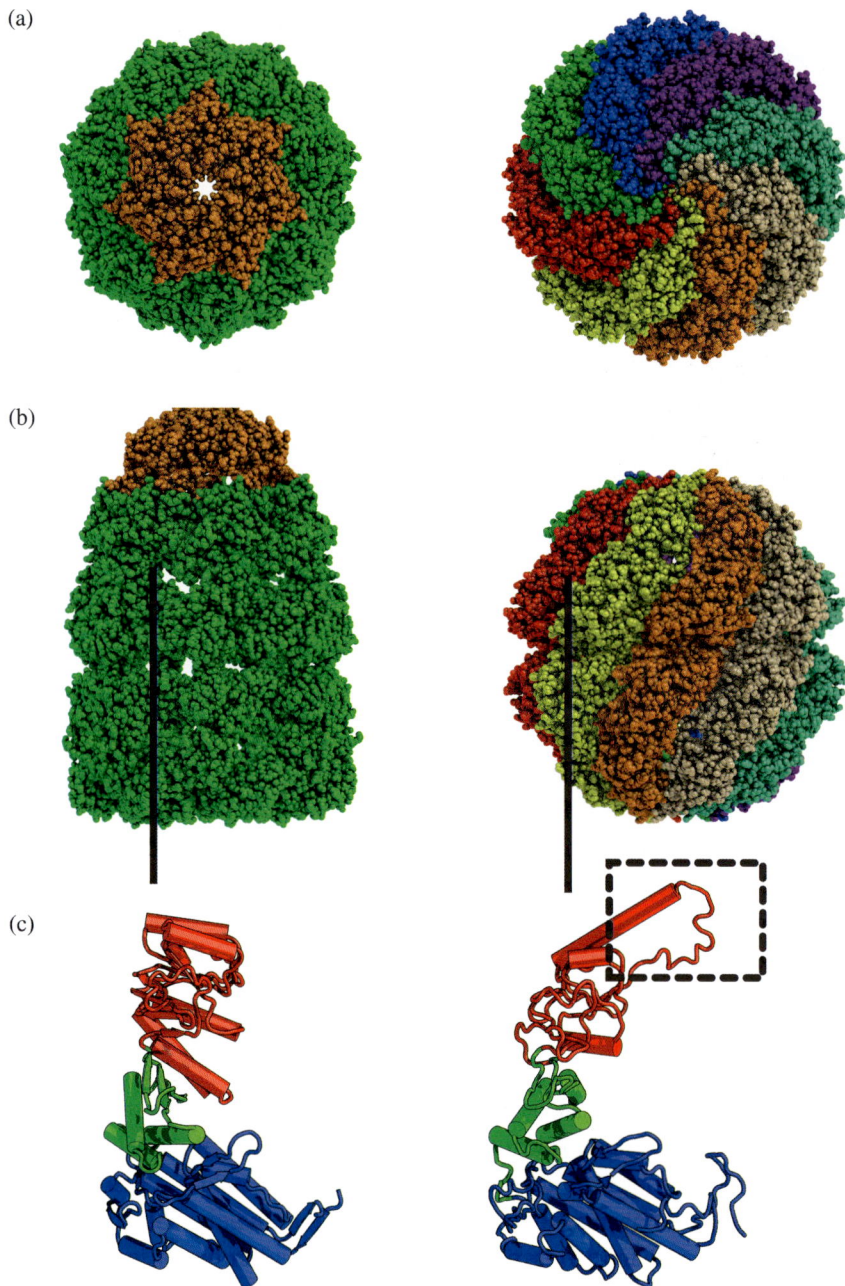

Figure 5.   Panel a shows a group 1 chaperonin (GroEL) on the left and a group 2 chaperonin (homology model of TRiC based on the thermosome) on the right (both viewed from the top). Panel b shows the same molecules viewed from the side. The molecules are colored to show the difference in subunits. GroEL has seven identical subunits per ring and a capping protein GroES with seven identical subunits. TRiC has eight different subunits per ring and no capping protein. Panel C shows an individual subunit from GroEL and an individual subunit from TRiC in the closed conformation. The dashed box highlights the extended apical domain found in group 2 chaperonins, which forms the built-in lid. It should be noted that the symmetric complex (i.e. two closed rings) shown here for group 2 chaperonins is an artifact of crystallization conditions (other lower resolution measures such as SAXS and Cryo-EM show the complex to be asymmetric much like the group 1 chaperonins).

prokaryotic chaperonin) with the thermosome (the archeal chaperonin) and TRiC (the eukaryotic chaperonin), we can explore this interesting question of cellular evolution. It turns out that not all chaperonins are created equal. GroEL, which is found only in prokaryotes (although a similar chaperonin is found in eukaryotic chloroplasts and mitochondria), is referred to as a group 1 chaperonin, while the cytosolic chaperonins of archea and eukaryotes are referred to as group 2 chaperonins (Figure 5).

There are a number of striking differences between group 1 and group 2 chaperonins. The most obvious is that the group 2 chaperonins do not require a co-protein (like GroES) to seal their cavities, but rather have an extended apical domain that acts as a built-in lid. The closing of a group 2 chaperonin resembles the closing of the aperture of a camera. Another difference is the number and diversity of subunits composing the cavity. The group 2 chaperonins generally have eight subunits per ring as opposed to the group 1 chaperonins, which have only seven. Additionally, the subunits can be homo-oligomers or hetero-oligimers. The chaperonin Mm-Cpn (found in the mesophillic archae *Methanococcus maripaludis*) is a homo-oligomer, while the thermosome (found in *Thermoplasma acidophilum*) is a hetero-oligomer (four alpha and four beta subunits per ring), and TRiC (found in eukaryotic cytosol) is a hetero-oligomer with eight different subunits per ring. This incredible diversity lends credibility to the assertion that group 2 chaperonins have evolved to deal with more specialized circumstances than those of group 1.[28]

TRiC is of particular interest considering that it has eight different subunits. The main difference between these subunits and those of GroEL comes in the apical domain, where the chaperonin interacts with substrate protein. It is known that the eukaryotic cytoskeletal proteins actin and tubulin are both obligate substrates of TRiC. When considering the immense importance of these substrates, it is reasonable to assume that one of TRiC's major tasks is to maintain these proteins in their folded conformations. But TRiC also interacts with a great number of other substrates as well. Recent experimental work concerning the binding motifs of a specific model protein to the various TRiC subunits has revealed that different peptide sequences can bind different subunits with varying degrees of affinity.[29] This seems to support the idea that TRiC has evolved to be a specialist in folding cytoskeletal proteins, but still retains some of its characteristics as a generalist as well. This evolutionary phenomenon was also demonstrated in a slightly different context with GroEL. Successive rounds of directed evolution selecting for GroEL's ability to fold the photoluminescent protein GFP has revealed that GroEL can be evolved to act as a specialist (much like a group 2 chaperonin).[30] But, this specialization in folding GFP comes at the cost of effectiveness in folding the other members in the diverse prokaryotic proteome.

5. Conclusions

In spite of our increasing body of knowledge regarding the structure, function, and mechanism of chaperonins, there is still a great amount that is unknown. How important are the different mechanisms of chaperonin action relative to each other for different substrate proteins and in different organisms? Is more information needed to fully understand the mechanisms of chaperonin-mediated protein folding (especially in the case of group 2 chaperonins)? What is it about the group 2 structures that make them more prevalent in extremophillic bacteria as well as the eukaryotic cytosol? With current theoretical, computational, and experimental techniques continuing to advance to meet the challenge of dealing with these large complex protein-folding machines, we can hope to have answers to these questions in the future.

Finally, it is useful to point out that the continuing study of chaperonins can have tremendous applied consequences in the fields of health and biotechnology. Evidence is beginning to build, supporting the idea that including chaperonins with *in vitro* protein preparations can stabilize poorly behaved protein complexes (such as RUBISCO and the VHL tumor suppressor) allowing them to be studied more easily.[31,32] Additionally, there currently are attempts underway to engineer chaperonins to bind different substrates, favor one folding mechanism over another, and even be used as tools of drug delivery. Finally, considering the fact that other molecular chaperones have become the targets of small molecule inhibitors (for example, the hsp90 chaperonin is emerging as a viable inhibition target for cancer therapy), it seems reasonable to begin to consider small molecule modulation of chaperonin activity as possible antimicrobial, antiviral, and anticancer treatments. Of course much work needs to be done in order to validate chaperonins as reasonable therapeutic targets, but such exciting possibilities are hopefully waiting on the horizon for those who study these fascinating protein-folding machines.

Suggested Additional Reading Materials

F. U. Hartl and M. Hayer-Hartl. 2002. Molecular chaperones in the cytosol: from nascent chain to folded protein. *Science* **295**, 1852–1858.

A. I. Jewett and J. E. Shea. 2009. Reconciling theories of chaperonin accelerated folding with experimental evidence. *Cell Mol Life Sci* **67(2)**, 255–276.

Z. Lin and H. S. Rye. 2006. GroEL-mediated protein folding: making the impossible, possible. *Crit Rev Biochem Mol Biol* **41**, 211–239.

D. Lucent, J. England and V. Pande. 2009. Inside the chaperonin toolbox: theoretical and computational models for chaperonin mechanism. *Phys Biol* **6**, 15003.

References

1. W. A. Fenton and A. L. Horwich. 2003. Chaperonin-mediated protein folding: Fate of substrate polypeptide. *Q Rev Biophys* **36**, 229–256.
2. F. U. Hartl and M. Hayer-Hartl. 2002. Molecular chaperones in the cytosol: From nascent chain to folded protein. *Science* **295**, 1852–1858.
3. Z. Lin and H. S. Rye. 2006. GroEL-mediated protein folding: Making the impossible, possible. *Crit Rev Biochem Mol Biol* **41**, 211–239.
4. M. J. Kerner, D. J. Naylor, Y. Ishihama, T. Maier, H. C. Chang, A. P. Stines, C. Georgopoulos, D. Frishman, M. Hayer-Hartl, M. Mann and F. U. Hartl. 2005. Proteome-wide analysis of chaperonin-dependent protein folding in *Escherichia coli*. *Cell* **122**, 209–220.
5. C. Spiess, A. S. Meyer, S. Reissmann and J. Frydman. 2004. Mechanism of the eukaryotic chaperonin: Protein folding in the chamber of secrets. *Trends Cell Biol* **14**, 598–604.
6. P. Genevaux, F. Keppel, F. Schwager, P. S. Langendijk-Genevaux, F. U. Hartl and C. Georgopoulos. 2004. *In vivo* analysis of the overlapping functions of DnaK and trigger factor. *EMBO Rep* **5**, 195–200.

7. O. Yifrach and A. Horovitz. 1995. Nested cooperativity in the ATPase activity of the oligomeric chaperonin GroEL. *Biochemistry* **34**, 5303–5308.

8. R. J. Ellis. 1994. Molecular chaperones. Opening and closing the Anfinsen cage. *Curr Biol* **4**, 633–635.

9. C. B. Anfinsen. 1973. Principles that govern the folding of protein chains. *Science* **181**, 223–230.

10. J. S. Weissman, Y. Kashi, W. A. Fenton and A. L. Horwich. 1994. GroEL-mediated protein folding proceeds by multiple rounds of binding and release of nonnative forms. *Cell* **78**, 693–702.

11. T. K. Chaudhuri, G. W. Farr, W. A. Fenton, S. Rospert and A. L. Horwich. 2001. GroEL/GroES-mediated folding of a protein too large to be encapsulated. *Cell* **107**, 235–246.

12. M. J. Todd, G. H. Lorimer and D. Thirumalai. 1996. Chaperonin-facilitated protein folding: Optimization of rate and yield by an iterative annealing mechanism. *Proc Natl Acad Sci U S A* **93**, 4030–4035.

13. M. R. Betancourt and D. Thirumalai. 1999. Exploring the kinetic requirements for enhancement of protein folding rates in the GroEL cavity. *J Mol Biol* **287**, 627–644.

14. D. Lucent, J. England and V. Pande. 2009. Inside the chaperonin toolbox: Theoretical and computational models for chaperonin mechanism. *Phys Biol* **6**, 015003.

15. A. Brinker, G. Pfeifer, M. J. Kerner, D. J. Naylor, F. U. Hartl and M. Hayer-Hartl. 2001. Dual function of protein confinement in chaperonin-assisted protein folding. *Cell* **107**, 223–233.

16. Z. Lin and H. S. Rye. 2004. Expansion and compression of a protein folding intermediate by GroEL. *Mol Cell* **16**, 23–34.

17. Y. C. Tang, H. C. Chang, A. Roeben, D. Wischnewski, N. Wischnewski, M. J. Kerner, F. U. Hartl and M. Hayer-Hartl. 2006. Structural features of the GroEL-GroES nano-cage required for rapid folding of encapsulated protein. *Cell* **125**, 903–914.

18. J. England, D. Lucent and V. Pande. 2008. Rattling the cage: Computational models of chaperonin-mediated protein folding. *Curr Opin Struct Biol* **18**(2), 163–169.

19. D. Lucent, V. Vishal and V. S. Pande. 2007. Protein folding under confinement: A role for solvent. *Proc Natl Acad Sci U S A* **104**, 10430–10434.

20. D. Chandler. 2005. Interfaces and the driving force of hydrophobic assembly. *Nature* **437**, 640–647.

21. J. Dzubiella and J. P. Hansen. 2003. Reduction of the hydrophobic attraction between charged solutes in water. *J Chem Phys* **119**, 12049–12052.

22. J. Dzubiella and J. P. Hansen. 2004. Competition of hydrophobic and Coulombic interactions between nanosized solutes. *J Chem Phys* **121**, 5514–5530.

23. J. L. England and V. S. Pande. 2008. Potential for modulation of the hydrophobic effect inside chaperonins. *Biophys J* **95**, 3391–3399.

24. D. K. Eggers and J. S. Valentine. 2001. Molecular confinement influences protein structure and enhances thermal protein stability. *Protein Sci* **10**, 250–261.

25. J. L. England, S. Park and V. S. Pande. 2008. Theory for an order-driven disruption of the liquid state in water. *J Chem Phys* **128**, 044503.

26. J. L. England, D. Lucent and V. Pande. 2008. A role for confined water in chaperonin function. *J Am Chem Soc* **130**, 11838–11839.

27. S. Selvaraj and M. M. Gromiha. 2003. Role of hydrophobic clusters and long-range contact networks in the folding of (alpha/beta) 8 barrel proteins. *Biophys J* **84**, 1919–1925.

28. M. G. Bigotti and A. R. Clarke. 2008. Chaperonins: The hunt for the Group II mechanism. *Arch Biochem Biophys* **474**, 331–339.

29. C. Spiess, E. J. Miller, A. J. McClellan and J. Frydman. 2006. Identification of the TRiC/CCT substrate binding sites uncovers the function of subunit diversity in eukaryotic chaperonins. *Mol Cell* **24**, 25–37.

30. J. D. Wang, C. Herman, K. A. Tipton, C. A. Gross and J. S. Weissman. 2002. Directed evolution of substrate-optimized GroEL/S chaperonins. *Cell* **111**, 1027–1039.

31. D. E. Feldman, V. Thulasiraman, R. G. Ferreyra and J. Frydman. 1999. Formation of the VHL-elongin BC tumor suppressor complex is mediated by the chaperonin TRiC. *Mol Cell* **4**, 1051–1061.

32. C. Liu, A. L. Young, A. Starling-Windhof, A. Bracher, S. Saschenbrecker, B. V. Rao, K. V. Rao, O. Berninghausen, T. Mielke, F. U. Hartl, R. Beckmann and M. Hayer-Hartl. Coupled chaperone action in folding and assembly of hexadecameric Rubisco. *Nature* **463**, 197–202.

Muscle and Myosin

6

Ronald S. Rock

1. A Structural Basis for Motion

Although muscle is a complex tissue where many proteins are required to organize the proper structure, the action of two main proteins drives muscle contraction. Since these two components are highly enriched in muscle, it is possible to extract and biochemically purify large quantities of muscle proteins from whole tissue using a straightforward series of buffer extraction steps.[1] This feature is a considerable advantage for those of us that work on contractility. Indeed, this system has become one of the darlings of biochemistry and biophysics (with around 12 000 PubMed citations that include "myosin" in the title) for exactly this reason. In this chapter, I will describe muscle in molecular terms, focusing on the structures and biochemistry of the central players, myosin and actin, and eventually building up to the organization of muscle tissue. However, keep in mind that, historically, many of the key ideas germinated in studies of muscle physiology, while the molecular picture developed later.

The first key player, known as actin, is a globular protein that polymerizes into a linear filament (Figure 1a). This form is called F-actin (filamentous actin), or sometimes a "thin filament". Although F-actin is found in all eukaryotic cells where its assembly, disassembly, and spatial organization is highly dynamic,[2,3] in the context of a muscle fiber, we can consider it to be a fairly static track, or a substrate for a muscle motor. The actin filament is helical, forming a 1-start left-handed helix. However, the resulting filament structure appears to have two long-pitch helical strands, wrapped around each other in a right-handed manner. These two helices cross over each other every 36 nm, using 13 actin monomers in that span.[4–6] The actin filament is polar, with distinct ends known as the barbed and pointed ends.

The second protein, myosin, is the engine of muscle contraction. In the simplest sense, myosin latches onto the actin filament and pulls, in much the same way that a person would pull on a rope. The pulling motion results from a rotation of subdomains within the myosin that is triggered upon actin binding.[7,8] Skeletal muscle myosin (also known as myosin II) is a large, 200 kD protein with an N-terminal catalytic domain (or motor domain) that binds to actin and to nucleotide, an alpha-helix that juts out from the motor domain and is supported by two smaller proteins known as the essential and regulatory light-chains (the lever arm), and a long, C-terminal coiled-coil (Figures 1b and 1c). The coiled-coil homodimerizes two of the heavy chains. Thus, the complete myosin "molecule" is a hexameric complex, with a total of two heavy-chains and four light-chains.[9]

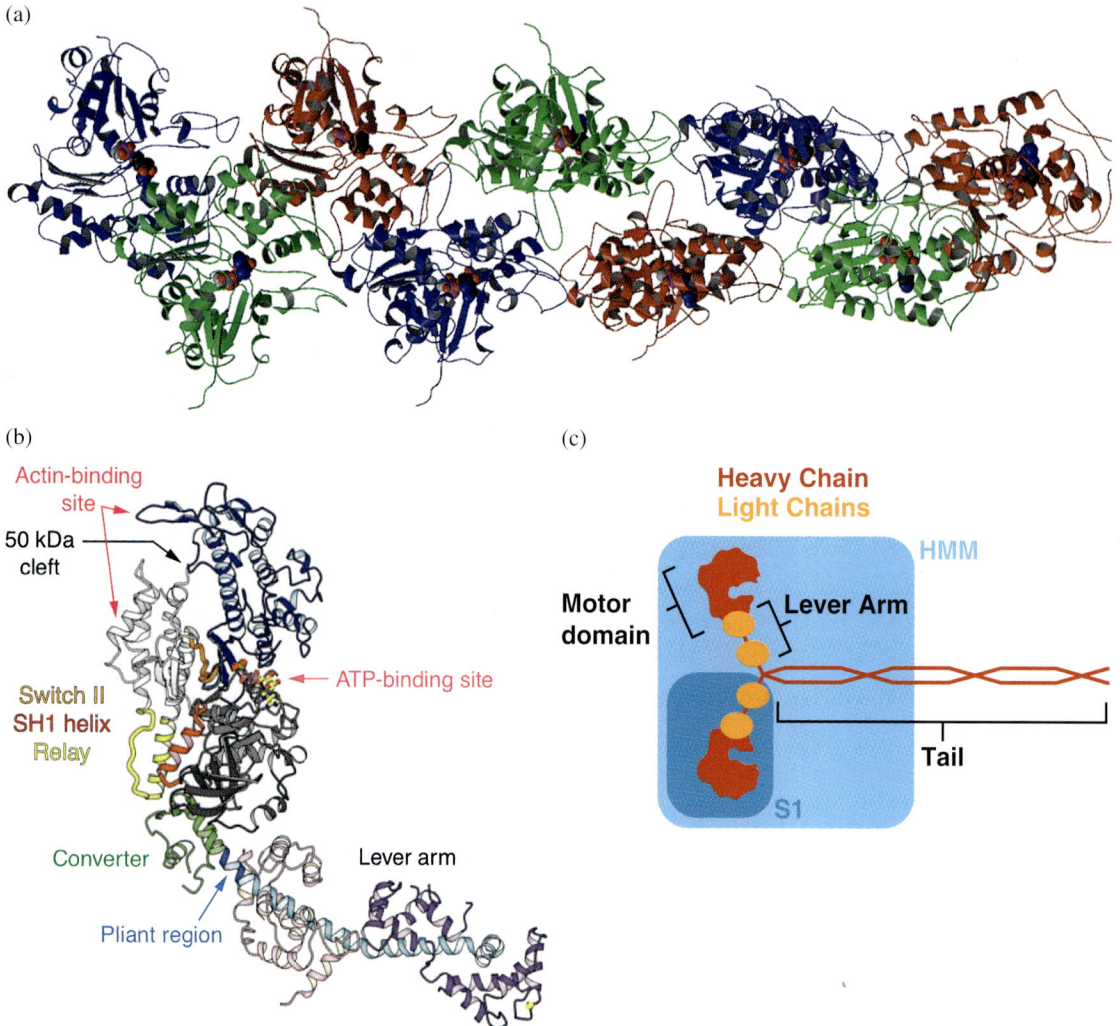

Figure 1. Actin and myosin structures. (a) F-actin is a helical polymer of actin monomers. Here, nine actin monomers are shown in alternating colors. The actin filament appears to have two strands (top and bottom) that wrap around each other to form two right-handed helices. The segment shown has a little more than a quarter turn of the helix. Cellular actin filaments are typically much longer than this, and may contain several hundred actin monomers in a single filament. This model was formed by fitting the individual actin monomers into the filament electron density obtained by electron microscopy. (Figure adapted with permission from Geeves and Holmes.)[17] (b) The crystal structure of the N-terminal fragment of chicken skeletal muscle myosin, known as subfragment-1 or S1. This molecule resembles a lobster claw, with a long cleft separating the digits (white and blue). Opening and closing of this cleft modulates the actin affinity at the sites located at the top of the molecule. Nucleotide binds to a pocket near the middle. Finally, a long alpha-helix extends from the "wrist" (green) at the C-terminal end of this domain. This alpha-helix is supported by two separate polypeptides, the "light chains" (light and dark purple), and is known as the "lever arm". (Figure adapted with permission from Houdusse and Sweeney.)[7] (c) Domain organization of a complete myosin molecule. Two myosin heavy chains dimerize through a coiled-coil that begins immediately after the lever arms. A second, somewhat longer proteolytic fragment of muscle myosin known as heavy meromyosin (HMM) is a stable dimer of heavy chains but does not form higher order assemblies as found in muscle sarcomeres.

Although our structural description of these players is still incomplete, we do have atomic structures of some of the players. The monomeric actin structure has been solved in various complexes with proteins and small molecules that block polymerization,[10–12] although the atomic structure of the complete filament still eludes us. The problem here is that purified actin will form filaments of various lengths, which prevents the growth of well-ordered crystals. Likewise, structures of myosin motors in various conformations have been solved,[13–15] although none while bound to actin filaments. Nevertheless, it is possible to assemble these individual structures to obtain a reasonable model of the events within a muscle fiber. One trick that has been quite useful is to dock the individual crystal structures within the low-resolution structures of muscle proteins that have been found by electron microscopy (EM).[4,6] These docking procedures are now highly refined, allowing us to infer some of the more subtle subdomain motions by using molecular dynamics to remodel the static crystal structures to fit the EM density.[16] The picture that has emerged is that the myosin light-chain binding domains form a mechanical lever that swings through approximately 70 degrees while myosin is attached to actin.[17,18] When myosin detaches from actin, the reverse swing occurs, priming the myosin for another round of tugging an actin filament. These working-stroke and recovery-stroke events make myosin similar, in some respects, to the two-stroke engines found in mopeds, lawnmowers, and model airplanes. This "swinging lever arm" or "tilting cross-bridge" hypothesis is the central notion that guides our thinking of how a muscle works (Figure 2).[19] Over time, more and more features of these models have been determined, so we now know how to correlate the structural models and the biochemical states of myosin to the ultimate motion within a muscle fiber.

2. Coupling of Motion to ATP Hydrolysis

Myosin binds and hydrolyses ATP, which it uses as a fuel source to produce mechanical work. Since the motor domain of myosin is small — about 10 nm long — it comes as no surprise that a myosin will hydrolyze multiple ATPs in a cyclic manner to drive motility over long distances. Much of our understanding of the myosin ATPase cycle comes from steady state and transient kinetic measures of myosin turnover. In the steady state reactions, myosin is mixed with ATP, and the production of the hydrolysis products, either ADP or P_i, is monitored over time through multiple hydrolysis cycles. The resulting ATPase rate gives information on the amount of time it takes to complete a single cycle. Note that we are considering the entire cycle here, including ATP binding, the hydrolysis event itself, phosphate release, and ADP release, and not just the single chemical event of ATP hydrolysis. Usually, we will be most interested in the ATPase rate at saturating concentrations of ATP, which mimic the situation in muscle tissue (1 mM ATP, 10 μM ADP). In this case, the ATP-binding event is essentially instantaneous and does not contribute to the overall rate.

In the absence of actin, the myosin ATPase rate is 0.1/s, or an average time of about 10 s for a single turnover. This rate without actin is known as the "basal ATPase rate". In then presence of saturating actin, the rate is stimulated at least 100-fold, yielding an "actin-activated ATPase rate" of around 10/s (or 100 ms to complete a turnover).[9,20] As we shall see, this actin-activation process is important for muscle efficiency; in a relaxed muscle, actin is essentially removed from the scene by other regulatory proteins, so the slow basal ATPase minimizes energy consumption in a relaxed muscle.

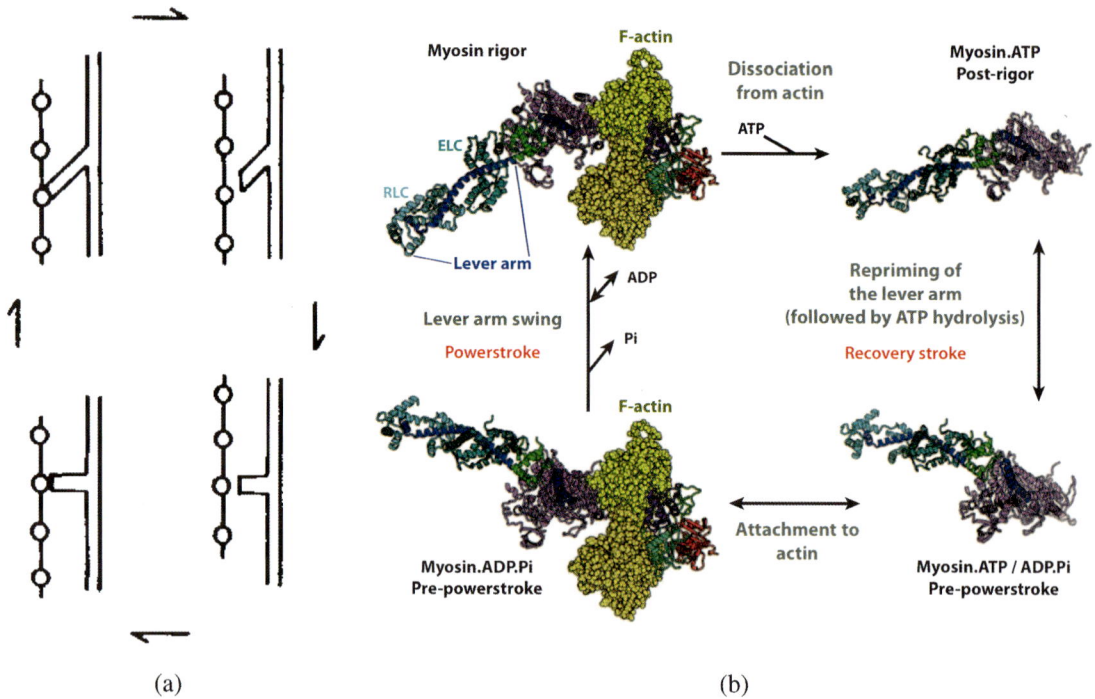

Figure 2. The swinging lever arm model. (a) The swinging lever arm model, as presented by Lymn and Taylor.[20] Actin is shown as a string of beads on the left, while myosin is shown as the rectangular structure to the right. A movement in the myosin while attached to actin propels the filament, while the reverse movement while detached resets the myosin for another cycle. These fours states are directly coupled to the nucleotide states of the myosin. (Figure adapted with permission from Lymn and Taylor.)[20] (b) The corresponding structural models and nucleotide states. Most of the motion of myosin occurs at the wrist where the lever arm meets the motor domain. Thus, the lever arm is a mechanical element that amplifies small movements within the motor domain. (Figure adapted with permission from Sweeney and Houdusse.)[8]

The goal of the transient kinetic measurements is twofold: to identify the most likely kinetic pathway and rate-limiting events in the hydrolysis cycle, and to provide some initial structural information about these events. The experimental design is to rapidly mix myosin or the actin-myosin complex (actomyosin) with nucleotides, so that all molecules start at a specific point in the ATPase cycle, and to follow the events that occur downstream of that point. Most of the transient kinetic measurements were performed in a series of landmark experiments performed nearly 40 years ago,[20–22] which were later refined as methods for following binding and release of nucleotide products improved over time. The essential features of the ATPase cycle are that (1) ATP binding stimulates release of myosin from actin, (2) hydrolysis occurs while myosin is detached, and (3) rebinding of myosin to actin stimulates the release of the hydrolysis products, ADP and P_i.[20]

When myosin is mixed with ATP, the hydrolysis products ADP and P_i are formed but remain bound, with only a slow release of P_i that matches the basal ATPase rate. However, if actomyosin is mixed with ATP, the first event is the dissociation of the actomyosin complex. The actomyosin complex is a large assembly that scatters a significant amount of light, thus dissociation of actomyosin may be followed through a decrease in this scattering signal. Lymn and Taylor found that actomyosin dissociation after mixing with ATP occurred far more rapidly than ATP hydrolysis, the

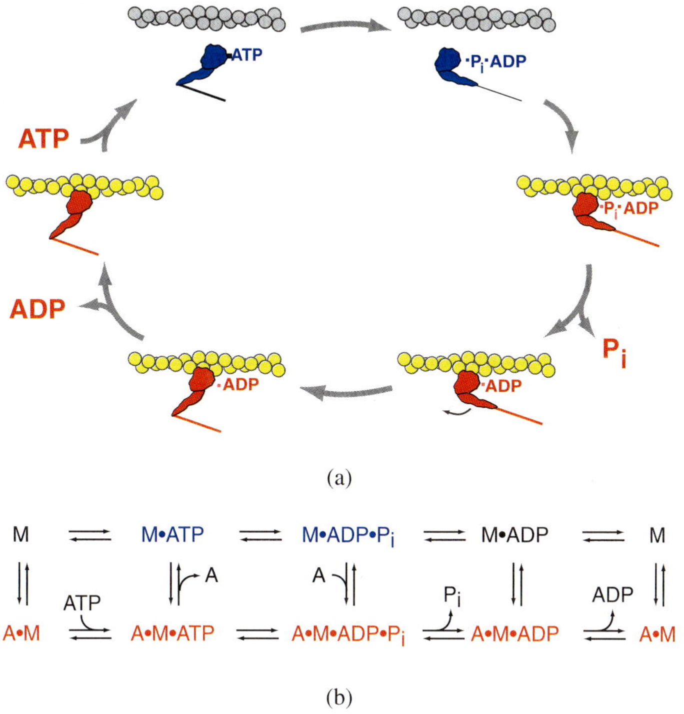

(a)

(b)

Figure 3. The myosin ATPase cycle. (a) The states of the myosin motor domain that are strongly attached to actin are shown in red. For skeletal muscle myosins, these states represent only about 2% of the total ATPase cycle time. The majority of the time is spent in the weakly associated states shown in blue. (b) A kinetic scheme listing the major biochemical intermediates. The most populated biochemical path is shown in colors that correspond to the states shown in (a). Note that some of these events may have intermediate species that are not shown. For example, rebinding of myosin to actin is thought to proceed via a weakly bound intermediate. (Figure adapted with permission from Murphy, Rock and Spudich.)[23]

latter monitored by the liberation of radiolabeled P_i before the reaction reaches steady state.[20] This observation requires that dissociation is the first event, followed by hydrolysis while myosin is free from actin.

To follow rebinding of myosin with bound nucleotide products, Lymn and Taylor rapidly mixed myosin, ATP, and then actin. The mixture was immediately applied to a gel filtration column to separate released products from the actomyosin complex.[20] They found an increase in hydrolysis products released from actomyosin, compared to myosin alone. This observation is consistent with actin stimulating the rate-limiting product release, which provides a direct mechanism for the actin activation of the ATPase activity over the basal level. Together, this set of experiments established the order of events that connect the biochemical and mechanical cycles of myosin (Figure 3).

3. *In Vitro* Motility Assays Reconstitute Motion from a Minimal Set of Components

Until now, we have considered only the chemical products of myosin, namely the turnover of ATP to ADP and P_i. However, motion itself is also one of the products of myosin motors, although one that is much more ephemeral and challenging to monitor. We expect that actin and myosin move relative to each other, but one cannot simply mix a quantity of purified actin, myosin, and ATP in a tube and measure the "total motion". Once the biochemical cycle had been determined, several groups started to work on reconstituting and directly observing motility. The ultimate goal was to determine the minimal set of components that result in directed motion; from there, the roles of the components would be further dissected. The basic approach was to use microscopy to follow the movement of small collections of motors. These efforts culminated in the first *in vitro* motility assays in the early 1980s, in what would turn out to be one of the most significant developments in muscle and motility research. In particular, these *in vitro* motility assays established that myosin, actin and ATP are the only components needed for motility.

The first *in vitro* motility assay was developed by Mike Sheetz and Jim Spudich,[24] and was truly a case of scientific serendipity. At the time, Spudich had a long daily commute in the San Francisco Bay area. To save some time, he brought home a microscope and some simple samples that he could examine in evenings, including an interesting algae, *Nitella axillaris*. The cells of *Nitella* are extremely large, about 1 mm × 40 mm long, and contain a parallel array of actin cables that line the chloroplasts. These actin cables are used in an actomyosin system that drives cytoplasmic streaming, which is essentially a circulatory system for the large cells to mix their contents. Spudich realized that the actin cables within *Nitella* were organized in a way that might support the motility of myosin-coated beads. Together with Sheetz, who was on a sabbatical in the Spudich lab to develop these sorts of assays, they developed a procedure to split open a *Nitella* cell using tiny scissors normally used for eye surgery. After attaching the opened cell on a microscope slide and gently washing away the cytoplasm, they then applied ATP along with small plastic beads, approximately 1 μm in diameter, that had been coated with skeletal muscle myosin. These beads moved unidirectionally down the actin cables at speeds of several microns per second, approaching the speeds estimated in muscle contraction (Figure 4a).[24] Because *Nitella* has neighboring zones of actin cables with opposite polarity, Sheetz and Spudich were even able to observe beads traveling in opposite directions along their respective zones within the same field of view. A fortunate finding of this assay is that the actin filaments determine the direction of motion, and that the specific organization of the myosin does not matter all that much: the myosin motor domains can apparently reorient so that they can bind to the actin and drive motility in one direction, set by the actin filament polarity.

However, there were clear drawbacks to this system. The dissection of *Nitella* took some skill and was labor-intensive. Furthermore, although the myosin was pure, the actin was not: it certainly contained other components that would hold together the bundles, perhaps along with other components of the cytoplasm that could affect motility. Soon after the *Nitella* work, Toshio Yanagida demonstrated that it was possible to see single actin filaments in a fluorescent microscope.[25] Normally, actin filaments are impossible to see; they are thin objects (approximately 8 nm in diameter) that are too small to scatter light. However, by coupling a fluorophore to phalloidin, a mushroom toxin that binds to actin, Yanagida's group was able to use fluorescence to image single actin filaments. Further development of *in vitro* motility assay took advantage of this system, ending in Steve Kron's and Jim Spudich's gliding filament assay.[26] In this assay geometry, the purified myosin motor was attached to the coverslip surface, usually through nonspecific adsorption to nitrocellulose-coated coverslips. Fluorescent actin filaments were then applied to the surface along

(a)

(b)

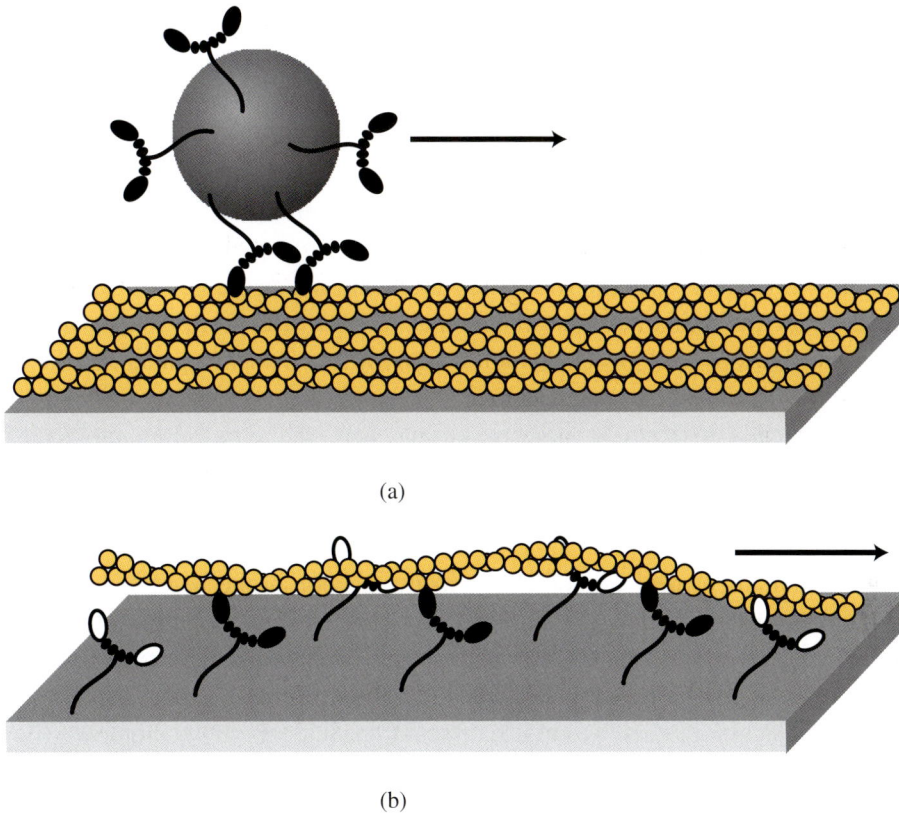

Figure 4. Two types of *in vitro* motility assays. (a) The *Nitella* assay. Parallel actin filament bundles from *Nitella* (orange) are immobilized on a glass surface. Myosin-coated beads are applied in a solution of ATP, and the beads are propelled by the myosin in one direction when they encounter the actin bundle. Note that parallel actin filaments are needed for this assay, since if the filaments were fixed to the surface with mixed polarity, then multiple myosin molecules would pull in different directions and compete. (b) The gliding filament assay. The principle is similar to the *Nitella* assay, although the components are inverted. Here, the myosin motor is fixed to the surface, and the actin filaments move. The actin filaments are visualized by fluorescence microscopy.

with ATP, resulting in smooth gliding of the filaments over the glass surface as they were propelled by the fixed myosin motors (Figure 4b). Soon after, this assay was combined with molecular genetics to demonstrate that the S1 fragment of myosin was the motor domain, which ruled out contraction of a "spring" in the distal portion of the myosin.[27] Later, the Spudich lab altered the length of the myosin II lever arm by expressing mutants with deletions in the light-chain binding domains. They showed that the velocity scaled with the length of the lever arm, suggesting that this structural element swings when myosin moves.[28] Due to its simplicity and power, this gliding filament *in vitro* assay is still routinely used in modern motility research.

4. The Stepsize Controversy and Single Molecule Motility

The early *in vitro* motility assays also generated their own share of puzzles. In particular, the early estimates of the myosin stepsize seemed to be unrealistically large. To take the simplest example, since skeletal muscle myosin moves at 3 μm/s, and its ATPase rate is 10/s, the myosin moves past 300 nm of actin during a single ATP turnover. That distance is much large than the dimensions of the myosin motor domain (~ 8 nm diameter) and seems unreasonable. More direct measures of the distance traveled per ATP centered on 100 nm, still far larger than expected.[29,30] These observations led to the proposal of unusual models, such as the suggestion that the free energy of ATP was somehow stored in the motor domain to be later parceled out over multiple discrete steps, somewhat like a wind-up toy.

Many of these arguments over the sliding distance centered on the number of motors that were active and engaged to the actin track, an estimate that varied widely from group to group. To tackle these sorts of questions, several groups began to develop more sophisticated motility assays that would be capable of detecting the steps generated by a single motor protein molecule. Although the field commonly calls these single-molecule assays, here, the term "single molecule" may also mean "single object" or "single-protein complex" such as the assembly of myosin heavy chains and light chains used here.

To see the steps of a single myosin motor, the fundamental challenge was to find a way to position a single actin filament near a single myosin using appropriate handles to gently hold the components in place. Too firm of a grip would prevent the myosin from moving once it was bound to the actin filament; too gentle, and the thermal motions of the sensor components would obscure the motion of the myosin. The solution, developed in at least three different labs in rapid succession for separate molecular motors, was to use optical traps to position the motors or filaments.[31–33] Optical traps[34,35] were developed by Arthur Ashkin and Steven Chu at Bell Labs in the mid 1980s, as an offshoot of efforts to develop laser-based cooling of atoms. These optical traps use a focused, infrared laser beam to hold on to micron-sized polystyrene or silica beads in an optical microscope. These optical traps can produce several piconewtons of force before the beads will escape the trap, which turns out to be an ideal force range for studying molecular motors.

To measure the single steps of myosin, Jeff Finer, Robert Simmons, and Jim Spudich developed a clever experimental geometry (Figure 5).[32] Instead of attaching the track to the coverslip surface, as was done by the Block lab in their studies of the kinesin motor protein,[31] the Spudich group attached beads to either end of an actin filament. This bead-filament-bead assembly, which they called a "dumbbell" (since it resembles a weightlifters dumbbell), was then stretched by moving the beads apart. They then moved the dumbbell over a raised surface platform that had a single myosin motor on it. To ensure that they were looking at single molecules, they reduced the surface density of myosin so that most tested surface platforms had no binding events. Thus, when they saw binding events, they could be confident that only a single motor was on that platform. They tracked the position of one of the beads to measure the size of the individual myosin steps. The key feature of this three-bead geometry is that the actin filament can be stretched before probing the myosin motors. If the filament were slack, it would be difficult to see motion in the end beads if a myosin grabbed and pulled on he middle of the filament.

The Spudich group measured a myosin stepsize of around 10 nm, comparable to the size of the myosin catalytic domain.[32] This stepsize was only an upper estimate, since their experimental design meant that they were likely to miss events with a near zero stepsize. Nevertheless, this short stepsize allowed them to rule out the more elaborate models involving travel of 100 nm per ATP molecule consumed. Soon after, the Molloy group improved these stepsize measurements

(a)

50 nm

50 nm

50 nm

1 sec

(b)

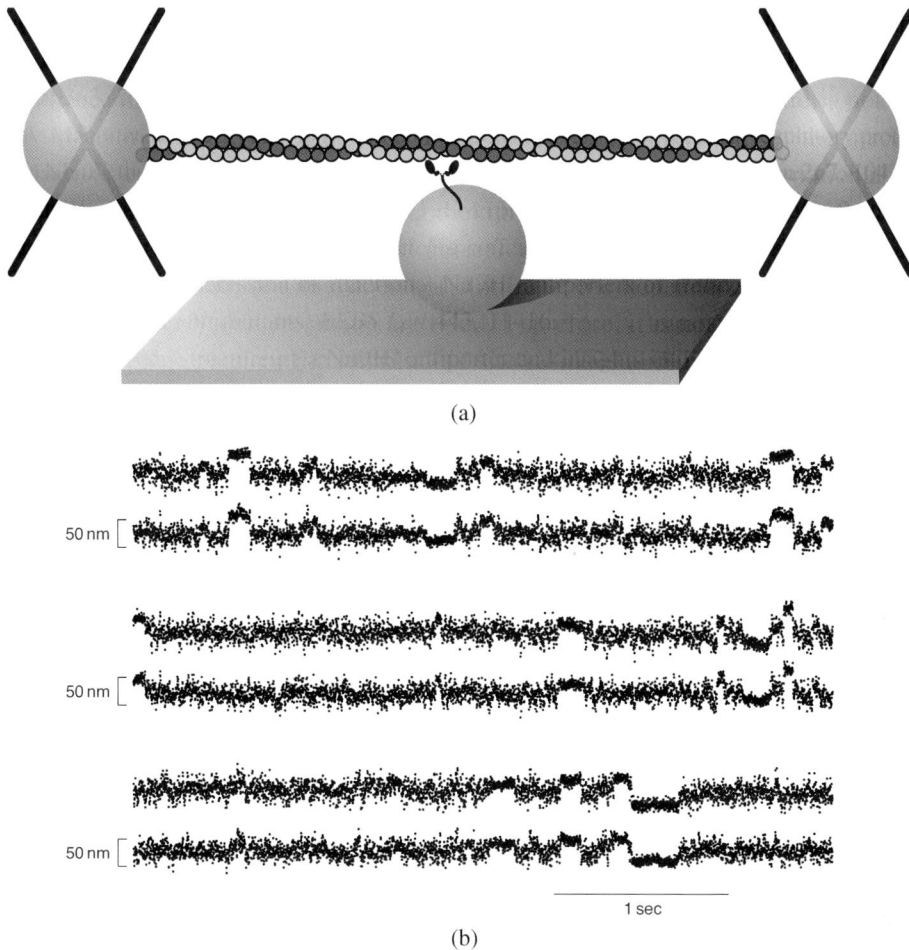

Figure 5. The three-bead optical trap assay to observe single stepping events. (a) In the three-bead geometry, poly-styrene beads held in two optical traps (illustrated by the "X"s) are attached to each end of an actin filament. The traps are moved apart to remove the slack in the filament. This trapped actin filament, called a "dumbbell", is then moved over a raised surface platform containing the myosin motor. (b) Binding events and displacements are recorded by optically tracking the position of the two beads. (Figure adapted with permission from Veigel *et al.*)[37] Note the drop in variance during the binding event. To ensure that data are recorded from single myosin molecules, the myosin is diluted so that observed binding events are rare. Under these conditions where most platforms have zero myosins molecules, the chance that two myosins are found on the same platform is even more remote, and can be calculated from the Poisson distribution.

through a detector with a higher bandwidth and an improved event detection scheme that allowed them to detect binding events independent of significant motion.[33] Since they were able to include events with near-zero stepsize, their estimate went down to 5 nm. Later refinements improved methods to deal with technical issues such as compliant connections between the actin and the bead, or between the motor and surface, which tend to absorb a portion of the myosin motion.[36,37] Most recently, the Higuchi lab developed a system to do away with the calculated corrections for system compliance in the three-bead assay, by tracking the position of the actin filament using a fluorescent quantum dot as a reporter. Their more accurate stepsize was 8 nm, which is consistent with the dimensions of the myosin motor domain and the putative angle over which the lever arm swings.[38]

To return to the initial question, given that myosin takes short steps in the 5–10 nm range, how does it manage to move at 3 μm/s if it takes a long time to process an ATP molecule? As is usually the case, the devil is in the details: the details of the myosin ATPase cycle. As discussed above, it turns out that myosin spends most of the ATP cycle time detached from actin. These detached myosin motors do not affect the sliding speed of an actin filament, since they do not "hold back" or produce any drag. Instead, what matters is the time myosin spends attached to actin, or the strongly-bound state time t_s, since the attached motors would be expected to produce drag after they complete their working stroke. The simplest model gives the sliding filament velocity as

$$V = d/t_s,$$

where d is the single molecule stepsize and t_s is the strongly bound state time of a few milliseconds. This short t_s, relative to the actin-activated ATPase cycle time, is why muscle myosins are known as low-duty ratio motors. The duty ratio is expressed as $R = t_s/t_{ATPase}$, and can range from zero (for motors that spend most of their life detached from tracks) to one (for motors that prefer to stay on their tracks).[39]

The overall molecular picture is that muscle myosin, as a low-duty ratio motor, has only fleeting interactions with its track. It comes on, steps while attached, and gets out of the way quickly. Note that in this model, the filament velocity does not depend on the actin-activated ATPase rate. For this low-duty ratio motor to work, it must have a large number of neighboring motors in the vicinity so that at least one myosin is attached at all times. Thus, at least 50 myosins must be near an actin filament to yield one attached myosin, on average.

This low-duty ratio form of motility allows for high filament sliding speed in muscle, since myosin does not have to wait for ATP hydrolysis when it can operate in a large ensemble.[39] This turns out to be an important advantage as muscle evolved from earlier molecular motors. As we shall see, the higher order organization of muscle makes this form of motility possible.

5. From Swinging Crossbridges to Sliding Filaments

Until now, I have described the molecular details that occur at the site of action, where myosin contacts the actin filament. However, the original idea of a rotating domain in myosin came from muscle physiology in the late 1960s, with an idea known as the "swinging crossbridge".[40]

In muscle, the fundamental repeating structural unit is known as the sarcomere (Figure 6). In the simplest terms, the sarcomere consists of two end plates, known as z-lines, where actin filaments (also known as thin filaments) are connected. The actin filaments project toward the center of the sarcomere with their pointed ends inward, so that myosin will tend to pull the two z-lines toward each other. Myosin sits in the center of the sarcomere in a large assembly known as a bipolar thick filament. The c-terminal portion of myosin forms a long, extended coiled coil. These coiled-coils self assemble in complex manner to produce an elongated object with myosin motor domains at each end, with roughly 300 myosin dimers on each half of the thick filament. This bipolar thick filament is ideally situated to pull the actin filaments from each end of the sarcomere together, thus shortening the distance between z-lines. In muscle, many sarcomeres are stacked

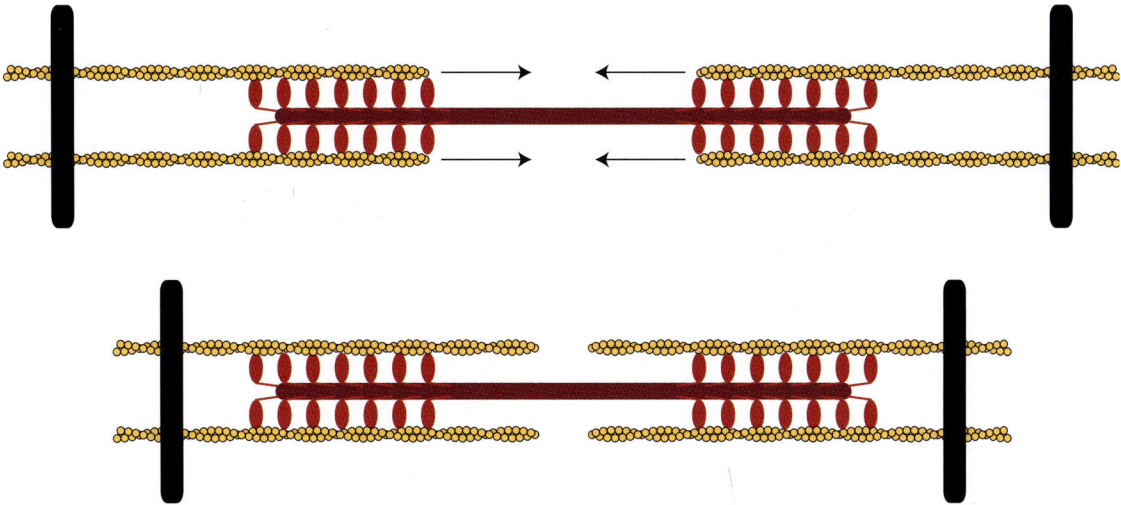

Figure 6. Sarcomere structure. In muscle, myosin (red) assembles through its coiled-coil C-terminal domain into an assembly of hundreds of molecules. The motor domains are found at each end of this elongated structure, known as a bipolar thick-filament. The motor domains form crossbridges between the thick filament and the actin filaments, also known as thin filaments (yellow). The thin filaments are anchored with their barbed ends at a structure known as the z-line (black). Thus, when the thin filament are activated, myosin can engage the actin an pull the z-lines in toward the center of the sarcomere. Only a single plane of the sarcomere is shown. In the full three-dimensional structure, the thin filaments are hexagonally packed around each thick filament. Moreover, many sarcomeres stack end-to-end in series. Here, only a small portion of the thin filaments from the next sarcomere is shown beyond the z-line.

end-to-end, producing a series mechanical element that greatly amplifies the overall movement. Thus, a 5–10 nm myosin working stroke can ultimately produce a centimeter-scale muscle contraction. This overall picture is necessarily simplified; a large number of other proteins are needed to assemble this structure and to ensure that all the components remain in the correct position.

In 1969, Hugh Huxley obtained EM images that showed "crossbridges", or connections between the actin thin filament and the myosin thick filament. When Huxley coupled these images with X-ray diffraction data on whole contracting muscle fibers, he found that the crossbridges tilted during muscle contraction.[40] Later, the Huxley group applied tension to intact muscle fibers, and used X-ray diffraction to observe the crossbridge tilting that follows changes in muscle tension.[41] More recently, the Goldman group used fluorescence polarization to follow probes attached to myosin lever arms, to watch the swinging crossbridges in real time in intact muscle fibers.[42,43] The swinging crossbridge model has been largely supported by the later molecular work on myosin fragments as discussed above (but see the work of the Yanagida group for an alternative view[44]).

6. The Regulation of Contraction

A relaxed skeletal muscle switches off this contractile assembly by blocking the myosin binding site on actin. These muscles must be able to contract rapidly, on demand. The regulatory system is based on a complex between a long coiled-coil protein, tropomyosin, an a calcium-sensing protein complex known at troponin. These two protein assemblies decorate the length of the actin filament, and in the absence of calcium, they block the myosin binding sites. Motor neurons stimulate the release of calcium from intracellular stores. This calcium then binds to troponin, resulting in a rapid shift of tropomyosin away from the key sites along actin.[45] Since the basal ATPase rate is low, the energy consumption in a relaxed muscle is also low. Myosin is maintained with bound nucleotide, poised to contract upon demand.

The situation in smooth muscle is somewhat different. These muscles, which line the gut and circulatory system, contract much more slowly which allows for a slower signaling system. These muscle tissues are instead regulated on the myosin molecule. A kinase phosphorylates the regulatory light-chain of smooth muscle myosin to switch on contraction, while a corresponding phosphatase switches off contraction.[46]

7. New Frontiers in Motility Research: Nonmuscle Myosins and the Evolution of Eukaryotic Motility

Although we are used to the idea of biological motion when we see muscle contraction in other animals, it turns out that muscle arrived late on the scene in evolutionary terms. In fact, ever since the first nonmuscle myosins were identified in *Acanthamoeba*,[47] it has become clear that myosin motors traffic organelles, transcripts, and other components. Moreover, myosin motors control and remodel cell adhesion and shape, especially in the final stages of cell division. Given myosin's diverse roles, it is perhaps unsurprising that nearly 40 distinct myosin families have now been identified.[48] Humans have forty distinct myosin genes, representing ten of these myosin families, and on average ten of these genes will be expressed in any given cell.[49] Adding to the myosin diversity, many of these human genes are alternatively spliced as well. In fact, the myosins are so diverse and ubiquitous, that their sequences have been used to reconstruct the evolutionary events that led to the earliest eukaryotes.[48,50]

Unlike the muscle myosins, most nonmuscle myosins make up only a small fraction of the total cellular mass and are therefore much more difficult to isolate. Although myosin V is reasonably abundant and can be isolated from brain tissue,[51] now most research groups express recombinant myosin V using insect cell expression systems. This fact alone presented an early barrier to nonmuscle myosin research. However, the additional effort here has yielded significant rewards. I will close by touching on the key features of a few of the dimeric myosins, myosin V, myosin VI, and myosin X, to give a sense of the diversity found in myosin function (Figure 7).

Myosin V was the first discovered "processive" myosin, meaning a myosin that takes many steps along actin before detaching.[52] This myosin walks somewhat like we walk, placing one foot (or motor domain) in front of the other. To do this, myosin V has two key adaptations, one structural, and one biochemical. Structurally, myosin V has long "legs" with six light chains in the lever arm, compared to the two found in muscle myosins.[51] Biochemically, myosin V has rate-limiting ADP release instead of rate-limiting actin binding.[53] This means that each motor domain of myosin V spends most of the time attached to actin, the feature of a high-duty ratio motor.

Myosin VI does something rather remarkable: backwards walking along an actin filament, toward the pointed end of actin.[54] Myosin VI achieves this reversal through a unique insertion of fifty residues in the lever arm domain that bends the lever arm nearly 180 degrees.[55] This bend means that the same rotation in the motor domain can produce a lever arm swing in the opposite

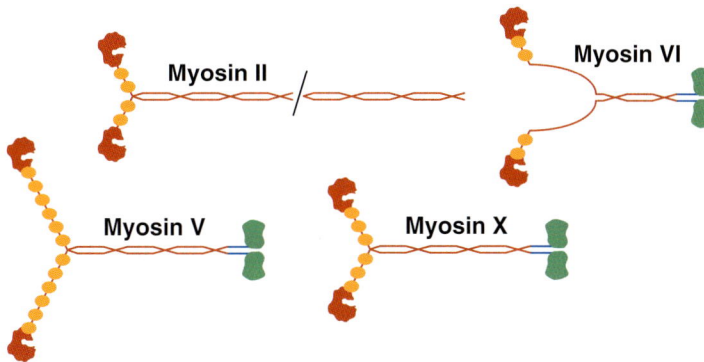

Figure 7. Domain structures of myosin V, myosin VI, and myosin X. Although these three nonmuscle myosins have similar overall motor domain structures to muscle myosin II, variable surface loops and other sequence variations adjust the biochemical rate constants and mechanical properties of these motor proteins. Beyond the motor domains, each motor has a distinct number of light-chains attached to each heavy-chain (orange), which adjusts the length of the lever arm. Finally, each motor has a set of C-terminal cargo-binding domains, which are used to dock to vesicles, cell surface receptors, and other cargo.

direction (although this is a bit of an oversimplification, since there are other structural differences in the pre-stroke structure of myosin VI[56]).

Finally, myosin X[57,58] is a processive myosin with a unique twist. It walks processively along bundles of actin, but walks poorly along single actin filaments.[59] To walk along bundles, myosin X seems to have structures that favor "straddling" two actin filaments. In the cell, myosin X is often found in regions of bundled actin, in particular at the tips of actin projections from the plasma membrane known as filopodia.[58] A preference for bundled actin begins to explain how myosin X finds the filopodial tips, and opens up further questions of how other myosins navigate to precise locations within the cell.

It is remarkable how Nature has tuned the molecular features of this common set of motors to carry out quite different functions within both muscle and nonmuscle cells. I imagine that we have yet to find many more surprising features, features that explain how myosins produce organized and mechanically responsive cells.

Suggested Additional Reading Material

J. Howard. 2001. Mechanics of Motor Proteins and the Cytoskeleton. Sinauer Associates, Sunderland, Massachusetts.

References

1. S. S. Margossian and S. Lowey. 1982. Preparation of myosin and its subfragments from rabbit skeletal muscle. *Methods Enzymol* **85 Pt B**, 55–71.
2. T. D. Pollard. 2003. The cytoskeleton, cellular motility and the reductionist agenda. *Nature* **422**, 741–745.
3. T. D. Pollard and J. A. Cooper. 2009. Actin, a central player in cell shape and movement. *Science* **326**, 1208–1212.

4. K. C. Holmes, D. Popp, W. Gebhard and W. Kabsch. 1990. Atomic model of the actin filament. *Nature* **347**, 44–49.

5. P. B. Moore, H. E. Huxley and D. J. DeRosier. 1970. Three-dimensional reconstruction of F-actin, thin filaments and decorated thin filaments. *J Mol Biol* **50**, 279–295.

6. T. Fujii, A. H. Iwane, T. Yanagida and K. Namba. 2010. Direct visualization of secondary structures of F-actin by electron cryomicroscopy. *Nature* **467**, 724–728.

7. A. Houdusse and H. L. Sweeney. 2001. Myosin motors: Missing structures and hidden springs. *Curr Opin Struct Biol* **11**, 182–194.

8. H. L. Sweeney and A. Houdusse. 2010. Structural and functional insights into the myosin motor mechanism. *Annu Rev Biophys* **39**, 539–557.

9. J. R. Sellers and H. V. Goodson. 1995. Motor proteins 2: Myosin. *Protein Profile* **2**, 1323–1423.

10. L. R. Otterbein, P. Graceffa and R. Dominguez. 2001. The crystal structure of uncomplexed actin in the ADP state. *Science* **293**, 708–711.

11. W. Kabsch, H. G. Mannherz, D. Suck, E. F. Pai and K. C. Holmes. 1990. Atomic structure of the actin: DNase I complex. *Nature* **347**, 37–44.

12. C. E. Schutt, J. C. Myslik, M. D. Rozycki, N. C. Goonesekere and U. Lindberg. 1993. The structure of crystalline profilin-beta-actin. *Nature* **365**, 810–816.

13. I. Rayment, W. R. Rypniewski, K. Schmidt-Base, R. Smith, D. R. Tomchick, M. M. Benning, D. A. Winkelmann, G. Wesenberg and H. M. Holden. 1993. Three-dimensional structure of myosin subfragment-1: A molecular motor. *Science* **261**, 50–58.

14. R. Dominguez, Y. Freyzon, K. M. Trybus and C. Cohen. 1998. Crystal structure of a vertebrate smooth muscle myosin motor domain and its complex with the essential light chain: Visualization of the pre-power stroke state. *Cell* **94**, 559–571.

15. A. Houdusse, V. N. Kalabokis, D. Himmel, A. G. Szent-Gyeorgyi and C. Cohen. 1999. Atomic structure of scallop myosin subfragment S1 complexed with MgADP: A novel conformation of the myosin head. *Cell* **97**, 459–470.

16. K. C. Holmes, I. Angert, F. J. Kull, W. Jahn and R. R. Schröder. 2003. Electron cryo-microscopy shows how strong binding of myosin to actin releases nucleotide. *Nature* **425**, 423–427.

17. M. A. Geeves and K. C. Holmes. 1999. Structural mechanism of muscle contraction. *Ann Rev Biochem* **68**, 687–728.

18. K. C. Holmes and M. A. Geeves. 2000. The structural basis of muscle contraction. *Philos Trans R Soc Lond, B, Biol Sci* **355**, 419–431.

19. A. F. Huxley. 1957. Muscle structure and theories of contraction. *Prog Biophys Biophys Chem* **7**, 255–318.

20. R. W. Lymn and E. W. Taylor. 1971. Mechanism of adenosine triphosphate hydrolysis by actomyosin. *Biochemistry* **10**, 4617–4624.

21. R. G. Yount, D. Ojala and D. Babcock. 1971. Interaction of P-N-P and P-C-P analogs of adenosine triphosphate with heavy meromyosin, myosin, and actomyosin. *Biochemistry* **10**, 2490–2496.

22. R. S. Goody and F. Eckstein. 1971. Thiophosphate analogs of nucleoside di- and triphosphates. *J Am Chem Soc* **93**, 6252–6257.

23. C. T. Murphy, R. S. Rock and J. A. Spudich. 2001. A myosin II mutation uncouples ATPase activity from motility and shortens step size. *Nat Cell Biol* **3**, 311–315.

24. M. P. Sheetz and J. A. Spudich. 1983. Movement of myosin-coated fluorescent beads on actin cables *in vitro*. *Nature* **303**, 31–35.

25. T. Yanagida, M. Nakase, K. Nishiyama and F. Oosawa. 1984. Direct observation of motion of single F-actin filaments in the presence of myosin. *Nature* **307**, 58–60.

26. S. J. Kron and J. A. Spudich. 1986. Fluorescent actin filaments move on myosin fixed to a glass surface. *Proc Natl Acad Sci U S A* **83**, 6272–6276.

27. Y. Y. Toyoshima, S. J. Kron, E. M. McNally, K. Niebling, C. Toyoshima and J. A. Spudich. 1987. Myosin subfragment-1 is sufficient to move actin filaments *in vitro*. *Nature* **328**, 536–539.

28. T. Q. Uyeda, P. D. Abramson and J. A. Spudich. 1996. The neck region of the myosin motor domain acts as a lever arm to generate movement. *Proc Natl Acad Sci U S A* **93**, 4459–4464.

29. T. Yanagida, T. Arata and F. Oosawa. 1985. Sliding distance of actin filament induced by a myosin cross-bridge during one ATP hydrolysis cycle. *Nature* **316**, 366–369.

30. Y. Harada, K. Sakurada, T. Aoki, D. D. Thomas and T. Yanagida. 1990. Mechanochemical coupling in actomyosin energy transduction studied by *in vitro* movement assay. *J Mol Biol* **216**, 49–68.

31. K. Svoboda, C. F. Schmidt, B. J. Schnapp and S. M. Block. 1993. Direct observation of kinesin stepping by optical trapping interferometry. *Nature* **365**, 721–727.

32. J. T. Finer, R. M. Simmons and J. A. Spudich. 1994. Single myosin molecule mechanics: Piconewton forces and nanometre steps. *Nature* **368**, 113–119.

33. J. E. Molloy, J. E. Burns, J. Kendrick-Jones, R. T. Tregear and D. C. White. 1995. Movement and force produced by a single myosin head. *Nature* **378**, 209–212.

34. J. Moffitt, Y. Chemla, S. Smith and C. Bustamante. 2008. Recent Advances in Optical Tweezers. *Ann Rev Biochem* **77**, 205–228.

35. K. C. Neuman and S. M. Block. 2004. Optical trapping. *Rev Sci Instrum* **75**, 2787–2809.

36. A. D. Mehta, J. T. Finer and J. A. Spudich. 1997. Detection of single-molecule interactions using correlated thermal diffusion. *Proc Natl Acad Sci U S A* **94**, 7927–7931.

37. C. Veigel, M. L. Bartoo, D. C. White, J. C. Sparrow and J. E. Molloy. 1998. The stiffness of rabbit skeletal actomyosin cross-bridges determined with an optical tweezers transducer. *Biophys J* **75**, 1424–1438.

38. M. Kaya and H. Higuchi. 2010. Nonlinear elasticity and an 8 nm working stroke of single myosin molecules in myofilaments. *Science* **329**, 686–689.

39. J. Howard. 1997. Molecular motors: Structural adaptations to cellular functions. *Nature* **389**, 561–567.

40. H. E. Huxley. 1969. The mechanism of muscular contraction. *Science* **164**, 1356–1365.

41. H. E. Huxley, A. R. Faruqi, M. Kress, J. Bordas and M. H. Koch. 1982. Time-resolved X-ray diffraction studies of the myosin layer-line reflections during muscle contraction. *J Mol Biol* **158**, 637–684.

42. M. Irving, St Claire Allen T. Sabido-David C. Craik J. S. Brandmeier B. Kendrick-Jones J. Corrie J. E. T. Trentham D. R. and Y. E. Goldman. 1995. Tilting of the light-chain region of myosin during step length changes and active force generation in skeletal muscle. *Nature* **375**, 688–691.

43. J. E. Corrie, B. D. Brandmeier, R. E. Ferguson, D. R. Trentham, J. Kendrick-Jones, S. C. Hopkins, U. A. van der Heide, Y. E. Goldman, C. Sabido-David, R. E. Dale, S. Criddle and M. Irving. 1999. Dynamic measurement of myosin light-chain-domain tilt and twist in muscle contraction. *Nature* **400**, 425–430.

44. K. Kitamura, M. Tokunaga, A. H. Iwane and T. Yanagida. 1999. A single myosin head moves along an actin filament with regular steps of 5.3 nanometres. *Nature* **397**, 129–134.

45. A. M. Gordon, E. Homsher and M. Regnier. 2000. Regulation of contraction in striated muscle. *Physiol Rev* **80**, 853–924.

46. R. B. Pearson, R. Jakes, M. John, J. Kendrick-Jones and B. E. Kemp. 1984. Phosphorylation site sequence of smooth muscle myosin light chain (Mr = 20 000). *FEBS Lett* **168**, 108–112.

47. T. D. Pollard and E. D. Korn. 1973. Acanthamoeba myosin. I. Isolation from Acanthamoeba castellanii of an enzyme similar to muscle myosin. *J Biol Chem* **248**, 4682–4690.

48. F. Odronitz and M. Kollmar. 2007. Drawing the tree of eukaryotic life based on the analysis of 2,269 manually annotated myosins from 328 species. *Genome Biol* **8**, R196.

49. J. S. Berg, B. C. Powell and R. E. Cheney. 2001. A millennial myosin census. *Mol Biol Cell* **12**, 780–794.

50. T. A. Richards and T. Cavalier-Smith. 2005. Myosin domain evolution and the primary divergence of eukaryotes. *Nature* **436**, 1113–1118.

51. R. E. Cheney, M. K. O'Shea, J. E. Heuser, M. V. Coelho, J. S. Wolenski, E. M. Espreafico, P. Forscher, R. E. Larson and M. S. Mooseker. 1993. Brain myosin-V is a two-headed unconventional myosin with motor activity. *Cell* **75**, 13–23.

52. A. D. Mehta, R. S. Rock, M. Rief, J. A. Spudich, M. S. Mooseker and R. E. Cheney. 1999. Myosin-V is a processive actin-based motor. *Nature* **400**, 590–593.

53. E. M. de la Cruz, A. L. Wells, S. S. Rosenfeld, E. M. Ostap and H. L. Sweeney. 1999. The kinetic mechanism of myosin V. *Proc Natl Acad Sci U S A* **96**, 13726–13731.

54. A. L. Wells, A. W. Lin, L. Q. Chen, D. Safer, S. M. Cain, T. Hasson, B. O. Carragher, R. A. Milligan and H. L. Sweeney. 1999. Myosin VI is an actin-based motor that moves backwards. *Nature* **401**, 505–508.

55. J. Ménétrey, A. Bahloul, A. L. Wells, C. M. Yengo, C. A. Morris, H. L. Sweeney and A. Houdusse. 2005. The structure of the myosin VI motor reveals the mechanism of directionality reversal. *Nature* **435**, 779–785.

56. J. Menetrey, P. Llinas, M. Mukherjea, H. L. Sweeney and A. Houdusse. 2007 Oct 19. The structural basis for the large powerstroke of myosin VI. *Cell* **131**, 300–308.

57. J. S. Berg, B. H. Derfler, C. M. Pennisi, D. P. Corey and R. E. Cheney. 2000. Myosin-X, a novel myosin with pleckstrin homology domains, associates with regions of dynamic actin. *J Cell Sci* **113**, 3439–3451.

58. J. S. Berg and R. E. Cheney. 2002. Myosin-X is an unconventional myosin that undergoes intrafilopodial motility. *Nat Cell Biol* **4**, 246–250.

59. S. Nagy, B. L. Ricca, M. F. Norstrom, D. S. Courson, C. M. Brawley, P. A. Smithback and R. S. Rock. 2008. A myosin motor that selects bundled actin for motility. *Proc Natl Acad Sci U S A* **105**, 9616–9620.

Protein Kinases

7

Phosphorylation Machines

Elaine E. Thompson, Susan S. Taylor and J. Andrew McCammon

1. Introduction

Protein kinases are enzymes that are essential to cellular function and survival, and central to many signal transduction pathways. Protein kinases regulate other proteins by covalently attaching a phosphate group from ATP (adenosine triphosphate) to the hydroxyl on a serine, threonine, or tyrosine side chain of a substrate protein. As many as 50% of all cellular proteins may be phosphorylated.[1] In mammals, there are two major classes of protein kinases, serine/threonine kinases and tyrosine kinases. Serine/threonine kinases transfer a phosphate group onto either a serine or threonine residue on the substrate protein. Tyrosine kinases transfer a phosphate group onto a tyrosine residue. There are also bacterial histidine kinases that transfer a phosphate group onto histidine. Kinases are enzymes that typically function as molecular machines that activate and inactivate in response to signals from within or outside the cell. Activation and inactivation is through movements of parts of the kinase molecule, and sometimes by association with other proteins.

The covalently bound, charged phosphate group added to the substrate protein by a kinase then acts as a reversible molecular switch. Residues that are phosphorylated are typically modified by a specific kinase. Specificity is achieved by the amino acid sequence around the residue that is phosphorylated, and sometimes additionally by tethering of the protein substrate to the kinase through a distal docking site. Many proteins are phosphorylated at more than one site, sometimes by different protein kinases. The phosphate group remains on the substrate protein until it is removed by a second enzyme, called a phosphatase. Edwin Krebs and Edmond Fisher were awarded the Nobel Prize in 1992 for their work discovering reversible protein phosphorylation and studying phosphorylase kinase phosphorylation of glycogen phosphorylase.[2,3] Phosphorylation can activate or inactivate proteins, or it can change their binding partners. An example of a protein that is activated by phosphorylation is cAMP-dependent protein kinase (PKA), whose activation will be discussed later in this chapter. A protein that is inactivated by phosphorylation is glycogen synthase, the protein that polymerizes UDP-glucose into glycogen in liver and muscle.[4]

Protein kinases are central to the signals for cellular growth, division, or migration.[5] Because of this, many tumor cells have mutations that either incorrectly activate or inactivate a kinase or its regulator protein. The importance of the kinase family in cancer was discovered through a virus that causes cancer, Rous sarcoma virus (RSV), named after Francis Peyton Rous. Rous was awarded a Nobel Prize in 1966 for his work in 1911 demonstrating that chicken sarcomas are virally transmitted. Scientists continued to study RSV and in the 1970s the oncogenic protein in RSV, named v-Src short for viral sarcoma protein, was determined to be a constitutively active protein tyrosine kinase.[6] At the same time, J. Michael Bishop and Harold E. Varmus discovered that there is a gene in chickens that is homologous to the retroviral v-Src, called c-Src for chicken-Src.[7,8] The discovery of a cellular homolog to viral oncogene changed the thinking about cancer, leading scientists to realize that normal cellular proteins, called proto-oncogenes, can mutate into cancer-causing genes. Bishop and Varmus were rewarded a Nobel Prize in 1989 for their discovery of c-Src, because of the impact it had on our understanding of oncogenesis. Kinases have since been extensively studied and as of 2009, ten cancer chemotherapeutic agents, including sunitinib and imatinib, have been developed to inhibit tyrosine kinases.[9]

As well as being regulatory molecules themselves, kinases are highly regulated, typically by phosphorylation, and often involved in a series of reactions called kinase cascades. Kinase cascades provide amplification of signals through the cell, so that the binding of a few molecules of a growth factor or neurotransmitter at the cell surface can result in the activation of multiple transcription factor proteins in a completely different part of the cell. Cascades also provide an opportunity for feedback regulation, so that an inappropriate signal does not commit the cell to do something drastic like die or divide. For this reason, cells can often overcome the effects of a single oncogenic mutation and still grow and divide normally. An example of a kinase cascade involved in cell growth and division is the mitogen-activated protein kinase (MAPK) pathway (Figure 1). The pathway has been studied extensively because the Src proto-oncogene is part of the pathway and because MAPK signaling is essential for cell growth and survival.[10] The pathway is started when a growth factor (or mitogen) binds to its receptor tyrosine kinase. Epidermal growth factor receptor (EGFR) is one of the receptors that signals through the MAPK pathway. Receptor tyrosine kinases all have a similar structure, with an extracellular domain that binds a growth factor, a single transmembrane helix that crosses the cell surface, and an intracellular tyrosine kinase domain.[11,12] When epidermal growth factor (EGF) binds to EGFR, the receptor undergoes a large conformational change and dimerizes. When the receptor dimerizes, the two kinase domains can phosphorylate each other on multiple tyrosine residues.[13] Inside the cell there is a complex of proteins, including the Ras proto-oncogene, which only binds to the phosphorylated form of the EGFR.[14] When the Ras complex binds to the EGFR, Ras becomes active and initiates a long kinase cascade of serine/threonine kinases. *The signal is as follows: Raf → MEK → MAPK.*[15]

MEK is short for mitogen-activated protein kinase/extracellular signal-regulated kinase kinase. It is the only enzyme that phosphorylates MAP kinase and is highly specific for its substrate.[10] Part of how the anthrax toxin kills cells is by proteolytic inactivation of MEK, essentially turning off the MAPK cascade.[16] MAPK has multiple intracellular targets, all of which eventually change mRNA transcription or translation in the nucleus and mediate a biological response to the initial EGF signal. EGF and other growth factors typically signal either cell maintenance or growth and division. MAPK directly activates the c-Myc transcription factor. c-Myc is another proto-oncogene and is estimated to be involved in transcription of 15% of

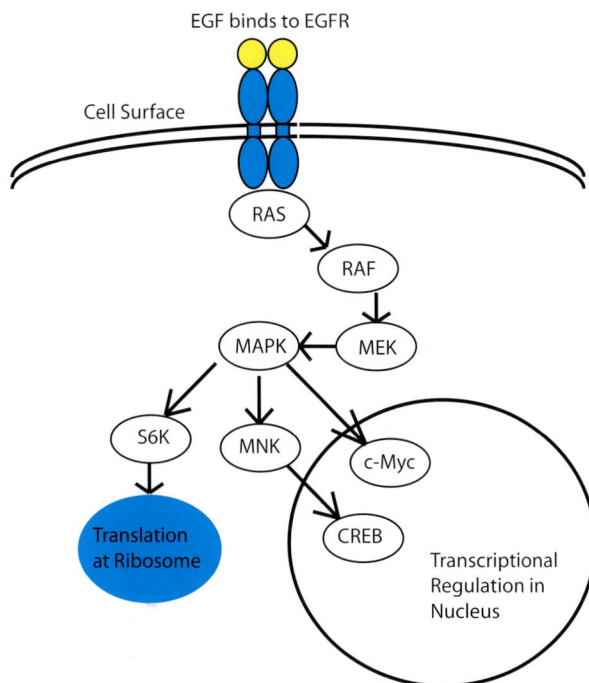

Figure 1. Diagram of MAP kinase pathway. The growth factor EGF binds to EGF receptor at the cell surface. Ras is activated and, in turn, activates the Raf kinase. Raf activates MAPK, which activates S6K and MNK. In turn, S6K regulates the ribosome to alter protein translation and the c-Myc and CREB nuclear transcription factors are activated and alter transcription. Some intermediate factors are omitted from this diagram for simplicity. For more details, see the review by Yarden and Silkowski.[18]

intracellular proteins.[17] MAPK also activates other kinases, including MAP kinase-interacting serine/threonine kinases (MNK), and 40S ribosomal protein S6 kinase (RSK or S6K). MNK activates a third transcription factor while RSK phosphorylates ribosomal protein S6 and changes protein translation at the ribosome.

Comparative studies of the kinase family have been undertaken both within and across species using sequence similarity. The human kinome has at least 518 putative kinase genes, with 244 mapped to known disease loci.[1] Kinases share a conserved catalytic core where the ATP and substrate bind and phosphoryl transfer occurs. The catalytic core may be the majority of the molecule, as with PKA or cyclin dependent protein kinase (CDK2), or it may be only a small part of a much larger molecule, as in the receptor tyrosine kinases. The conserved catalytic core consists of two lobes. The N-terminal lobe or N-lobe is made primarily of five antiparallel β-strands and a prominent helix, called αC or the C-helix. There is a highly conserved loop between strands β1 and β2 called the glycine loop. It has a glycine- rich motif (GXGXφG), where φ is typically tyrosine or phenylalanine that forms part of the ATP binding site.[19] The C-terminal lobe, or C-lobe, is larger and formed primarily from α-helices. The substrate ATP binds in a pocket formed by the two lobes and the glycine loop, called the active site cleft, and the substrate peptide binds on the surface of the kinase in a groove formed between the lobes. The substrate phosphorylation site, called the P site, is positioned very close to the γ-phosphate

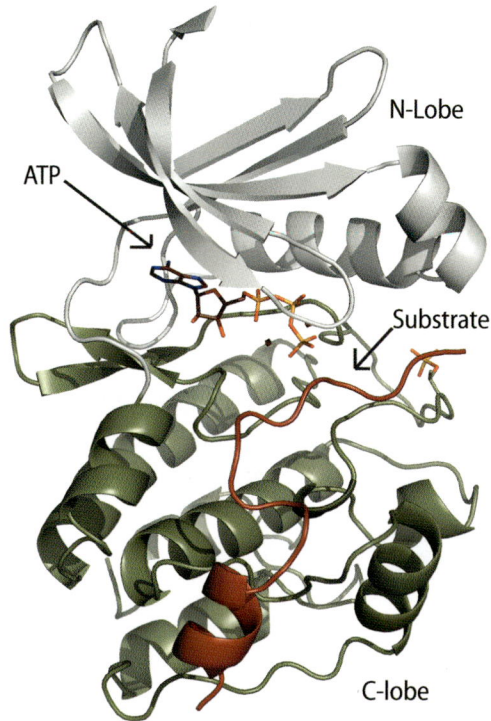

Figure 2. Catalytic core of protein kinase A. The N-lobe is shown in grey and the C-lobe in tan. ATP is bound between the N and C lobes. The red peptide is a pseudosubstrate peptide bound to the peptide substrate site.

of ATP. The active site cleft is deep, and shields the ATP and substrate from hydrolysis by the surrounding water. Figure 2 shows a ribbon diagram of the the catalytic core of PKA, with the N-lobe in grey, the C-lobe in tan, ATP in the binding pocket, and a pseudosubstrate peptide in the active site.

Kinases have both an active, catalytically competent state, and an inactive state where they do not bind ATP or phosphorylate substrate. Some kinases are always active and are sequestered by companion proteins, while others are reversibly phosphorylated. An example of a kinase that is sequestered is PKA. PKA phosphorylates multiple targets in response to intracellular increases in cAMP (3′-5′-cyclic adenosine monophate), an important second messenger molecule. The catalytic subunit has a structurally conserved catalytic core that is present in all mammalian kinases,[20] an N-terminal tail of 40 amino acids, and a C-terminal tail of 50 residues. The cAMP binding domains are on the regulatory subunit. Upon cAMP binding to the regulatory subunit, the catalytic subunit dissociates from the regulatory subunit and phosphorylates target proteins. When the intracellular cAMP concentration falls, the regulatory subunit binds to the kinase again, sequestering it. Src kinase is an example of a kinase that is rapidly phosphorylated. As well as the catalytic subdomain, c-Src has two other subdomains called SH2 and SH3 (short for Src-homology 2 and 3). SH2 domains are very common in intracellular signaling proteins, and their function is to bind phosphotyrosine.[21] SH3 domains bind to polyproline motifs.[22] Detailed computational models have been very important in elucidating the activation mechanism of Src.

Computer modeling by Roux, confirmed by Kuriyan with site-directed mutagenesis in yeast, showed that when Src is inactive, its SH2 domain is bound to phosphotyrosine 527 on the C-terminal tail. The SH3 domain binds to a polyproline motif between the kinase domain and the SH2 domain, snap-locking the kinase molecule in an inactive state.[23] The oncogenic mutation in viral Src is a truncation that removes this critical tyrosine residue.[24] Further computer modeling by Roux has shown that when Src is activated, either by dephosphorylation of Tyr527, a receptor binding to the SH3 domain, or by another molecule binding to the SH2 domain, the SH2 and SH3 domains move away from the kinase.[25] The activation loop in seen to partly unfold in the simulation, exposing Tyr416 to phosphorylation by another Src kinase. Src becomes fully active when the phosphate is removed from Tyr527 and Tyr416 in the activation loop is phosphorylated instead. Without a phosphate on Tyr527, the SH2 and SH3 domains can no longer lock in place and they become fully mobile.

Kinase activation involves a series of structural changes within the protein that stabilize the ATP-binding site and form the substrate recognition site. There are four motifs that are important in kinase activation:

1. The activation loop, phosphorylation of which activates many kinases.
2. The aspartate-phenylalanine-glycine (DFG) loop, which forms part of the ATP binding site.
3. The C-helix, which changes position upon activation, allowing the formation of a lysine-glutamate salt bridge inside the kinase.
4. The kinase spines, two sets of hydrophobic residues that run from the F-helix in the C-lobe through the active site to the N-lobe.

Figure 3 shows a diagram of PKA activation. The inactive kinase is on the right. Thr197 in the activation loop is not phosphorylated, the C-helix is moved away from the kinase core, which allows Glu91 to move out of the active site cleft. The kinase spine is not assembled, and catalytic

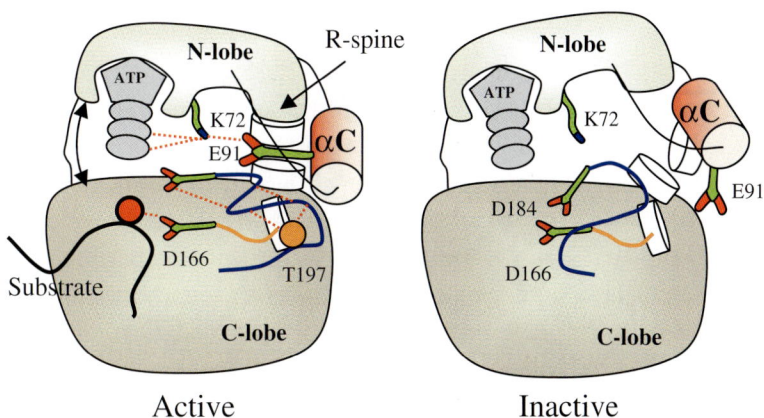

Figure 3. Protein kinase A subdomains and conformations. The two main conformations of PKA are illustrated. Phosphorylation of Thr197 and positioning of the C-helix to move Glu91 into the active site cleft activates the kinase. In the inactive form, Thr197 is not phosphorylated and the C-helix is moved away from the protein, removing Glu91 from the active site cleft. (Figure courtesy of A. Kornev, adapted by Thompson.)

residues are moved out of position for catalysis. The left-hand diagram shows the active kinase. When the C-helix is in the active position, a conserved set of amino acids lines up between the F-helix and the active site cleft, forming a regulatory spine that stabilizes the kinase.[26] Thr197 on the activation loop is phosphorylated and the C-helix is positioned close enough to the enzyme to move Glu91 into the active site cleft to form a salt bridge with Lys72. The conserved Asp184 from the DFG motif is moved into the active site cleft, and the catalytic loop (in blue) fold is stabilized by a hydrogen bond between Arg156 and the Tyr197 phosphate. When the enzyme is active, it is still flexible. There is subtle interdomain movement between the N- and C-lobes. It is unclear how far the enzyme has to open to release ADP and bind ATP between each round of catalysis but some movement is likely necessary.[27,28] In the next sections, the parts of the active kinase will be examined individually.

2. Activation Loop

Activation loop phosphorylation is required for activation of many kinases. Some kinases are only phosphorylated on one site in the activation loop, while others are phosphorylated in multiple places. The phosphorylation on the activation loop is a mechanism for controlling kinase activity. Figure 4 shows shows the location of the activation loop on the active PKA catalytic core, residues 191–197 colored in red. When the activation loop of a kinase is phosphorylated, it folds down and away from the active site. There is a pair of insulin receptor kinase (IRK) structures that are particularly interesting because they demonstrate auto-inhibition.[29] In insulin receptor kinase, the unphosphorylated activation loop folds up into the active site and prevents a substrate protein from binding to the kinase at all. Phosphorylation adds negative charges to the activation loop, and interaction with multiple positively charged amino acids on the C-lobe holds the activation loop out of the substrate binding site and helps position key catalytic residues.[30] Similar autoinhibition has been shown in fibroblast growth factor receptor and cyclin.[29]

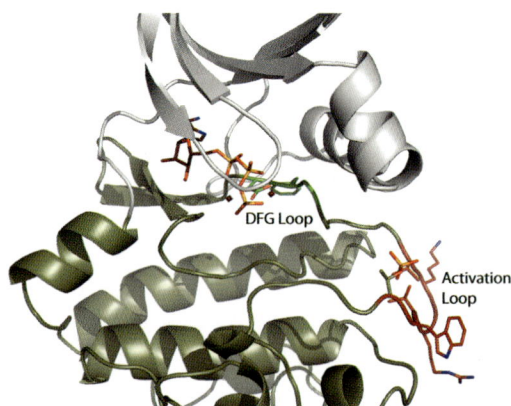

Figure 4.   Location of the activation loop and DFG motif on protein kinase A. The activation loop is shown in red. It is phosphorylated and held away from the substrate binding site. The DFG loop is shown in green, immediately next to the bound ATP molecule in the center of the kinase.

3. DFG Loop

The DFG loop is named for its amino acid sequence, aspartate-phenylalanine-glycine. It is highly conserved in mammalian kinases, and the negatively charged DFG aspartate side chain is critical for ATP binding. The DFG loop on PKA is also pictured in Figure 4, showing the close proximity of the aspartate to the two magnesium ions that bind to the kinase with the ATP. The carboxylic acid is negatively charged and forms a salt bridge with the positively charged magnesium ions.[31] In some kinases, the DFG loop can undergo a large motion, where the bulky phenylalanine changes position and the aspartate points away from the binding site. This conformation, called DFG-out, is inactive because the kinase cannot bind ATP properly, and (as will be discussed later), the DFG phenylalanine is also not positioned correctly to form the kinase spine. Changes in DFG loop positioning are initiated and maintained by the phosphorylation of both the adjacent activation loop and the regulatory subdomains that bind to the kinase core.

The DFG loop motion has been particularly well studied in c-Abl, the cellular form of the Abelson leukemia virus tyrosine kinase. Chronic myelogenous leukemia is caused by a chromosomal translocation called the Philadelphia chromosome that fuses a gene called BCR with c-Abl.[32] The resulting kinase is constitutively active because it lacks the regulatory domains that would keep the DFG loop flipped out. When another oncogene like Ras or p53 is also mutated, the leukemia becomes dangerous.[33] In the past, the only cure was chemotherapy and a bone marrow transplant. An Abl inhibitor called imatinib mesylate (Gleevec) has completely changed the treatment of the disease, putting the cancer into remission for most people who take it.[34] When imatinib was co-crystallized with c-Abl, it was found to bind to the DFG-out conformation of the kinase. The drug thus traps the kinase into its inhibited conformation.[35] Figure 5 shows two different crystal structures of c-Abl kinase, with the DFG loop highlighted. The top DFG-in structure has the DFG loop in orange.[36] There is an Aurora kinase inhibitor in the ATP site and the aspartate is oriented towards the ATP-binding site. A hydrogen bond between the DFG aspartate and the kinase inhibitor is shown as a dotted red line. While this structure is DFG-in, the phenylalanine is twisted out of position to complete the kinase spine so the kinase would not be able to phosphorylate. A fully active structure of Abl has not yet been crystallized. The DFG-out, imatinib-bound structure, has the DFG loop highlighted in blue.[35] The DFG aspartate and phenylalanine have rotated around the protein backbone so that the aspartate is pointed away from the ATP binding site and the phenylalanine is pulled away from the spine. A kinase in this conformation cannot be activated until the DFG flips in. As is suggested by the hydrogen bond in the DFG-in structure, computational modeling by Shaw suggested that imatinib binding is pH dependent, which was experimentally confirmed.[37]

4. C-helix

Another part of the kinase that moves on activation is the C-helix. The C-helix has a highly conserved glutamic acid residue, Glu91, shown in the cartoon in figure three. When the C-helix is in its active conformation, the glutamic acid residue forms an ion pair with a highly conserved lysine, Lys72 in PKA. This lysine residue also binds to ATP, and Taylor demonstrated that chemical modification of this critical amino acid results in a permanently inactive

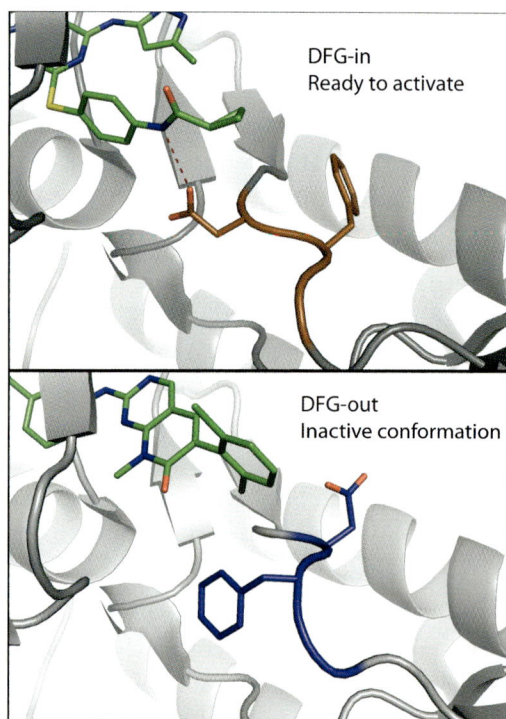

Figure 5. DFG-in (top) and DFG-out (bottom) conformations of c-Abl. In the top figure, the aspartate is oriented towards the ATP-binding site, and is hydrogen bonded to the inhibitor crystallized with the protein. In the bottom panel, imatinib is bound to the kinase and the DFG loop has moved, pointing the aspartate away from the ATP-binding site and moving the phenylalanine away from the kinase spine.

kinase molecule.[38,39] Similar to the DFG-loop aspartate, correct positioning of this lysine by a salt bridge to the C-helix glutamic acid is essential for catalytic activity. Most of the kinases mentioned thus far are regulated by C-helix positioning, including cyclin-dependent protein kinase (CDK), c-Src, IRK, PKA, and death-associated protein kinase (DAPK), and ERK2. If protein kinase cores are studied using software designed to detect pockets,[40] pockets around the C-helix are typically detected. These pockets are binding sites for hydrophobic protein-protein interactions that serve to anchor the C-helix against the rest of the kinase core. The C-helix can either be regulated by subdomains of the kinase, or by separate proteins. Many kinases have short tails either C-terminal or N-terminal to the conserved core and these tails often function to interact with the C-helix.

The C-helices from PKA, ERK2, CDK, and DAPK are pictured in Figure 6. Software was run to compute pockets on the protein surface to highlight areas of interaction and the computed pockets are shown in gold.[40] These are all serine/threonine kinases and show similar C-helix anchoring. In PKA, the C-helix is anchored by both the C-terminal and N-terminal tails. Deletion of the PKA tails will inactivate the kinase. In ERK2, a long extension of the C-terminal tail wraps around the C-helix and holds it against the kinase core. In DAPK, the short N-terminal tail folds over the N-lobe and into a pocket on the N-lobe side of the C-helix. CDK is activated by binding to cyclin, a cell-cycle signaling protein. Part of the interaction with cyclin is at the C-helix of CDK. A

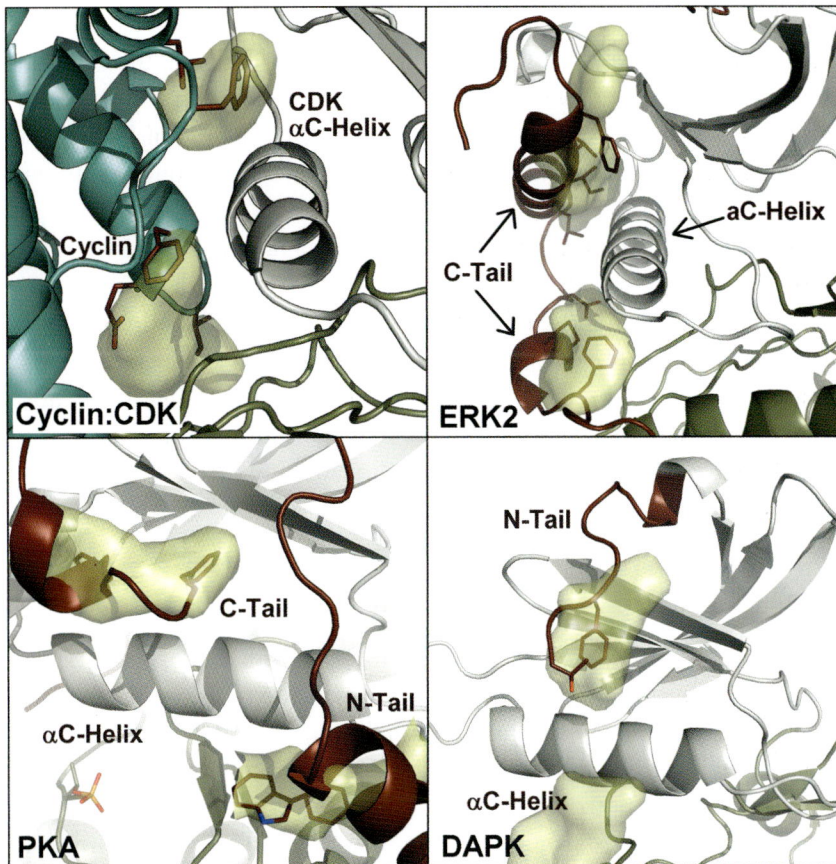

Figure 6. C-helix pockets and amino acids filling them. The computed pockets are in gold, N-lobe in grey, and amino acids outside the kinase core are shown in red; cyclin is in teal. ERK2 and CDK are in a different orientation to show the phenylalanine rings clearly.

common feature in the amino acids packed against the N-lobe surface of the C-helix is the presence of at least one bulky phenylalanine residue.

Tyrosine kinases have slightly different mechanisms for C-helix positioning and activation. As mentioned earlier, c-Src is autoinhibited by its SH2 and SH3 domains. To inactivate c-Src, the C-tail is phosphorylated, and the phosphate binds to its own SH2 domain. The SH2 and SH3 domains fold against the catalytic subunit and hold the C-helix away from the kinase core to prevent the formation of the glutamic acid-lysine bridge. Instead, there is a salt bridge between Arg409 in the activation loop and the conserved C-helix glutamic acid, numbered 310 in Src. The conserved lysine is also pulled out of position, bound to Asp404 and there is a bulky tryptophan residue, Trp260, in the linker between the core and SH2 domain that sits alongside the C-helix, stabilizing the inactive position.[41] When the SH2 and SH3 domains bind other proteins and the C-tail tyrosine is dephosphorylated, the "snap lock" breaks loose, and the whole kinase becomes more flexible.[23] As was mentioned earlier, computer models show that once the catalytic core is free to move, the C-helix rotates and the conserved glutamic acid-lysine bridge is formed. As the C-helix rotates, the salt bridge to the activation loop is broken, allowing the activation loop to partly unfold and expose Tyr416 for phosphorylation.[42] Once Tyr416 is phosphorylated, it forms a strong salt bridge with the Arg409 that was previously bound to the C-helix glutamic acid and the kinase is locked into its active position.[43] Post has shown with computational models that the shifting salt bridges inside c-Src help to pull the kinase in and out of its active conformation, and can briefly stabilize the kinase so that it has some catalytic activity even before Tyr416 is phoshporylated.[44]

In contrast to Src, c-Fes, another tyrosine kinase, is actually activated when its SH2 domain binds to a phosphotyrosine.[45] In this case, the SH2 domain becomes neatly ordered around the charged phosphate and presses the C-helix into place. In contrast to the hydrophobic interactions on the serine/threonine kinases, c-Fes has charged interactions at the interface between the C-helix and the SH2 domain. c-Fes has been tested with different substrates and the ideal substrate has a phosphotyrosine ten residues away from the site that c-Fes will phosphorylate. The SH2 domain binds to the upstream phosphotyrosine, the kinase activates, and then it adds another phosphate to a tyrosine ten residues away.

5. Kinase Spines

Thus far, the focus has been on sets of continuous amino acids that define separated motifs. These motifs converge at the active site to either bind ATP or mediate catalysis. Kinases also have sets of amino acids that run through the core from the F-helix in the middle of the C-lobe, through the ATP-binding site, to the β sheets in the N-lobe. These amino acids are not a continuous sequence like the DFG motif or activation loop, but rather a spatially conserved set of noncontinuous residues. They have been termed "spines" because they form a flexible but firm network through the kinase molecule that acts much like a vertebrate spine. The spines were discovered by Kornev using a computer program to search for residues with highly conserved backbone positions across the kinase family.[26]

The kinase spines are pictured on PKA in Figure 7. There are two spines, called the R-spine and C-spine for regulatory and catalytic spine. Both spines must be formed for the protein to be catalytically active. All of the amino acids in the spines are hydrophobic, and must be close enough to form noncovalent hydrophobic bonds with each other for the kinase to be active. Hydrophobic bonds are strong enough to stabilize the protein, but less rigid than salt bridges, allowing the kinase enough flexibility to bind and release substrates. The C-spine, shown in blue, is the larger of the two and it includes the adenine from the ATP molecule. It is termed catalytic because the adenine ring from the ATP completes the spine so that the motions of the N- and C-lobes are optimized for catalysis.[46] In PKA, it is made of Met231 and Lys227 from the F-helix, Met128 from the D-helix, Leu-Leu-Ile 172–174, and Ala70 and Val57 from the N-lobe. Leu173 and Val57 are in contact with the ATP adenine.[47] The second spine is shown in orange in the figure. It is made of Tyr164, Phe185, Leu95, and Leu106.[48] Phe185 is the phenylalanine from the DFG motif discussed earlier. Not only does movement of the DFG loop remove a crucial aspartate from the ATP-binding site, but it disrupts the R-spine of the kinase. Leu95 is on the important C-helix, one turn away from Glu91. Its position in the R-spine helps lock the C-helix into position from inside the kinase and ensures that Glu91 will be in the correct position to form the salt bridge with Lys72.

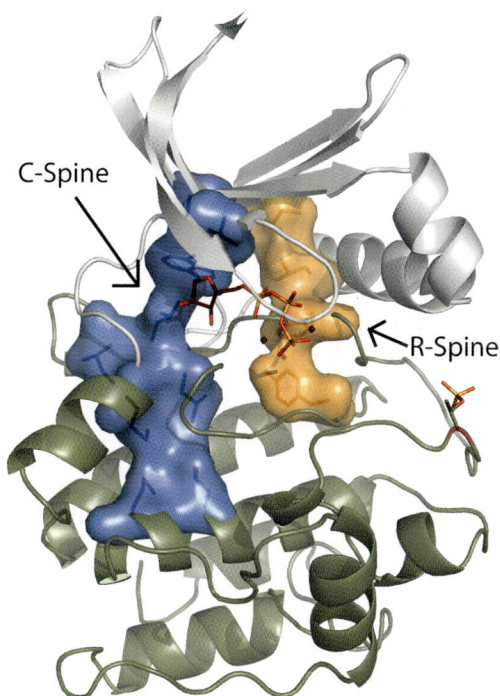

Figure 7. Kinase spines. The C-spine is pictured in blue and the R-spine is pictured in orange. Notice how the spines run through the kinase, helping to hold the ATP molecule in position.

The molecular basis for the DFG flip mentioned earlier relates to the R-spine. Not only is the DFG aspartate positioned to accept the magnesium of ATP in the active conformation, but the DFG phenylalanine completes the regulatory spine so that the N- and C-lobes are linked and poised for catalysis. The importance of the spine motif was demonstrated in Src and Abl when Daley and Kuriyan solved the crystal structure of an oncogenic, constitutively active chicken Src kinase with a Thr315 mutation.[49] Thr315 is referred to as the "gatekeeper" residue because it allows drugs to have access to the deep, hydrophobic pocket that results when the DFG loop flips out. Mutation of this residue to isoleucine in BCR-Abl is the most common form of Gleevec resistance. Daley showed that mutation of this threonine to a bulky, hydrophobic residue not only physically blocks Gleevec by removing the gatekeeping function and blocking the drug from binding, but it stabilizes the R-spine in its active conformation. The stabilization of the R-spine leads to abnormal kinase activation and is part of the oncogenesis process.

6. Processive Phosphorylation

Thus far, we have examined kinases that bind a protein, phosphorylate a single residue, and release the substrate. There is a different kind of phosphorylation, called processive phosphorylation, where a kinase works its way along a substrate protein phosphorylating multiple serine residues before releasing the substrate.[50] An example of processive phosphorylation is the phosphorylation of ASF/SF2 by SR protein kinase 1 (SRPK).[51] ASF/SF2 is a member of the serine-rich (SR) protein family and is part of the spliceosome, the assembly where cells splice mRNA. Members of the SR protein family have lengthy sequences of alternating serine and arginine residues, called RS domains. The cytosolic SRPK progressively phosphorylates a series of between 10 and 12 of the RS domain serine residues, creating hyperphosphorylated ASF/SF2. The hyperphosphorylation is essential for ASF/SF2 to be relocated to the nucleus where it is further phosphorylated by the nuclear Cdc2-like kinase/serine-threonine-tyrosine (CLK/STY).[51]

SRPK has a very high affinity interaction with its ASF/SF2 substrate, which contrasts with kinases that phosphorylate a single residue and quickly release the phosphorylated protein. SRPK binds to its ASF/SF2 substrate in three places. SRPK has a negatively charged groove for the unphosphorylated RS domain, a positively charged pocket for the phosphoserine it creates, and a binding site for another domain of ASF/SF2 called RRM2. The electrostatic interactions with the RS sequence create a very tight complex between kinase and substrate. The RS domain binds to the negatively charged area on the kinase C-lobe first, and then the RRM2 domain binds. RRM2 does not bind well to the kinase alone, but in the absence of the RRM2 domain, only about five serine residues are phosphorylated before SRPK releases its substrate. RRM2 is important in helping SRPK stay associated with ASF/SF2 and plays a role in substrate positioning. When ATP binds to the complex of ASF/SF2 and SRPK, a helix in the RRM2 domain partly unfolds into the kinase active site and helps slide the next RS serine into position to be phosphorylated.[51] As the kinase phosphorlyates more serine residues and there is less unphosphorylated RS domain to bind in the negatively charged groove on the kinase surface, the RRM2 domain also helps stabilize the kinase/substrate complex. Once the 10 to 12 serine residues have been phosphorylated, the ASF/SF2 is released. SRPK is also atypical in that it is a constitutively active kinase that is very robust to inactivation.[52] SRPK is able to compensate for the deletion of a domain that helps position the activation loop, and even for mutations directly in the activation loop. SRPK appears to be organized in an unusually stable and active conformation that may be related to its high efficiency and processivity. The sliding, processive phosphorylation mechanism of SRPK is an example of a molecular machine.

7. Conclusion

The activation loop, DFG motif, C-helix and its anchor, and kinase spines all work together to form a catalytically active kinase protein. Typically, the active core is also further stabilized by interactions with the domains and/or linkers that flank the core. The protein kinases are highly regulated and the different ways of activating and inactivating a kinase core allow the highly conserved structure to respond to different extracellular signals and then relay that signal into a biological response. Autoinhibition is a common mechanism of inactivation, and transcriptionally regulated binding partners, like cyclin, allow the cell to turn kinases on and off with a separate protein. In the case of PKA, inhibition is released by a second messenger, cAMP, binding to the inhibitory R-subunit and promoting its dissociation. Kinases like the EGF receptor kinase are transiently activated by EGF binding at the cell surface. Intracellular signals are amplified and transmitted through the cell by kinase cascades, which allow a small signal at the cell surface to have broad effects on transcription and translation. The cascades also provide flexibility, to be certain that essential cellular processes like growth and division are carried out correctly in response to extracellular signals. Kinases can phosphorylate proteins at single sites, or processively phosphorylate a substrate, adding multiple phosphate groups to a region of a protein. The wide variety of mechanisms that activate a common molecular core show the evolutionary importance and breadth of the protein kinase family.

Acknowledgments

Work in the McCammon group is supported in part by NIH, NSF, HHMI, CTBP, NBCR, SDSC, and TeraGrid resources. Work in the Taylor group is supported in part by NIH and HHMI resources.

Suggested Additional Reading Materials

L. N. Johnson. 2009. Protein kinase inhibitors: Contributions from structure to clinical compounds *Q Rev Biophys* **42**, 1–40.

A. P. Kornev and S. S. Taylor. 2010. Defining the conserved internal architecture of a protein kinase *Biochim Biophys Acta* **1804**, 440–444.

M. A. Lemmon and J. Schlessinger. 2010. Cell signaling by receptor tyrosine kinases. *Cell* **141**, 1117–1134.

References

1. G. Manning, D. B. Whyte, R. Martinez, T. Hunter and S. Sudarsanam. 2002. The protein kinase complement of the human genome. *Science* **298**, 1912–1934.

2. E. H. Fischer and E. G. Krebs. 1955. Conversion of phosphorylase b to phosphorylase a in muscle extracts. *J Biol Chem* **216**, 121–132.

3. E. G. Krebs, D. J. Graves and E. H. Fischer. 1959. Factors affecting the activity of muscle phosphorylase b kinase. *J Biol Chem* **234**, 2867–2873.

4. D. L. Friedman and J. Larner. 1963. Studies on UDPG-alpha-glucan transglucosylase. III. Interconversion of two forms of muscle UDPG-alpha-glucan transglucosylase by a phosphorylation-dephosphorylation reaction sequence. *Biochemistry* **2**, 669–675.

5. E. G. Krebs and E. H. Fischer. 1955. Phosphorylase activity of skeletal muscle extracts. *J Biol Chem* **216**, 113–120.

6. M. S. Collett and R. L. Erikson. 1978. Protein kinase activity associated with the avian sarcoma virus src gene product *Proc Natl Acad Sci U S A* **75**, 2021–2024.

7. D. Stehelin, H. E. Varmus, J. M. Bishop and P. K. Vogt. 1976. DNA related to the transforming gene(s) of avian sarcoma viruses is present in normal avian DNA. *Nature* **260**, 170–173.

8. D. Sheiness and J. M. Bishop. 1979. DNA and RNA from uninfected vertebrate cells contain nucleotide sequences related to the putative transforming gene of avian myelocytomatosis virus. *J Virol* **31**, 514–521.

9. L. N. Johnson. 2009. Protein kinase inhibitors: Contributions from structure to clinical compounds. *Q Rev Biophys* **42**, 1–40.

10. Z. Chen, T. B. Gibson, F. Robinson, L. Silvestro, G. Pearson, B. Xu, A. Wright, C. Vanderbilt and M. H. Cobb. 2001. MAP kinases. *Chem Rev* **101**, 2449–2476.

11. N. Jura, N. F. Endres, K. Engel, S. Deindl, R. Das, M. H. Lamers, D. E. Wemmer, X. Zhang and J. Kuriyan. 2009. Mechanism for activation of the EGF receptor catalytic domain by the juxtamembrane segment. *Cell* **137**, 1293–1307.

12. J. Brown, E. Y. Jones and B. E. Forbes. 2009. Keeping IGF-II under control: Lessons from the IGF-II-IGF2R crystal structure. *Trends Biochem Sci* **34**, 612–619.

13. J. Schlessinger. 2002. Ligand-induced, Receptor-mediated dimerization and activation of EGF receptor. *Cell* **110**, 669–672.

14. M. A. Lemmon and J. Schlessinger. 2010. Cell signaling by receptor tyrosine kinases. *Cell* **141**, 1117–1134.

15. G. Pearson, F. Robinson, T. Beers Gibson, B. E. Xu, M. Karandikar, K. Berman and M. H. Cobb. 2001. Mitogen-activated protein (MAP) kinase pathways: Regulation and physiological functions. *Endocr Rev* **22**, 153–183.

16. N. S. Duesbery, C. P. Webb, S. H. Leppla, V. M. Gordon, K. R. Klimpel, T. D. Copeland, N. G. Ahn, M. K. Oskarsson, K. Fukasawa, K. D. Paull and G. F. Vande Woude. 1998. Proteolytic inactivation of MAP-kinase-kinase by anthrax lethal factor. *Science* **280**, 734–737.

17. J. Gearhart, E. E. Pashos and M. K. Prasad. 2007. Pluripotency redux — advances in stem-cell research. *N Engl J Med* **357**, 1469–1472.

18. Y. Yarden and M. X. Sliwkowski. 2001. Untangling the ErbB signalling network. *Nat Rev Mol Cell Biol.* **2**, 127–137.

19. M. Huse and J. Kuriyan. 2002. The conformational plasticity of protein kinases. *Cell* **109**, 275–282.

20. S. K. Hanks and T. Hunter. 1995. Protein kinases 6. The eukaryotic protein kinase superfamily: Kinase (catalytic) domain structure and classification. *FASEB J* **9**, 576–596.

21. M. F. Moran, C. A. Koch, D. Anderson, C. Ellis, L. England, G. S. Martin and T. Pawson. 1990. Src homology region 2 domains direct protein-protein interactions in signal transduction. *Proc Natl Acad Sci U S A* **87**, 8622–8626.

22. K. Alexandropoulos, G. Cheng and D. Baltimore. 1995. Proline-rich sequences that bind to Src homology 3 domains with individual specificities. *Proc Natl Acad Sci U S A* **92**, 3110–3114.

23. M. A. Young, S. Gonfloni, G. Superti-Furga, B. Roux and J. Kuriyan. 2001. Dynamic coupling between the SH2 and SH3 domains of c-Src and Hck underlies their inactivation by C-Terminal tyrosine phosphorylation. *Cell* **105**, 115–126.

24. J. Kuriyan and D. Cowburn. 1997. Modular peptide recognition domains in eucaryotic signaling. *Annu Rev Biophys Biomol Struct* **26**, 259–288.

25. N. K. Banavali and B. Roux. 2009. Flexibility and charge asymmetry in the activation loop of Src tyrosine kinases. *Proteins* **74**, 378–389.

26. A. P. Kornev, N. M. Haste, S. S. Taylor and L. F. Eyck. 2006. Surface comparison of active and inactive protein kinases identifies a conserved activation mechanism. *Proc Natl Acad Sci U S A* **103**, 17783–17788.

27. D. A. Johnson, P. Akamine, E. Radzio-Andzelm, M. Madhusudan and S. S. Taylor. 2001. Dynamics of cAMP-dependent protein kinase. *Chem Rev* **101**, 2243–2270.

28. I. V. Khavrutskii, B. Grant, S. S. Taylor and J. A. McCammon. 2009. A transition path ensemble study reveals a linchpin role for Mg(2+) during rate-limiting ADP release from protein kinase A. *Biochemistry* **48**, 11532–11545.

29. J. A. Adams. 2003. Activation loop phosphorylation and catalysis in protein kinases: is there functional evidence for the autoinhibitor model? *Biochemistry* **42**, 601–607.

30. J. W. Orr and A. C. Newton. 1994. Requirement for negative charge on "activation loop" of protein kinase C *J Biol Chem* **269**, 27715–27718.

31. J. Zheng, D. R. Knighton, L. F. ten Eyck, R. Karlsson, N. Xuong, S. S. Taylor and J. M. Sowadski. 1993. Crystal structure of the catalytic subunit of cAMP-dependent protein kinase complexed with MgATP and peptide inhibitor. *Biochemistry* **32**, 2154–2161.

32. R. Kurzrock, W. S. Kloetzer, M. Talpaz, M. Blick, R. Walters, R. B. Arlinghaus and J. U. Gutterman. 1987. Identification of molecular variants of p210bcr-abl in chronic myelogenous leukemia *Blood* **70**, 233–236.

33. G. Gaidano, A. Guerrasio, A. Serra, G. Rege-Cambrin and G. Saglio. 1994. Molecular mechanisms of tumor progression in chronic myeloproliferative disorders. *Leukemia* **8**(Suppl. 1), S27–S29.

34. D. Pytel, T. Sliwinski, T. Poplawski, D. Ferriola and I. Majsterek. 2009. Tyrosine kinase blockers: New hope for successful cancer therapy. *Anticancer Agents Med Chem* **9**, 66–76.

35. B. Nagar, O. Hantschel, M. A. Young, K. Scheffzek, D. Veach, W. Bornmann, B. Clarkson, G. Superti-Furga and J. Kuriyan. 2003. Structural basis for the autoinhibition of c-Abl tyrosine kinase. *Cell* **112**, 859–871.

36. M. A. Young, N. P. Shah, L. H. Chao, M. Seeliger, Z. V. Milanov, W. H. Biggs 3rd, D. K. Treiber, H. K. Patel, P. P. Zarrinkar, D. J. Lockhart, C. L. Sawyers and J. Kuriyan. 2006. Structure of the kinase domain of an imatinib-resistant Abl mutant in complex with the Aurora kinase inhibitor VX-680. *Cancer Res* **66**, 1007–1014.

37. Y. Shan, M. A. Seeliger, M. P. Eastwood, F. Frank, H. Xu, M. O. Jensen, R. O. Dror, J. Kuriyan and D. E. Shaw. 2009. A conserved protonation-dependent switch controls drug binding in the Abl kinase. *Proc Natl Acad Sci U S A* **106**, 139–144.

38. M. J. Zoller, N. C. Nelson and S. S. Taylor. 1981. Affinity labeling of cAMP-dependent protein kinase with p-fluorosulfonylbenzoyl adenosine. Covalent modification of lysine 71. *J Biol Chem* **256**, 10837–10842.

39. J. A. Buechler and S. S. Taylor 1989. Dicyclohexylcarbodiimide cross-links two conserved residues, Asp-184 and Lys-72, at the active site of the catalytic subunit of cAMP-dependent protein kinase. *Biochemistry* **28**, 2065–2070.

40. E. E. Thompson, A. P. Kornev, N. Kannan, C. Kim, L. F. Ten Eyck and S. S. Taylor. 2009. Comparative surface geometry of the protein kinase family. *Protein Sci* **18**, 2016–2026.

41. M. LaFevre-Bernt, F. Sicheri, A. Pico, M. Porter, J. Kuriyan and W. T. Miller. 1998. Intramolecular regulatory interactions in the Src family kinase Hck probed by mutagenesis of a conserved tryptophan residue. *J Biol Chem* **273**, 32129–32134.

42. S. Yang, N. K. Banavali and B. Roux. 2009. Mapping the conformational transition in Src activation by cumulating the information from multiple molecular dynamics trajectories. *Proc Natl Acad Sci U S A* **106**, 3776–3781.

43. E. Ozkirimli and C. B. Post. 2006. Src kinase activation: A switched electrostatic network. *Protein Sci* **15**, 1051–1062.

44. E. Ozkirimli, S. S. Yadav, W. T. Miller and C. B. Post. 2008. An electrostatic network and long-range regulation of Src kinases. *Protein Sci.* **17**, 1871–1880.

45. P. Filippakopoulos, M. Kofler, O. Hantschel, G. D. Gish, F. Grebien, E. Salah, P. Neudecker, L. E. Kay, B. E. Turk, G. Superti-Furga, T. Pawson and S. Knapp. 2008. Structural coupling of SH2-kinase domains links Fes and Abl substrate recognition and kinase activation. *Cell* **134**, 793–803.

46. L. R. Masterson, C. Cheng, T. Yu, M. Tonelli, A. Kornev, S. S. Taylor and G. Veglia. 2010. Dynamics connect substrate recognition to catalysis in protein kinase A. *Nat Chem Biol* **6**, 821–828.

47. A. P. Kornev, S. S. Taylor and L. F. Ten Eyck. 2008. A helix scaffold for the assembly of active protein kinases. *Proc Natl Acad Sci U S A* **105**, 14377–14382.

48. A. P. Kornev and S. S. Taylor. 2010. Defining the conserved internal architecture of a protein kinase *Biochim Biophys Acta* **1804**, 440–444.

49. M. Azam, M. A. Seeliger, N. S. Gray, J. Kuriyan and G. Q. Daley. 2008. Activation of tyrosine kinases by mutation of the gatekeeper threonine. *Nat Struct Mol Biol* **15**, 1109–1118.

50. P. Patwardhan and W. T. Miller. 2007. Processive phosphorylation: Mechanism and biological importance. *Cell Signal* **19**, 2218–2226.

51. N. Huynh, C. T. Ma, N. Giang, J. Hagopian, J. Ngo, J. Adams and G. Ghosh. 2009. Allosteric interactions direct binding and phosphorylation of ASF/SF2 by SRPK1. *Biochemistry* **48**, 11432–11440.

52. J. C. Ngo, J. Gullingsrud, K. Giang, M. J. Yeh, X. D. Fu, J. A. Adams, J. A. McCammon and G. Ghosh. 2007. SR protein kinase 1 is resilient to inactivation. *Structure* **15**, 123–133.

Computational Studies of Na^+/H^+ Antiporter

8

Structure, Dynamics and Function

Assaf Ganoth, Raphael Alhadeff and Isaiah T. Arkin

Living cells are crucially dependent on processes that regulate their intracellular salinity, acidity and resultant volume. The activity of Na^+/H^+ antiporters and exchangers, discovered in the mid-70s of the last century, has been implicated to play essential roles in these processes. These integral membrane proteins are ubiquitous in the biological world and may be referred to as highly efficient nanomachines, characterized by specificity, modular selectivity and an ability to actively transport protons and Na^+ with no apparent leakage. Since Na^+/H^+ antiporters have been extensively studied and are one of the fastest active transporters known, they make suitable targets for computational analysis. Therefore, this chapter deals with the Na^+/H^+ antiporters and exchangers from a computational perspective and reviews the current literature in the field.

1. Introduction

Sodium ions and protons play an integral role in the functioning of living cells, whether directly (e.g. through activation of proteins) or indirectly (e.g. through osmotic pressure affecting cell volume). As such, the importance of their regulation cannot be overstated. Living organisms control the content of their cells by separating them from the surroundings using hydrophobic membranes. However, it is clear that full separation cannot be attained nor desired as certain molecules from the outside are needed inside the cells for them to survive and similarly secretion of waste products is essential. Toward that end, cells have developed transport mechanisms, allowing them to actively, or passively, exchange matter with their surrounding.

In this chapter, we will focus on one example of such a transport system, the Na^+/H^+ antiporter family. Antiporters (a.k.a. exchangers) are active transporters that rely on the electrochemical gradient of one substrate (in this case, protons) to transport another substrate (Na^+) against its electrochemical gradient, in opposite directions through the membrane. Na^+/H^+ antiporters can be found in practically all cells, ranging from extreme archaea to higher eukaryotes including humans and plants.[1]

Two comprehensive books dealing with Na^+/H^+ antiporters have been published during the last decade. The first covers a wide spectrum of topics ranging from the molecular basis of the Na^+/H^+

antiporting and the role of the Na$^+$/H$^+$ antiporters in diseases.[2] The second presents a broad review about basic properties, distribution, function, regulation, current methods and the use of inhibitors and ligands for the study of the Na$^+$/H$^+$ antiporters.[3] For an inclusive detailed background about the proteins, the reader is referred to the aforementioned books, whereas the current chapter focuses on computational studies of the Na$^+$/H$^+$ antiporter. The most studied member of the family is the *Escherichia coli* Na$^+$/H$^+$ antiporters A (NhaA) which is the major theme of this chapter; however, due to the clinical importance of the human orthologs, work has been done on these members as well, and will be discussed towards the end of the chapter.

The existence of Na$^+$/H$^+$ antiporting has first been reported in 1974 by West and Mitchell.[4] The protein was yet uncharacterized but work was already underway, mainly by the Schuldiner group (e.g. Refs. 5,6). It was biochemically discovered that the antiporter is electrogenic, with a stoichiometry of two protons for every single Na$^+$ transported.[7] NhaA's sequence was published in 1988 by Schuldiner, Padan and co-workers,[8] allowing genetic studies to be conducted. Mutational analyses and the search for indispensable residues were performed, and while many mutations lower the antiporter's apparent K_M, three conserved aspartates, D-133, D-163 and D-164 were found to be critical for function.[9]

As mentioned above, the antiporter has a role in regulation, and is therefore expected to be regulated by the ions' involved concentrations. Indeed, both Na$^+$ and pH tightly regulate NhaA, but on different levels. Na$^+$ regulation comes into play by upregulating NhaR, an initiator that activates transcription of NhaA.[10] This allows the bacterium to express NhaA only when internal Na$^+$ becomes high, and expression will decrease as the sodium stress subsides. The regulation by pH is done by directly activating and deactivating the antiporter, whereby NhaA was found to be inactive at pH 6.5 and lower, and its activity was found to rise up to 2000-fold when pH increases to 8.[11] This two-faced regulation allows the NhaA to function when sodium levels are too high, but deactivates it if pH is decreased beyond the physiological range, probably to prevent the antiporter from harming the cell. At high external pH values, NhaA cannot function due to a lack of electrochemical energy on which it depends. Finally, it is important to note that low activity at acidic pH is not an inherent function of the Na$^+$/H$^+$ antiporters since a homolog of NhaA from *Helicobacter pylori* was found to function under acidic conditions.[12]

In 2005, Michel and co-workers solved the structure of NhaA in a landmark study[13] and paved the road to computational analysis. Since, the mechanism of the protein has been extensively looked at, model structures of family members were constructed and the search for insights on pH-sensing was intensified.

Ion selectivity of membrane proteins, mainly channels, has been the focus of many computational and experimental scientists alike.[14–17] The mechanism behind the ability of proteins to discriminate between different ions remained elusive in many cases; however, recent technological and algorithmic improvements have given tentative solutions to this enigma (e.g. Ref. 18). NhaA is selective as well; whereas *Escherichia coli* NhaA can transport only Na$^+$ and Li$^+$, the *Vibrio parahaemolyticus* homolog was found to transport K$^+$ ions as well.[19] This example is probably not unique, and shows first hand that the selectivity filter of NhaA is adjustable. As such, the *Vibrio parahaemolyticus* NhaA model structure[20] might serve as a platform to study the selectivity of the antiporter further.[21]

We begin by discussing the solved structure of *Escherichia coli* NhaA, with all the important characteristics that are addressed in the various studies. We will then present computational work done on NhaA to better understand its mechanism of action, and its ability to sense the surrounding pH. We will shortly review the *Vibrio parahaemolyticus* NhaA model structure and its potential interest and lastly switch to the human NhaA counterparts (NHE1 and NHA2). These bear little, if any, sequence homology to NhaA but have an important clinical role, causing damage during various cardiac procedures; specific inhibitors for them could prove to be beneficial.[22]

2. Structure of the Na⁺/H⁺ Antiporter

Structural data for NhaA have been collected earlier using Cryo-EM,[23,24] but only after Michel and co-workers obtained crystals of *Escherichia coli* NhaA was the 3D structure revealed in atomic resolution.[13] This step was a milestone in the way of understanding the protein better, and it is not surprising that all computational studies done on the protein were conducted quickly thereafter (e.g. Refs. 25, 26).

Crystals were grown and the phasing problem was solved using single-wavelength anomalous dispersion. The final structure was determined to 3.45 Å resolution, with an R_{free} of 31.7%. It should be noted that the protein was crystallized at pH 4, where it is fully inactive.[11] This issue is well known and is taken under consideration when studying the NhaA.

The protein consists of 388 amino acids, with both termini exposed to the cytoplasm; the electron density, however, only appears clearly for residues 9–384. The overall structure of NhaA is composed of 12 transmembrane segments (TMSs), and at the time solved, described a novel architecture (Figure 1). Of high interest is the so-called TMS IV/XI assembly, consisting of two pairs of oppositely oriented TMSs, built up of a short helix, an unfolded short coil, and another short helix (referred to as TMS IVc and IVp for cytoplasmic and periplasmic side short helices, respectively). This arrangement creates a core in which the polar ends from all helices do not neutralize each other's charge, and was described to be energetically unfavorable.[13] It is generally accepted to be the place where the ions bind and the conformation changes take place, at least in part (for review, see Ref. 27). With that, this arrangement is not unique, and a similar one has been reported before.[28] The helical dipoles mentioned are compensated by the presence of D-133 and K-300 near the positive and negative helix termini, respectively. Due to this role, both residues are thought to be found at their charged states at physiologically relevant pHs.

The surface of the protein is quite asymmetric, showing a smooth face parallel to the membrane at the periplasmic side; while the cytoplasmic side has loops and helices protruding from the membrane, creating a rough surface. Moreover, on the periplasmic side, a β-hairpin is found, also parallel to the membrane, that was shown to play a part in the dimerization of the protein.[29] However, recent studies of NhaA homologs from different bacteria have shown a noticeable minority of sequences that lack this β-hairpin, raising interesting questions about the role of the β-hairpin and the importance of the dimerization.[20]

The protein, being an antiporter, must have means of transferring its substrates, and naturally exhibits funnels elongating from the surrounding bulk to the core of the protein. Two such vestibules are seen, one from the cytoplasmic side and one from the periplasmic side (Figure 1). Both funnels roughly converge to the putative ion binding site but are disconnected by a hydrophobic barrier. The cytoplasmic funnel was calculated to be negatively charged, and was suggested to serve as an ion lining passage leading to the putative binding site.[13] D-163 as well as D-164 were

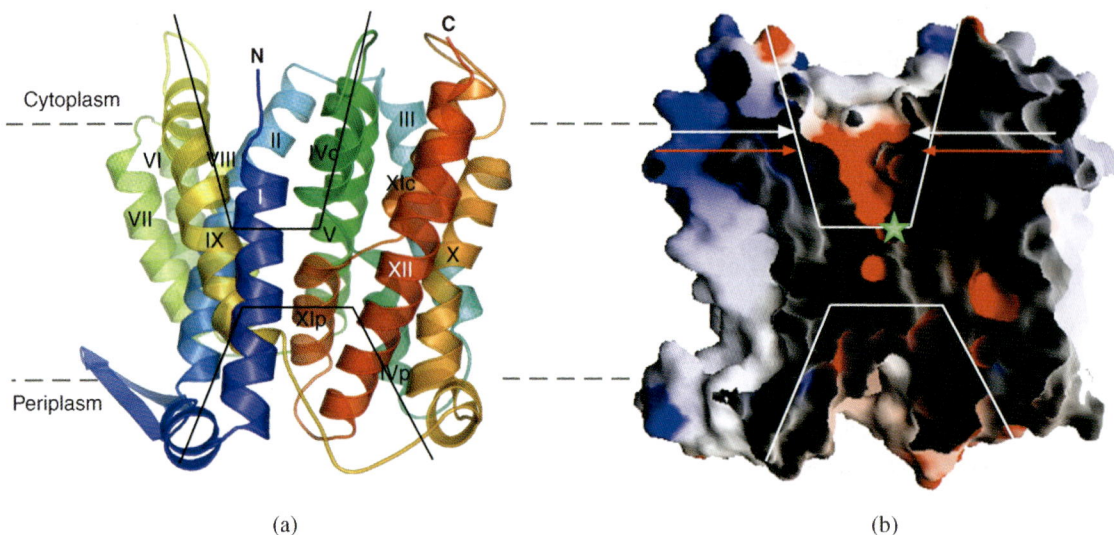

(a)   (b)

Figure 1. General architecture of NhaA antiporter. (a) Ribbon representation of the crystal structure of NhaA, viewed parallel to the membrane (broken lines). The transmembrane segments (TMSs) are labeled with roman numerals. N and C indicate the N- and C-terminus, respectively. Cytoplasmic and periplasmic funnels are marked by continuous black lines; c or p denotes helices in the cytoplasmic or periplasmic sides, respectively. (b) Cross-section through the antiporter normal to the membrane, with the front part removed. The cytoplasmic and periplasmic funnels are marked by white lines. Down to the white arrow, hydrated cations are potentially accessible into the cytoplasmic funnel. Below the red arrow, only non-hydrated Na^+ or Li^+ can potentially enter. The location of D-164 is marked by a green star. (Figure adapted from Padan *et al.*)[52]

found to be highly conserved and indispensable,[9] and of the two, D-164 is exposed to the cavity, making it a more likely candidate for the ion binding site.

The mechanistic model suggested by Michel and co-workers in their work corresponds to the alternate accessibility model proposed by Jardetzky nearly four decades ago.[30] The presence of both funnels and their separation is in line with that hypothesis, and the authors suggest that the TMS's IV/XI assembly is responsible for the delicate (and fast) movement allowing the alternate accessibility of the binding site. The reasoning behind that is given to the fact that D-164 is placed on a rod-shaped, supposedly rigid helix (TMS V) while the assembly is composed of many components who could potentially move, and are expected to be well affected by charge changes during the process of antiporting.

The solved structure of the NhaA[13] enabled subsequent biochemical and functional studies of the protein and further proteins of this family. Its unique features, exposed in atomic details, made way for further understanding as presented at the following subsections.

3. The Dynamics of the Na^+/H^+ Antiporter

3.1 *The influence of protonation states on the NhaA*

Michel and co-workers conducted molecular dynamics (MD) simulations on the NhaA hoping to find insights into the pH activation dynamics.[31] Relying on experimental data that show a separation in space between the antiporting center and the pH-sensor,[32] they looked at the protein's movement after short MD trajectories (4 ns) under different protonation states (mimicking active and inactive pH conditions, referred to as pH 8 and pH 4 in the article, respectively) and compared those to simulations done using an NhaA variant, G338S, which was shown experimentally to have an altered pH-sensitivity profile and an increased K_M.[33] The protonation states of the specific residues in the simulations were calculated using the multi-conformation continuum electrostatics (mcce) technique,[34] yet due to structural considerations (see above), D-133 was chosen to be deprotonated although its pKa calculated value predicted it should be otherwise.

The study shows that, as opposed to the rest of the protein, TMSs X and IVp exhibit a unique pH-dependent behavior (the latter supported by experiment[35]), both exhibiting higher fluctuations at the simulated higher pH. Michel and co-workers report that TMS X is, in fact, not a rigid helix, but two rigid shorter helices connected by a hinge; both its bend and wobble angles, characteristics of such a hinge, are different at the two pH states. This helical phenomenon has been suggested and seen before in other proteins[36] and might be attributed to a particular function.

The work reports that, at the pH 4 simulation, there is no connectivity between the cytoplasmic and periplasmic funnels, in line with Arkin and co-workers.[26] However, in the pH 8 simulation, they observe a continuous hydrogen bond network, refers to as "water molecules diffuse into the 'hydrophobic' barrier".[31]

Lastly, these findings were tested on a G338S variant, and as expected, TMS X of the protein behaved as in the pH 4 simulation of the WT, in both conditions. Michel and co-workers suggest a mechanism for the pH independence of the variant. In the simulation, a stable water chain is formed between S-338 and K-300 (on TMS X), preventing TMS X to move as it should for the pH-activation. Notably, other residues that might play a role in the pH response have been identified elsewhere, such as D-65.[37]

3.2 Mechanism of Na$^+$/H$^+$ antiporting

Arkin and co-workers[26] have used the crystal structure of NhaA and put it through elongated MD simulations to try and elucidate its mechanism of function. Considering that D-163 and D-164 were the only carboxylic residues that were shown to be absolutely indispensable for activity,[9] simulations were done under the assumption that they play a critical role in the mechanism and, furthermore that their protonation states dictate its steps. The study consisted of simulating NhaA under the four possible protonation state combinations of D-163 and D-164 while having a Na$^+$ ion adjacent to either of the aspartates (eight states in all). The study revealed that the Na$^+$ binding site is most likely not D-163 based on the fact that in the simulations, regardless of the protonation states of D-163 and D-164, the Na$^+$ ion became trapped when placed near D-163, and no water could penetrate the protein's cavities in those cases.

On the other hand, when the Na$^+$ was placed near D-164 in its deprotonated state, the ion was quickly hydrated by water molecules. However, the most interesting findings were obtained when the ion was placed adjacent to a protonated D-164; in these instances not only did the ion exit the protein, but it did so in a differential manner, according to the protonation state of D-163, denoted in the article as the accessibility-control site. Taking the different simulations into account, a model

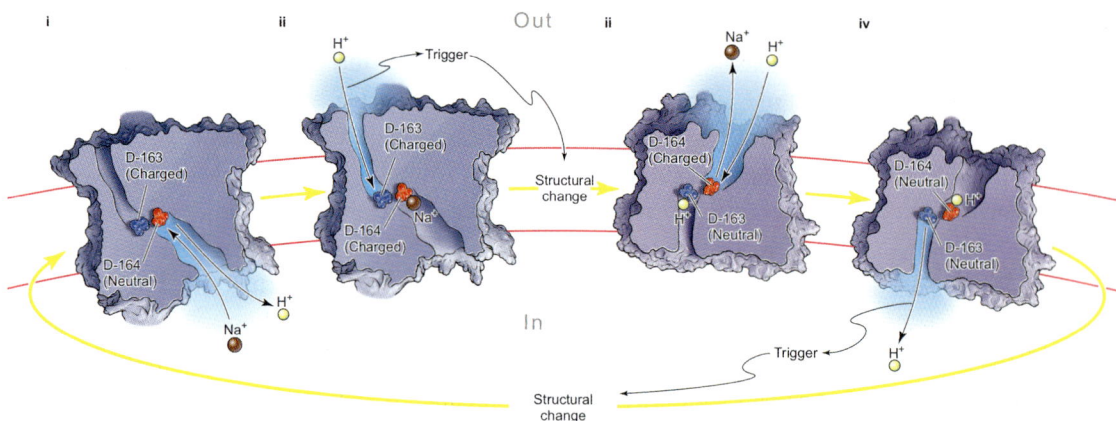

Figure 2. Schematic representation of the transport model of NhaA. The carboxylic group of D-163 in the accessibility-control site is colored blue, and D-164 in the Na$^+$-binding site is colored red. (Figure adapted from Arkin *et al.*)[26]

for the mechanism is suggested (Figure 2) and functions as follows: (i) D-163 is deprotonated, resulting in cytoplasmic accessibility of the Na$^+$ binding site. D-164 exchanges a proton for a cyto-plasmic Na$^+$ ion. (ii) D-163 becomes protonated by a periplasmic proton, resulting in the accessibility changing and the Na$^+$ binding site pointing toward the periplasm. (iii) D-164 exchanges a proton for a Na$^+$ ion with the periplasm. (iv) D-163 deprotonates, losing its proton to the cyto-plasm, resulting in another conformational change, leading back to step (i) for another cycle thereafter. As one can see, the mechanism presented accommodates for the known stoichiometry of one Na$^+$ to two H$^+$. Also worth noting is that the model abides to the alternate accessibility mecha-nism hypothesized by Jardetzky[30] back in the 60's, and is in line with Michel and co-workers above.

The work presents a suggestion for how the alternate accessibility is achieved. As D-163 becomes protonated, it shifts its hydrogen bond network from the amide hydrogens of M-105 and T-132 to the carboxyl of M-105. This shift carries with it a movement, which causes helix V to tilt, orienting D-164 as a consequence.

Lastly, and based on pKa calculations done by Michel and co-workers,[25] Arkin and co-workers conducted more simulations, this time exploring the effect of protonating candidate aspartic-acid residues to serve as the pH sensor. The reason being that previous experimental analyses have shown that histidine residues (the likely candidates as pH sensor when the pKa of the protein is around pH 7) play no role in pH sensing.[38,39] Arkin and co-workers hypothesized that the pH sensor, if indeed an aspartic-acid residue, would cause the protein to enter the conformation that most resembles the crys-tal structure that was obtained at pH 4.[13] MD simulations were done for all aspartic-acid residues with high calculated pKa's but only in the case of protonated D-133 has the protein adopted a structure more similar to the X-ray structure. This is supported by a mechanistic explanation, arguing that D-133 serves as a charge compensation for the TMS IV/XI assembly (see above) and being proto-nated it can no longer neutralize the dipoles meeting in the TMS intersection, causing electrostatic-induced movement. This movement caused (in the simulations) the binding site to point towards the protein lumen (similar to the crystal structure), making it unable to bind substrate.

Finally, it is important to point out that it is improbable that a single residue acts as the sole pH sensor. The reason is that three orders of magnitude change in protein activity across one and a half pH units most likely arises from more than one protonation event. Hence, it is likely that the change arises from three protonation events that act cooperatively to control the protein function. This cooperativity might be the reason why so far the full identity of the pH sensor has yet to be described, in that single point mutation of an ionizable residue were not able to completely abolish the pH sensitivity of the protein.

3.3 *The unique Na⁺/H⁺ electrostatic organization*

In a complementary study, Michel and co-workers computationally investigated the electrostatic interactions in the NhaA structure in its closed conformation. They examined the effect of pH values on the protonation states of titratable residues in NhaA.[25] Their method of research was the mcce technique,[34] which is a hybrid method combining continuum electrostatics with molecular mechanics. It allows multiple positions of side chain rotamers, hydroxyls, and water hydrogens in the calculation of the pH dependence of the ionization equilibria of titratable groups. Explicit conformers for polar and ionizable side chains, the cofactor, and buried water molecules were added subsequently.

Four clusters of electrostatically tightly interacting residues in the protein, consisting of 18 titratable groups altogether, have been identified (Figure 3). These clusters are spread in the protein along its cross-membrane axis: cluster I is located at the N-terminal side of the cytoplasmic orifice funnel's and contains residues of the proposed pH sensor, including E-78, E-252, E-253, and H-256; cluster II is located on the opposite side of the cytoplasmic funnel; cluster III, formed mainly by residues of the putative ion-binding site, including D-163 and D-164, and is located at the middle of the membrane; cluster IV is at the rim of the periplasmic funnel. H-256 provides an electrostatic connection between clusters I and II, and D-133 connects clusters I and III. The residues of these four clusters are responsible for the electrostatic properties of NhaA and for the structural changes upon activation by raising the pH.

The interpretation and the analysis of the computational results were performed by the aid of experimental characterization of site-directed mutants and cys-replacement studies. These biochemical studies were mainly conducted by Padan and co-workers (such as Refs. 40, 41) and by a subsequent study of Michel and co-workers.[42] The extreme pKa values of the clustered residues suggested no change in the protonation state of these residues within the physiological pH range, unless the structure of NhaA alters. E-78, with a calculated pKa near the physiological range, is suggested to trigger such pH-dependent structural changes. Long-range interactions between the pH sensor and D-133 have a significant effect on the electrostatic potential of the antiporter. Analysis of the X-ray data[13] suggests that D-163 and D-164 contribute to the Na⁺ binding site. Biochemical data obtained with NhaA activated at alkaline pH show that these conserved residues are essential for the activity of the antiporter.[9] The calculations of the model predicted that D-163 and D-164 continue to hold their protons in the acid-locked conformation of NhaA. The identification of protein-bound water molecules allowed the examination of the initial hydrogen-bonded network of NhaA that was found to be between residues D-163 and K-300 via four explicit water molecules.

The inclusion of water molecules altered the pKa values of several individual titratable residues. The most significant change of a pKa value was found for K-300. The deprotonation of

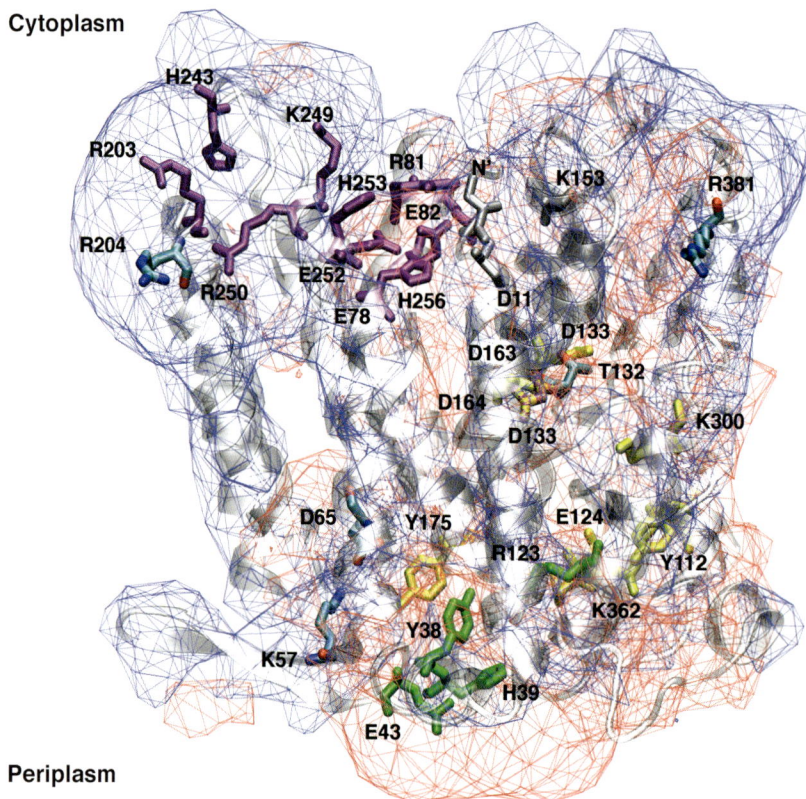

Figure 3. Clusters of strongly electrostatically interaction ionizable groups for NhaA antiporter. Zones of negative and positive potential are colored red and blue, respectively. Four clusters of interacting residues are shown in gray, green, magenta, and yellow. Additionally, the isolated sites K-57 and D-65, and the conserved T-132, R-204, and R-381 are included. (Figure modified from Olkhova *et al.*)[25]

D-65 at pH 8 is accompanied by a conformational reorientation of its side chain to the cytoplasmic side. Furthermore, analysis of the results indicated that K-57 and D-65, whose functional importance has been demonstrated by site-directed mutagenesis, form an isolated charge pair.

To assess the mechanistic basis underlying the significant roles of D-163 and D-164 in ion binding, and pH activation of NhaA, Michel and co-workers[42] also studied the effects of the substituted residues in the variants D163E, D163N, D164E, and D164N on the electrostatic properties of NhaA and compared the pKa values and mean field interactions energies of the titratable groups in the variants with those of the wild type antiporter. Their results assigned different roles to D-163 and D-164 in the electrostatic interaction between them and the residues of the pH sensor. They concluded that D-163 influences the long-range electrostatic interaction between residues of the pH sensor, the ion-binding site and helix X; whereas D-164 is an important element for triggering the pH-induced structural reorganization of the antiporter. The site-directed mutagenesis experimental results confirmed the interaction between the electrostatically coupled clusters comprising the pH sensor and the cation-binding site emphasizing the crucial role of this interaction in the pH activation of NhaA.

It is suggested that the cross-membrane arrangement of polar residues in the antiporter proteins facilitates the access of the ionic substrates to the central group, their binding and their transport.

Most interestingly, based on functional analysis of mutants, a similar arrangement of polar conserved residues has been suggested for Vc-NhaD, a *Vibrio cholera* antiporter that is not homologous to *Escherichia coli* NhaA (in sequence) and of which the structure is not known.[43]

A word of caution should be mentioned about the possible limitations of the mcce computational approach when implemented to the NhaA protein. pH-induced conformational changes by deprotonation/protonation of residues forming the ion binding site could make the NhaA structure, which was determined at pH 4, inappropriate for calculations at high pH. The side chain flexibility implemented in mcce reduces, but does not eliminate, the dependence on the initial structure since the positions of the backbone atoms are fixed. Therefore, it is not possible to define whether or not it affects the pKa values of the ion binding site residues and the pH sensor residues.[42]

The studies of Michel and co-workers provide valuable information on the activation of the antiporter and the role of individual amino acids, and exemplifies the power of combining structural, computational and biochemical approaches. Additionally, they offer clues for future analyses of the mechanism of activation and transport as well as for the interpretation or design of experimental work.

4. Models for the Na$^+$/H$^+$ Proteins

Genome sequencing efforts discovered genes encoding putative Na$^+$/H$^+$ antiporters that were classified as members of the monovalent cation/proton antiporters (CPA) super-family according to the nomenclature of Transport Classification Database (http://www.tcdb.org/).[44] Eleven human CPAs have been identified and assigned to specific subfamilies of the monovalent CPA family with orthologs throughout the biological world.[45] Two subfamilies of this superfamily, named CPA1 and CPA2, have been identified as the Na$^+$/H$^+$ exchanger (NHE)[46] and the NHA,[45] both have orthologs from bacteria to human. The NHE subfamily includes nine human isoforms (NHE1 through NHE9) that are well characterized and believed to be electroneutral.[46] It was found that they play critical roles in human heart, kidney, brain, uterus, skeletal muscle and stomach (for review see Ref. 47). In plants, NHE orthologs have been characterized in several species such as *Oryza sativa*[48] and *Arabidopsis thaliana*.[49] The NHA subfamily, which was named NHA on the basis of its sequence similarity to fungal NHA genes and bacterial NhaA genes, is widely spread both in eukaryotes[50,51] and prokaryotes.[52] This subfamily includes NhaA proteins of bacteria, and the human NHA1 and NHA2.

Study of membrane proteins (as the CPAs) is particularly challenging due to the fact that they are embedded in the membrane, and therefore need to be solubilized by detergents and then reconstituted back into the membrane. Other than the X-ray structure of the *Escherichia coli* NhaA,[13] structures of other prokaryotic or eukaryotic Na$^+$/H$^+$ proteins have yet to be solved. Due to the scarcity of structural data, models to few of these proteins have been recently constructed.[20,53,54]

4.1 *Model structure of the vibrio parahaemolyticus NhaA*

Vibrio parahaemolyticus is a slightly halophilic gram-negative marine bacterium that naturally inhabits coastal waters and can cause illness resulting from eating contaminated shellfish, or, less frequently, through open wound exposure to sea water. In the United States, *Vibrio parahaemolyticus* is the most common *Vibrio* species isolated from humans, as well as the most frequent cause of *Vibrio*-associated gastroenteritis.[55,56]

Vibrio parahaemolyticus is of special interest not only because of its illness causing potential but also due to its broad antiporter selectivity. In contrast to the *Escherichia coli* NhaA, the *Vibrio parahaemolyticus* NhaA is able to use not only Na[+] or Li[+] as its substrate ion but also K[+].[19] However, the structure of the *Vibrio parahaemolyticus* NhaA is unknown and the biochemical data about it are limited. Understanding its structure-function relationship will help to elucidate the selectivity mechanism, which might shed light into other selective systems as well.

Arkin and co-workers[20] utilized a combination of computational methodologies in order to construct a structural model for the Na[+]/H[+] antiporter from *Vibrio parahaemolyticus*. First, they performed a wide spectrum multiple sequence alignment analysis, constructed a phylogenetic tree of the bacteria domain and mapped the *Vibrio parahaemolyticus* NhaA. Based on the results of the phylogenetic study, they applied a homology modeling procedure by using the *Escherichia coli* NhaA as a template in order to generate an initial model for the protein's structure. Given that the NhaA of *Vibrio parahaemolyticus* shares a high identity (59%) and similarity (74%) with the NhaA of *Escherichia coli* (E value = $7 \cdot 10^{-122}$), the NhaA of *Escherichia coli* was selected as a template for the homology modeling process. The extensive sequence resemblance between the proteins ensures a high success of the modeling process. However, the β-hairpin region existing in the *Escherichia coli* NhaA appears to be absent in the *Vibrio parahaemolyticus* NhaA structure, based on multiple sequence alignment and a subsequent phylogenetic evolutionary analysis. The obtained model was evaluated for its quality from structural, geometrical and evolutionary perspectives.

Then, the authors both fine tuned and provided credence to the structure by performing atomic-level MD simulations. As a comparison, they simulated the *Escherichia coli* NhaA antiporter that was used as a template for the homology modeling process. Arkin and co-workers analyzed several main features for both proteins (e.g. secondary structure and electrostatic potential). Representative structures of the *Vibrio parahaemolyticus* NhaA, superimposed with the X-ray structure of the *Escherichia coli* NhaA, are presented on Figure 4. The two structures highly resemble each other, and the *Vibrio parahaemolyticus* NhaA displays all the distinguishing characteristics of the *Escherichia coli* protein, except the aforementioned missing β-hairpin region.

Although the similarity between the target protein and the template is striking, one must take into account that this work presents a model which should ultimately be validated by structural means. Nevertheless, it is stressed that the model for the *Vibrio parahaemolyticus* NhaA can still be considered a useful working tool to aid theoretical and experimental studies. Better understanding of the *Vibrio parahaemolyticus* NhaA structure-function may assist in designing selective blockers of NhaA to help in fighting *Vibrio*-associated infections such as gastroenteritis. Moreover, the suggested model might prove useful in provide insight into the selectivity basis of the NhaA.[21]

4.2 *Model structure of NHE1 — A human Na⁺/H⁺ exchanger*

NHE1 is the most widely expressed isoform among the family of the mammalian Na[+]/H[+] exchangers (for recent review see Ref. 57). Following allosteric activation by intracellular acidification, NHE1 exchanges extracellular Na[+] for intracellular protons with Na[+]:H[+] stoichiometry of 1:1.[46] NHE1 is associated with a plethora of physiological and pathological conditions including hypoxia

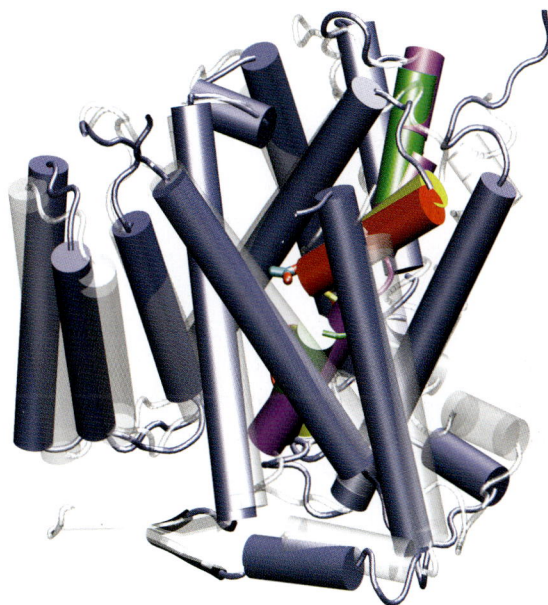

Figure 4. Schematic diagrams of a representative MD-derived structure of the *Vibrio parahaemolyticus* NhaA (blue) superimposed to the X-ray structure (PDB entry 1ZCD) of the *Escherichia coli* NhaA (white). TMSs IV and XI of the *Vibrio parahaemolyticus* NhaA are colored with red and green and these of the *Escherichia coli* with yellow and purple, respectively. Highlighted in black is the location on the *Vibrio parahaemolyticus* where the β-hairpin of *Escherichia coli* is present. (Figure modified from Ganoth *et al.*)[20]

and hypertension as well as cancer, heart, blood vessels, central nervous system, vascular, gastric, and kidney diseases (for review see Refs. 47, 58). For example, NHE plays a key role in damage to the mammalian myocardium that occurs during ischemia and reperfusion and is involved in hypertrophy of the myocardium,[59] and hence its potential inhibitors should reduce heart hypertrophy and ischemia-reperfusion damage.[60] Moreover, NHE1 is a key component in carcinogenesis and tumor progression processes by reversing the pH gradient in malignant cells, a phenomenon known as malignant acidosis, which is a crucial stage in oncogenic transformation.[61] NHE1 is inhibited by amiloride and its derivatives (such as eniporide and cariporide) and by benzoylguanidine type compounds,[62] and, accordingly, pre-clinical studies and clinical trials were conducted.[63]

Using the crystal structure of the NhaA antiporter from *Escherichia coli* as a template,[13] Ben-Tal and co-workers built a model for the three-dimensional structure of human Na⁺/H⁺ exchanger 1 (NHE1).[53] The bacterial *Escherichia coli* NhaA and eukaryotic Na⁺/H⁺ exchangers play similar roles in controlling pH and electrolyte homeostasis, and have been suggested to share a common evolutionary origin and possibly also have a similar structural fold despite lack of any sequence homology.[45,47] Thus, Ben-Tal and co-workers' working hypothesis was that the *Escherichia coli* NhaA can be utilized as a template to predict the structure of the TMSs of NHE1. However, the proteins share very low sequence identity of about 10%, and hence Ben-Tal and co-workers used the homology modeling technique in combination with a fold-recognition approach in order to obtain a putative three-dimensional model of NHE1 and propose novel mechanistic details for the protein's function.

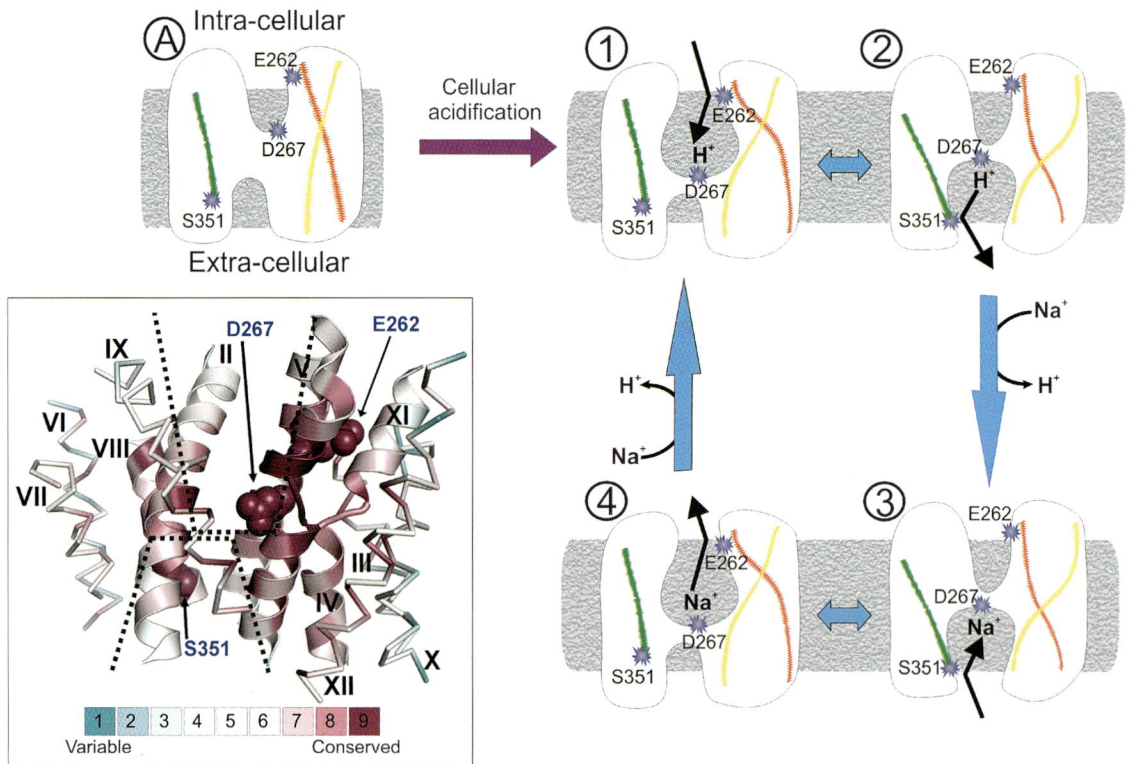

Figure 5. A suggested Na⁺/H⁺ exchange mechanism of NHE1. Alternating-access mechanism for Na⁺/H⁺ exchange in NHE1. State A represents an inactive conformation, and the exchange cycle is presented by states 1–4. Residues are marked by purple stars, and TMSs IV and XI are indicated in red and yellow, respectively. The boxed figure depicts the model structure of the TMSs of NHE1 in its inactive conformation. TMSs and residues suggested to be important for function are represented by colored ribbons and spheres, respectively. The evolutionary conservation profile is mapped on the model. The amino acids are colored by their conservation grades using the color-coding bar. TMS I was omitted for clarity. The cavities are indicated by dashed lines. (Figure modified from Landau *et al.*)[53]

The suggested model for NHE1 enables correlation between essential residues in NHE1 and their molecular role in the function of the transporter and proposes an alternating-access mechanism for Na⁺/H⁺ exchange in NHE (Figure 5). Overall, the mechanism, which involves consecutive transformations between pairs of conformations that are at chemical equilibrium, is driven by the concentration gradients of Na⁺ ions or protons across the membrane. The cation-transport path is formed by two discontinuous funnels comprised of TMSs II, IV, V and IX at the cytoplasmic side and TMSs II, VIII and XI at the extracellular side. TMS II and VIII are involved in the cation path and cation translocation, respectively.

Main residues that are thought to be implicated in the exchange mechanism are highlighted on Figure 5, which also presents the evolutionary conservation of the protein. Upon activation by intracellular acidification, a proton, possibly attracted by E-262 (TMS V), enters the cytoplasmic funnel and binds to D-267 (TMS V) located at the bottom of the funnel (state 1 in Figure 5). Conformational changes, induced by the TMS IV/XI assembly (which might include rotation of TMS VIII by ~180°), then shield the proton from the cytoplasm. Alternatively, an external path opens to the extracellular matrix (state 2 in Figure 5), which is enriched with Na^+. Na^+ ion can consequently compete with the proton for binding to the extracellular site, possibly mediated by S-351 (TMS VIII) (state 3 in Figure 5). Binding of Na^+ ion results in conformational changes, which shield the Na^+ cation from the extracellular matrix and open the path to the cytoplasm (state 4 in Figure 5). The Na^+ can then be released and replaced by a proton, again in accordance with the chemical gradient of these cations (state 1 in Figure 5), and the cycle continues.

Importantly, the suggested model was mainly derived from information pertaining to the bacterial antiporter[13] and is supported by phylogenetic[45] and data available for NHE1 and other eukaryotic Na^+/H^+ exchanges, specifically pertaining to the protein core TMSs (II, IV, V, VIII and XI). These central segments are evolutionarily conserved and include essential residues in the NHE1 and *Escherichia coli* NhaA proteins. Moreover, both transporters display a cluster of titratable residues in the center of the conserved protein core that are essential and involved in conformational changes and cation translocation. Yet, due to the low sequence similarity between the template (*Escherichia coli* NhaA) and the target (NHE1) proteins, this model has to be carefully assessed and taken only as an initial preliminary rough prediction. Even Ben-Tal and co-workers note that the location of the peripheral TMSs (I, III, VI, VII, IX, X and XII) in the model structure might be approximate, and that the conformations of the extra-membranal loops are tentative.

Nevertheless, this model structure, particularly when integrated with experimental data, can be used to propose testable hypotheses that will ultimately shed light on function and regulatory mechanisms. Given the urgent need for NHE1 inhibitors for myocardium-related diseases,[60] the model may also pave the way to structure-based drug design, yielding additional NHE1 inhibitors of clinical benefit. Additional structural data, e.g. from high-resolution cryo-EM and X-ray crystallography, are needed to further explore the structure of NHE1 and improve the model.

4.3 *Model structure of NHA2 — A human Na⁺/H⁺ antiporter*

NHA2, a mammalian Na^+/H^+ exchanger with a transcript bearing more similarity to prokaryotic than known eukaryotic Na^+/H^+ proteins, was recently characterized by Moe and co-workers.[51] Its expression and distribution are still debatable: one study found that NHA2 is restricted to the distal convoluted tubule in the kidney,[51] while another study concluded that it is selectively expressed in osteoclasts,[64] whereas a third study described its ubiquitous distribution.[22]

Ben-Tal and co-workers applied a reverse approach to study the mechanism and function of NHA2.[54] Usually, investigation of membrane proteins begins with accumulation of biochemical and physiological data that ultimately validate structural information. Yet, Ben-Tal and co-workers first built a structural model of NHA2, and then obtained molecular insights on its function based on the model. As in their previous study regarding the NHE1 model structure,[53] they encountered the same problem of extremely low sequence similarity between the template (*Escherichia coli* NhaA) that was used for the homology modeling process and the target (NHA2) proteins. Hence, they optimized the pairwise alignment between human NHA2 and NhaA by combining fold recognition, profile-to-profile alignment, and hydrophobicity analysis. The new alignment was used to produce a three-dimensional model of NHA2 that was supported by evolutionary conservation analysis. The model guided design of mutations that revealed new key residues for function. Together, experimental and structural data identified novel structural attributes of NHA2 likely to contribute to a mechanism of antiporting distinct from the previously characterized NhaA- and NHE1- type transporters.

The suggested model structure for NHA2 is shown at Figure 6, where helices and specific residues that participate at the proposed NHA2 antiporting mechanism are marked. Notably, partial dipoles at the TMS IV/XI assembly of NHA2 are seen at the model. These dipoles can be compensated by polar residues (Gln, Ser, Thr, Cys), in the absence of oppositely charged residues present in both NhaA and NHE1. In addition, conserved negative residue (E-215) in the vicinity of the TMS IV/XI assembly and essential charged residue (D-278) in TMS V were identified. R-187, a unique and conserved residue, was implicated to be located on TMS II where it has the potential of stabilizing the oppositely charged D-279 in TMS V. The DD motif on TMS V in NHA2 was clearly recognized to be functionally important as the same characterized motif at NhaA. S-362 and K-460 were proposed to line the translocation pathways similarly to their counterparts at NhaA.

An apparent limitation to the presented approach is the lack of a suitable template protein for the homology modeling process. Although Ben-Tal and co-workers address this issue, it is more appropriate to refer to the suggested structure as a tentative rough evaluation rather than an atomic resolution structure. However, due the potential physiological role that NHA2 may play in hypertension,[22] the suggested model is highly valuable as a start point for further studies. For example, the model can be used to assess sequence variants in the human population that may be associated with hypertension and other diseases in following studies.

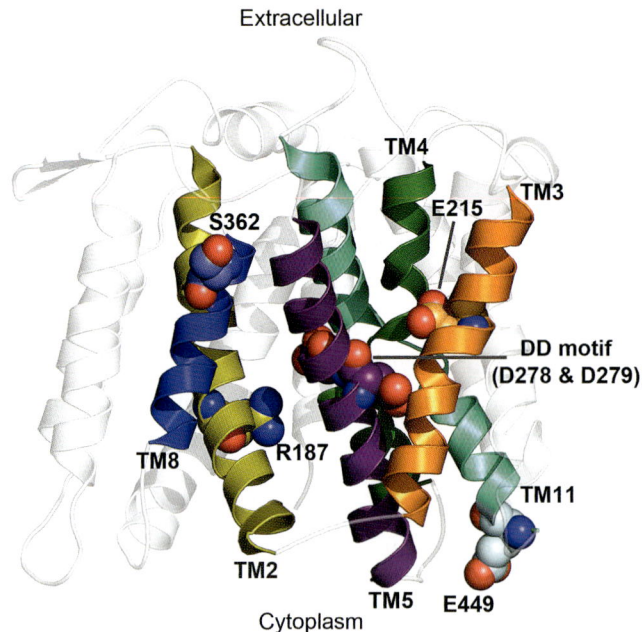

Figure 6. Functional residues mapped on the structural architecture of NHA2. The model structure of NHA2 is shown as transparent ribbons, whereas helices implicated in the function mechanism are marked. Helices IX and XII, along with the extra-membrane loops, were omitted for clarity. Specific residues suggested to participate in transport are shown as spheres and highlighted. (Figure adapted from Schushan *et al.*)[54]

5. Concluding Remarks

Na$^+$/H$^+$ transporters and exchangers are a large family of integral membrane proteins. They have substantial roles in regulating intracellular pH level, Na$^+$ content and cell volume. Despite their importance and profound distribution, many aspects concerning their structure and function are still an enigma. Therefore, computational methodologies have been used to study them in an atomic detail in order to get insights about their structure and function. In this chapter, we surveyed the current body of studies obtained by computational techniques regarding these proteins: Arkin and co-workers proposed a mechanism for the NhaA antiporting mechanism;[26] Michel and co-workers investigated the electrostatic interactions at the NhaA structure,[25] followed the effects of substituted residues at NhaA on its pH activation mechanism,[42] and conducted MD simulations of NhaA in order to find insights into its pH activation dynamics.[31] And, finally, possible models for the *Vibrio parahaemolyticus* NhaA[20] and the human NHE1[53] and NHA2[54] were constructed.

A fundamental appreciation for how biological macromolecules, such as proteins, may work requires knowledge of their structure. Computer modeling and simulations have become increasingly popular in the last few decades as tools to investigate not only structures of proteins but their dynamics and function as well.[65] Accordingly, the studies discussed at the current chapter cover few of the main themes and methodologies applied nowadays at the field of computational and molecular biophysics: structural bioinformatics, molecular dynamics simulations, homology modeling, computational phylogenetic analyses, and the mcce approach that combines continuum electrostatics and molecular mechanics. Taken together, these methods are well established for modeling, characterization of proteins' motions and dynamics, and provide valuable insights into the function of bimolecular systems at spatial and temporal scales that are difficult to access experimentally.[66] Either at the presence (as for the *Escherichia coli* NhaA) or the absence (as for the *Vibrio parahaemolyticus* NhaA, and the eukaryotic NHE1 and NHA2) of high resolution atomic structures, computational techniques are powerful means to gain better understanding of structure-function relationship and will hopefully soon be used as a common practice for computational drug design.

We argue that the Na$^+$/H$^+$ transporters and exchangers can be viewed as cellular nano-machines. At full activity, the translocation ion cycle of the *Escherichia coli* NhaA may reach up to $9 \cdot 10^4$ min^{-1} which makes it one of the fastest active transporters known.[11] Additionally, they are highly regulated by transcriptional and activation levels, and possess a high-fidelity selectivity filter. Their atomic characteristics, as identified and analyzed at the bunch of the presented studies at this chapter, only start to expose the fascinating molecular machinery of the Na$^+$/H$^+$ proteins. Evidently, these proteins can be considered as amazing examples for nature-made, highly efficient, agile, structural nano-machines.

Besides gaining basic scientific knowledge on the Na$^+$/H$^+$ transporters and exchangers, computational means will ultimately be used to screen and suggest potent inhibitors since these proteins are potentially efficient drug targets. Up to now, rational design of drugs for the Na$^+$/H$^+$ proteins has not been reported, and therefore having detailed insights about the proteins structure-function as well as blueprint structures of *Vibrio parahaemolyticus* NhaA, NHE1 and NHA2 provide a basis for a functional analysis of them. An improved understanding of the function may aid design selective blockers of the proteins to help in fighting a variety of physiological and pathological conditions.

Furthermore, the presented data in this chapter can be proved useful in virtual screening or *de novo* inhibitor design for discovery of new lead compounds. Due to the potentially high clinical importance of the functional data and the crucial epidemiological need for NhaA, NHE1 and NHA2 inhibitors, we hope that the current summary will be considered useful not only from a mechanistic perspective but from a pharmaceutical point of view as well and will stimulate further studies.

Acknowledgments

This work was supported in part by grants from The Lady Davis Fellowship Trust and The Valazzi-Pikovsky Fellowship Fund (to A.G.), the Rudin Fellowship Trust (to R.A.) and the Israeli Science Foundation (784/01, 1249/05, 1581/08 to I.T.A.). I.T.A. is the Arthur Lejwa Professor of Structural Biochemistry at the Hebrew University of Jerusalem.

Suggested Additional Reading Materials

The following review articles are suggested for those who wish to expand their knowledge in the field:

C. Ganea and K. Fendler. Bacterial transporters: Charge translocation and mechanism. 2009. *Biochim Biophys Acta* **1787(6)**, 706–713.

E. Padan, L. Kozachkov, K. Herz and A. Rimon. NhaA crystal structure: Functional-structural insights. 2009. *J Exp Biol* **212(Pt 11)**, 1593–1603.

E. Screpanti and C. Hunte. Discontinuous membrane helices in transport proteins and their correlation with function. 2007. *J Struct Biol* **159(2)**, 261–267.

E. R. Slepkov, J. K. Rainey, B. D. Sykes and L. Fliegel. Structural and functional analysis of the Na$^+$/H$^+$ exchanger. 2007. *Biochem J* **401(3)**, 623–633.

References

1. E. Padan, M. Venturi, Y. Gerchman and N. Dover. 2001. Na$^+$/H$^+$ antiporters. *Biochim Biophys Acta* **1505**, 144–157.

2. M. Karmazyn, M. Avkiran and L. Fliegel. 2003. *The Sodium-Hydrogen Exchanger: From Molecule to its Role in Disease*. Springer.

3. S. Grinstein. (ed.) 2005. *Na$^+$/H$^+$ Exchange*. CRC Press Inc.

4. I. C. West and P. Mitchell. 1974. Proton/sodium ion antiport in *Escherichia coli*. *Biochem J* **144**, 87–90.

5. D. Zilberstein, V. Agmon, S. Schuldiner and E. Padan. 1982. The sodium/proton antiporter is part of the pH homeostasis mechanism in *Escherichia coli*. *J Biol Chem* **257**, 3687–3691.

6. S. Schuldiner and H. Fishkes. 1978. Sodium-proton antiport in isolated membrane vesicles of *Escherichia coli*. *Biochemistry* **17**, 706–711.

7. D. Taglicht, E. Padan and S. Schuldiner. 1993. Proton-sodium stoichiometry of NhaA, an electrogenic antiporter from *Escherichia coli*. *J Biol Chem* **268**, 5382–5387.

8. R. Karpel, Y. Olami, D. Taglicht, S. Schuldiner and E. Padan. 1988. Sequencing of the gene ant which affects the Na$^+$/H$^+$ antiporter activity in *Escherichia coli*. *J Biol Chem* **263**, 10408–10414.

9. H. Inoue, T. Noumi, T. Tsuchiya and H. Kanazawa. 1995. Essential aspartic acid residues, Asp-133, Asp-163 and Asp-164, in the transmembrane helices of a Na⁺/H⁺ antiporter (NhaA) from *Escherichia coli*. *FEBS Lett* **363**, 264–268.

10. O. Rahav-Manor *et al.* 1992. Nhar, a protein homologous to a family of bacterial regulatory proteins (LysR), regulates NhaA, the sodium proton antiporter gene in *Escherichia coli*. *J Biol Chem* **267**, 10433–10438.

11. D. Taglicht, E. Padan and S. Schuldiner. 1991. Overproduction and purification of a functional Na⁺/H⁺ antiporter coded by nhaA (*ant*) from *Escherichia coli*. *J Biol Chem* **266**, 11289–11294.

12. H. Inoue *et al.* 1999. Expression of functional Na⁺/H⁺ antiporters of *Helicobacter pylori* in antiporter-deficient *Escherichia coli* mutants. *FEBS Lett* **443**, 11–16.

13. C. Hunte *et al.* 2005. Structure of a Na⁺/H⁺ antiporter and insights into mechanism of action and regulation by pH. *Nature* **435**, 1197–1202.

14. O. S. Andersen, R. E. N. Koeppe and B. Roux. 2005. Gramicidin channels. *IEEE Trans Nanobioscience* **4**, 10–20.

15. A. Yamashita, S. K. Singh, T. Kawate, Y. Jin and E. Gouaux. 2005. Crystal structure of a bacterial homologue of Na⁺/Cl⁻ dependent neurotransmitter transporters. *Nature* **437**, 215–223.

16. S. Y. Noskov and B. Roux. 2008. Control of ion selectivity in LeuT: Two Na⁺ binding sites with two different mechanisms. *J Mol Biol* **377**, 804–818.

17. J. S. Hub. H. Grubmuller and B. L. Groot. 2009. Dynamics and energetics of permeation through aquaporins. What do we learn from molecular dynamics simulations? *Handb Exp Pharmacol* **190**, 57–76.

18. S. Y. Noskov and B. Roux. 2006. Ion selectivity in potassium channels. *Biophys Chem* **124**, 279–291.

19. M. V. Radchenko *et al.* 2006. Cloning, functional expression and primary characterization of *Vibrio parahaemolyticus* K⁺/H⁺ antiporter genes in *Escherichia coli*. *Mol Microbiol* **59**, 651–663.

20. A. Ganoth, R. Alhadeff and I. T. Arkin. Computational study of the Na⁺/H⁺ antiporter from *Vibrio parahaemolyticus*. *J Mol Model* in press.

21. R. Alhadeff, A. Ganoth and I. T. Arkin. Molecular basis of Na⁺/H⁺ antiporter selectivity. Submitted.

22. M. Xiang, M. Feng, S. Muend and R. Rao. 2007. A human Na⁺/H⁺ antiporter sharing evolutionary origins with bacterial NhaA may be a candidate gene for essential hypertension. *Proc Natl Acad Sci U S A* **104**, 18677–18681.

23. K. A. Williams, U. Geldmacher-Kaufer, E. Padan, S. Schuldiner and W. Kühlbrandt. 1999. Projection structure of NhaA, a secondary transporter from *Escherichia coli*, at 4.0 Å resolution. *EMBO J* **18**, 3558–3563.

24. K. A. Williams. 2000. Three-dimensional structure of the ion-coupled transport protein NhaA. *Nature* **403**, 112–115.

25. E. Olkhova, C. Hunte, E. Screpanti, E. Padan and H. Michel. 2006. Multiconformation continuum electrostatics analysis of the NhaA Na⁺/H⁺ antiporter of *Escherichia coli* with functional implications. *Proc Natl Acad Sci U S A* **103**, 2629–2634.

26. I. T. Arkin. *et al.* 2007. Mechanism of Na⁺/H⁺ antiporting. *Science* **317**, 799–803.

27. E. Padan. 2008. The enlightening encounter between structure and function in the NhaA Na⁺-H⁺ antiporter. *Trends Biochem Sci* **33**, 435–443.

28. R. Dutzler, E. B. Campbell, M. Cadene, B. T. Chait and R. MacKinnon. 2002. X-ray structure of a ClC chloride channel at 3.0 Å reveals the molecular basis of anion selectivity. *Nature* **415**, 287–294.

29. K. Herz, A. Rimon, G. Jeschke and E. Padan. 2009. Beta-sheet-dependent dimerization is essential for the stability of NhaA Na⁺/H⁺ antiporter. *J Biol Chem* **284**, 6337–6347.

30. O. Jardetzky. 1966. Simple allosteric model for membrane pumps. *Nature* **211**, 969–970.

31. E. Olkhova, E. Padan and H. Michel. 2007. The influence of protonation states on the dynamics of the NhaA antiporter from *Escherichia coli*. *Biophys J* **92**, 3784–3791.

32. E. Padan. *et al.* 2004. NhaA of *Escherichia coli*, as a model of a pH-regulated Na^+/H^+ antiporter. *Biochim Biophys Acta* **1658**, 2–13.

33. A. Rimon, Y. Gerchman, Z. Kariv and E. Padan. 1998. A point mutation (G338S) and its suppressor mutations affect both the pH response of the NhaA-Na^+/H^+ antiporter as well as the growth phenotype of *Escherichia coli*. *J Biol Chem* **273**, 26470–26476.

34. M. R. Gunner and E. Alexov. 2000. A pragmatic approach to structure based calculation of coupled proton and electron transfer in proteins. *Biochim Biophys Acta* **1458**, 63–87.

35. M. Appel, D. Hizlan, K. R. Vinothkumar, C. Ziegler and W. Kühlbrandt. 2009. Conformations of NhaA, the Na^+/H^+ exchanger from *Escherichia coli*, in the pH-activated and ion-translocating states. *J Mol Biol* **388**, 659–672.

36. H. Luecke. 2000. Atomic resolution structures of bacteriorhodopsin photocycle intermediates: The role of discrete water molecules in the function of this light-driven ion pump. *Biochim Biophys Acta* **1460**, 133–156.

37. K. Herz, A. Rimon, E. Olkhova, L. Kozachkov and E. Padan. 2010. Transmembrane segment II of NhaA Na^+/H^+ antiporter lines the cation passage, and Asp65 is critical for pH activation of the antiporter. *J Biol Chem* **285**, 2211–2220.

38. Y. Gerchman. *et al.* 1993. Histidine-226 is part of the pH sensor of NhaA, a Na^+/H^+ antiporter in *Escherichia coli*. *Proc Natl Acad Sci U S A* **90**, 1212–1216.

39. A. Rimon, Y. Gerchman, Y. Olami, S. Schuldiner and E. Padan. 1995. Replacements of histidine 226 of NhaA-Na^+/H^+ antiporter of *Escherichia coli*. Cysteine (H226C) or serine (H226S) retain both normal activity and pH sensitivity, aspartate (H226D) shifts the pH profile toward basic pH, and alanine (H226A) inactivates the carrier at all pH values. *J Biol Chem* **270**, 26813–26817.

40. L. Galili, A. Rothman, L. Kozachkov, A. Rimon and E. Padan. 2002. Trans membrane domain IV is involved in ion transport activity and pH regulation of the NhaA-Na^+/H^+ antiporter of *Escherichia coli*. *Biochemistry* **41**, 609–617.

41. Y. Gerchman, A. Rimon and E. Padan. 1999. A pH-dependent conformational change of NhaA Na^+/H^+ antiporter of *Escherichia coli* involves loop VIII–IX, plays a role in the pH response of the protein, and is maintained by the pure protein in dodecyl maltoside. *J Biol Chem* **274**, 24617–24624.

42. E. Olkhova, L. Kozachkov, E. Padan and H. Michel. 2009. Combined computational and biochemical study reveals the importance of electrostatic interactions between the "pH sensor" and the cation binding site of the sodium/proton antiporter NhaA of *Escherichia coli*. *Proteins* **76**, 548–559.

43. R. Habibian *et al.* 2005. Functional analysis of conserved polar residues in Vc-NhaD, Na^+/H^+ antiporter of *Vibrio cholerae*. *J Biol Chem* **280**, 39637–39643.

44. M. H. Saier, Jr., C. V. Tran and R. D. Barabote. 2006. TCDB: The Transporter Classification Database for membrane transport protein analyses and information. *Nucleic Acids Res* **34**, D181–D186.

45. C. L. Brett, M. Donowitz and R. Rao. 2005. Evolutionary origins of eukaryotic sodium/proton exchangers. *Am J Physiol Cell Physiol* **288**, C223–C239.

46. J. Orlowski and S. Grinstein. 2004. Diversity of the mammalian sodium/proton exchanger SLC9 gene family. *Pflugers Arch* **447**, 549–565.

47. E. R. Slepkov, J. K. Rainey, B. D. Sykes and L. Fliegel. 2007. Structural and functional analysis of the Na$^+$/H$^+$ exchanger. *Biochem J* **401**, 623–633.

48. A. Fukuda, A. Nakamura and Y. Tanaka. 1999. Molecular cloning and expression of the Na$^+$/H$^+$ exchanger gene in *Oryza sativa*. *Biochim Biophys Acta* **1446**, 149–155.

49. Y. Sato and M. Sakaguchi. 2005. Topogenic properties of transmembrane segments of *Arabidopsis thaliana* NHX1 reveal a common topology model of the Na$^+$/H$^+$ exchanger family. *J Biochem* **138**, 425–431.

50. L. Pham, P. Purcell, L. Morse, P. Stashenko and R. A. Battaglino. 2007. Expression analysis of nha-oc/NHA2: A novel gene selectively expressed in osteoclasts. *Gene Expr Patterns* **7**, 846–851.

51. D. G. Fuster. *et al.* 2008. Characterization of the sodium/hydrogen exchanger NHA2. *J Am Soc Nephrol* **19**, 1547–1556.

52. E. Padan, L. Kozachkov, K. Herz and A. Rimon. 2009. NhaA crystal structure: Functional-structural insights. *J Exp Biol* **212**, 1593–1603.

53. M. Landau, K. Herz, E. Padan and N. Ben-Tal. 2007. Model structure of the Na$^+$/H$^+$ exchanger 1 (NHE1): Functional and clinical implications. *J Biol Chem* **282**, 37854–37863.

54. M. Schushan. *et al.* 2010. Model-guided mutagenesis drives functional studies of human NHA2, implicated in hypertension. *J Mol Biol* **396**, 1181–1196.

55. L. McCarter. 1999. The multiple identities of *Vibrio parahaemolyticus*. *J Mol Microbiol Biotechnol* **1**, 51–57.

56. Y.-C. Su and C. Liu. 2007. *Vibrio parahaemolyticus*: A concern of seafood safety. *Food Microbiol* **24**, 549–558.

57. G. Kemp, H. Young and L. Fliegel. 2008. Structure and function of the human Na$^+$/H$^+$ exchanger isoform 1. *Channels (Austin)* **2**.

58. M. E. Malo and L. Fliegel. 2006. Physiological role and regulation of the Na$^+$/H$^+$ exchanger. *Can J Physiol Pharmacol* **84**, 1081–1095.

59. M. Karmazyn, M. Sawyer and L. Fliegel. 2005. The Na$^+$/H$^+$ exchanger: A target for cardiac therapeutic intervention. *Curr Drug Targets Cardiovasc Haematol Disord* **5**, 323–335.

60. L. Fliegel. 2009. Regulation of the Na$^+$/H$^+$ exchanger in the healthy and diseased myocardium. *Expert Opin Ther Targets* **13**, 55–68.

61. R. A. Cardone, V. Casavola and S. J. Reshkin. 2005. The role of disturbed pH dynamics and the Na$^+$/H$^+$ exchanger in metastasis. *Nat Rev Cancer* **5**, 786–795.

62. S. F. Pedersen, S. A. King, E. B. Nygaard, R. R. Rigor and P. M. Cala. 2007. NHE1 inhibition by amiloride- and benzoylguanidine-type compounds. Inhibitor binding loci deduced from chimeras of NHE1 homologues with endogenous differences in inhibitor sensitivity. *J Biol Chem* **282**, 19716–19727.

63. B. Masereel, L. Pochet and D. Laeckmann. 2003. An overview of inhibitors of Na$^+$/H$^+$ exchanger. *Eur J Med Chem* **38**, 547–554.

64. R. A. Battaglino. *et al.* 2008. NHA-oc/NHA2: A mitochondrial cation-proton antiporter selectively expressed in osteoclasts. *Bone* **42**, 180–192.

65. M. Karplus and J. Kuriyan. 2005. Molecular dynamics and protein function. *Proc Natl Acad Sci U S A* **102**, 6679–6685.

66. J. L. Klepeis, K. Lindorff-Larsen, R. O. Dror and D. E. Shaw. 2009. Long-timescale molecular dynamics simulations of protein structure and function. *Curr Opin Struct Biol* **19**, 120–127.

Membrane Transporters

Molecular Machines Coupling Cellular Energy to Vectorial Transport Across the Membrane

9

Zhijian Huang, Saher A. Shaikh, Po-Chao Wen, Giray Enkavi,
Jing Li and Emad Tajkhorshid

1. Introduction

Membrane transport is one of the most fundamental processes in all living cells. The transport of most molecular species across the cellular membrane is mediated by highly specialized membrane proteins broadly known as membrane channels and transporters. These proteins empower the cell to implement selectivity measures, and, in the case of membrane transporters, facilitate the translocation of the substrates against their electrochemical gradients, a process which relies on delicate coupling mechanisms between substrate transport and various sources of energy in the cell furnished by these molecular machines. Examples of such sources of cellular energy include ATP, membrane electric potential, and established electrochemical gradients for other species.

Specific protein conformational changes of membrane transport proteins are at the heart of the mechanisms employed for efficient and selective transport of materials across the membrane. For instance, the opening and closing of membrane channels (gating) involve large-scale protein conformational changes that are regulated by various signals in the cell, including ligand binding (ligand-gated channels), membrane electrical potential (voltage-gated channels), or mechanical stress (mechanosensitive channels). Once opened in response to such stimuli, membrane channels allow passive diffusion of their substrates through selective pores from one side of the membrane to the other, down their concentration gradient. Active membrane transporters, on the other hand, couple various sources of cellular energy to vectorial translocation of their substrates, often against the chemical gradient (*pumping*). While the energy is provided by ATP in *primary* transporters, *secondary* transporters couple substrate transport to co-transport (*symport* or *antiport*) of ionic species (most prominently H$^+$ and Na$^+$ ions). The energy-coupling mechanism relies on a much more complex set of molecular events within the transporter engaging many of its structural elements. As such, membrane transporters are significantly slower and mechanistically more complex than channels. The transport cycle in membrane transporters is, thus, composed of a large number of steps that are mediated by significant protein conformational changes whose nature and magnitude are largely unknown and often difficult to characterize experimentally.

The dynamics of membrane transporters is highly relevant to their function. Relevant structural motions of membrane transporters span a wide spectrum of scales, ranging from localized side-chain conformational changes, to loop flipping motions, and up to extensive subdomain/domain structural transitions. Similar motions in channels constitute the mechanism of gating (open/closed transition in response to signal), while the process of substrate conduction itself usually does not involve significant protein motions. In contrast, in active transporters, protein conformational changes of various forms and magnitudes are an integral part of the substrate translocation process. Molecular dynamics (MD) simulation offers a method with sufficient temporal and spatial resolutions to characterize functionally relevant molecular events in proteins. Employing MD simulations, we have described several functionally relevant motions in membrane channels and transporters.[1–7]

For transporters, a general mechanism named the "alternating-access mechanism" has been proposed.[8] This model ensures that the substrate is only accessible from one side of the membrane at any given time, thus, preventing the formation of a *leak*, i.e. a channel-like structure that would allow free diffusion of the substrate or other molecular species, during the transport cycle. The alternating-access model requires at least two major conformational states of the protein, namely, the inward-facing (IF) and outward-facing (OF) states, whose inter-conversion switches the substrate accessibility from one side of the membrane to the other (Figure 1). The complete transport cycle naturally involves many other intermediate states; for instance, it is known for many transporters that both the IF and OF states can exist in either open (IF-o and OF-o) or occluded (IF-occ and OF-occ) states. Due to various technical difficulties, for most transporters only one of these major conformational state has been structurally characterized. The other state(s), and the conformational changes involved in their transitions, therefore, have to be studied using other methodologies that would yield a dynamical description of these complex proteins.

The complexity and time scale of the function of membrane transporters pose a great challenge also to computational studies. Simulation of membrane transporters requires the inclusion of the embedding lipid bilayer, water and ions explicitly in the system, as often these elements directly participate in the mechanism. This often results in large system sizes (100 000–500 000 atoms), requiring large computational resources to simulate. More importantly, characterizing the complete transport cycle in transporters would require simulations on the order of at least μs-ms time scales, which are currently not possible. Despite these technical limitations, we have demonstrated that extended, large-scale MD simulations of membrane transporters can be very effective in describing key molecular events and processes involved in their function.[1–7,9–14] These studies show that MD simulations, once designed and analyzed carefully, can indeed significantly advance our understanding of the molecular mechanisms of energy coupling and transport phenomena in membrane transporters.

In this chapter, we use five systems, viz. the glutamate transporter (GlT), an ATP-binding cassette transporter (ABCT), glycerol-3-phosphate transporter (GlpT), leucine transporter (LeuT), and Na$^+$-coupled galactose transporter (vSGLT), to showcase the power of MD simulations in capturing key molecular events in membrane transporters (Figure 2). In all the systems, the simulation setup uses experimentally solved, atomic-resolution protein structures, while water,

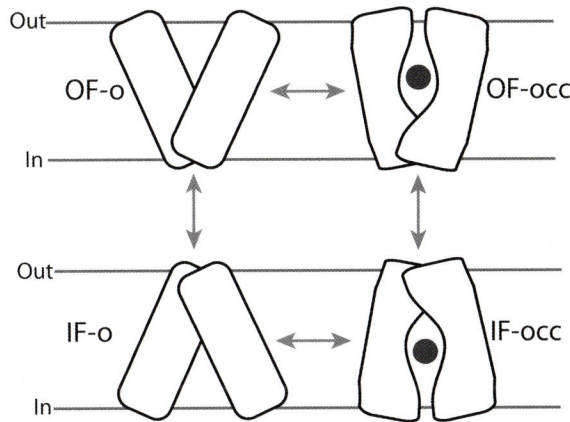

Figure 1. The alternate access model proposed for the transporter function, including two major open states, outward facing-open (OF-o) and inward facing-open (IF-o), and two intermediate substrate-occluded states (OF-occ and IF-occ). "Out" and "In" represent the extracellular/periplasmic and cytoplasmic side of the membrane, respectively. The access to the substrate (dark circle) from the two sides of the membrane is controlled by protein conformational changes of different nature and magnitudes.

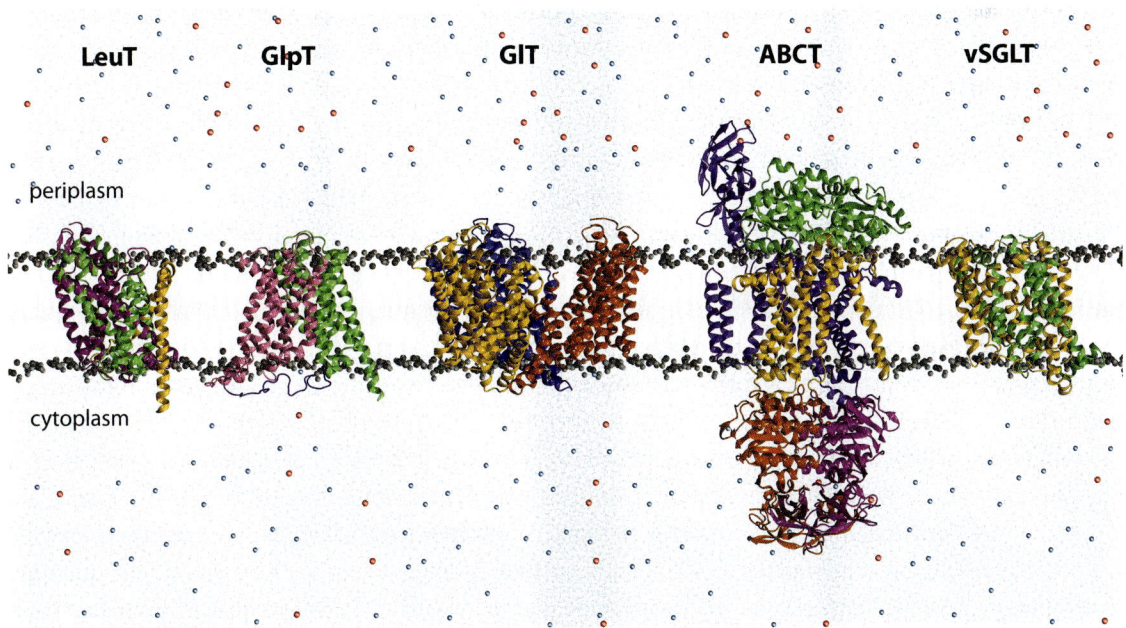

Figure 2. The five membrane transporters discussed in this review — LeuT, GlpT, GlT, ABCT, and vSGLT. The proteins are shown embedded in a lipid bilayer (grey), solvated by explicit water (blue) and ions (Na$^+$, blue, Cl$^-$, red), as a representation of the typical simulation setup used for each system.

lipid (membrane), and ions are added by modeling. The program NAMD,[15] an efficiently parallelized code for MD simulations of biomolecules, was used for all the simulations. The initial equilibration period typically ranges between 1–5 nanoseconds (ns), while the production simulations are carried out for 50–200 ns. As will be demonstrated, on these time scales, we have been able to capture not only fast motions such as water and ion diffusion, side chain reorientations, and loop movements, but also functionally relevant larger-scale domain motions of the studied transporters.

2. ATP-Driven Transport in ABC Transporters

ATP-binding cassette (ABC) transporters are the most abundant primary transporters in living cells.[16] This superfamily consists of both ABC-exporters and ABC-importers, which are involved in a large number of key processes in different organisms. For example, many ABC-importers are responsible for nutrient uptake in prokaryotic cells, while the human genome encodes a large number of ABC-exporters responsible for material secretion and xenobiotic removal. Many pathological conditions are directly related to ABC transporters, e.g. cystic fibrosis, Tangier disease, adrenoleukodystrophy, as well as multidrug resistance in cancer and bacterial cells.

Despite opposite directions of transport (import vs. export) achieved by different transporters in this family, all ABC transporters utilize a universal mechanism — ATP binding and hydrolysis — to facilitate the substrate translocation.[17–19] Mechanistically, the transport process is a result of concerted motion of different units within the protein, typically two cytoplasmic nucleotide binding domains (NBDs) and two transmembrane domains (also known as the permease domains). ATP binding and hydrolysis take place at the two NBDs, while the substrate translocation occurs within the permease domains.

Almost a dozen of full ABC transporter complexes have been crystallized in recent years (recently reviewed by Kos and Ford[19]). The conformations captured in these crystal structures are highly dependent on the presence of the nucleotides in the NBDs. Whether importers or exporters, all structures obtained with ATP, or with ATP analogs, show an opening of the transmembrane domains at the periplasmic/extracellular side, whereas those co-crystallized with ADP or under nucleotide-free conditions generally exhibit a cytoplasmic opening in the transmembrane domains (Figure 3a). Based on these conformations, a general transport mechanism for ABC transporters has been proposed: the permease domains are able to open toward either side of the membrane, depending on the nucleotide species present in the NBDs; while the NBDs bind and hydrolyze ATP in each transport cycle, the permease domains shift between the IF-o and the OF-o states, thus switching the accessibility of the substrate from one side of the membrane to the other (Figure 3b).

The key to this transport mechanism is the large conformational changes induced by ATP binding and hydrolysis, which originate in the NBDs and then propagate into the transmembrane domains to facilitate the substrate transport. The nucleotide dependence of the NBD conformations has been captured in great details in numerous crystal structures of isolated NBDs (recently reviewed by Moussatova *et al.*[20]). While the closed NBD dimers are exclusively available with ATP or ATP analogs, an ADP-bound or nucleotide-free NBD is usually captured as separate monomers or as an open dimer. It is therefore suggested that the effect of ATP binding and hydrolysis is to change the dimerization state of the NBDs: ATP binding promotes dimer formation, whereas its hydrolysis results in dimer opening. These conformational changes are then propagated from the NBDs to the transmembrane domains hence inducing structural transitions required for substrate translocation.

We have investigated the mechanism of NBD dimer opening induced by ATP hydrolysis with MD simulations,[5] using simulations performed on the crystal structure of isolated NBDs of the

(a)

(b)

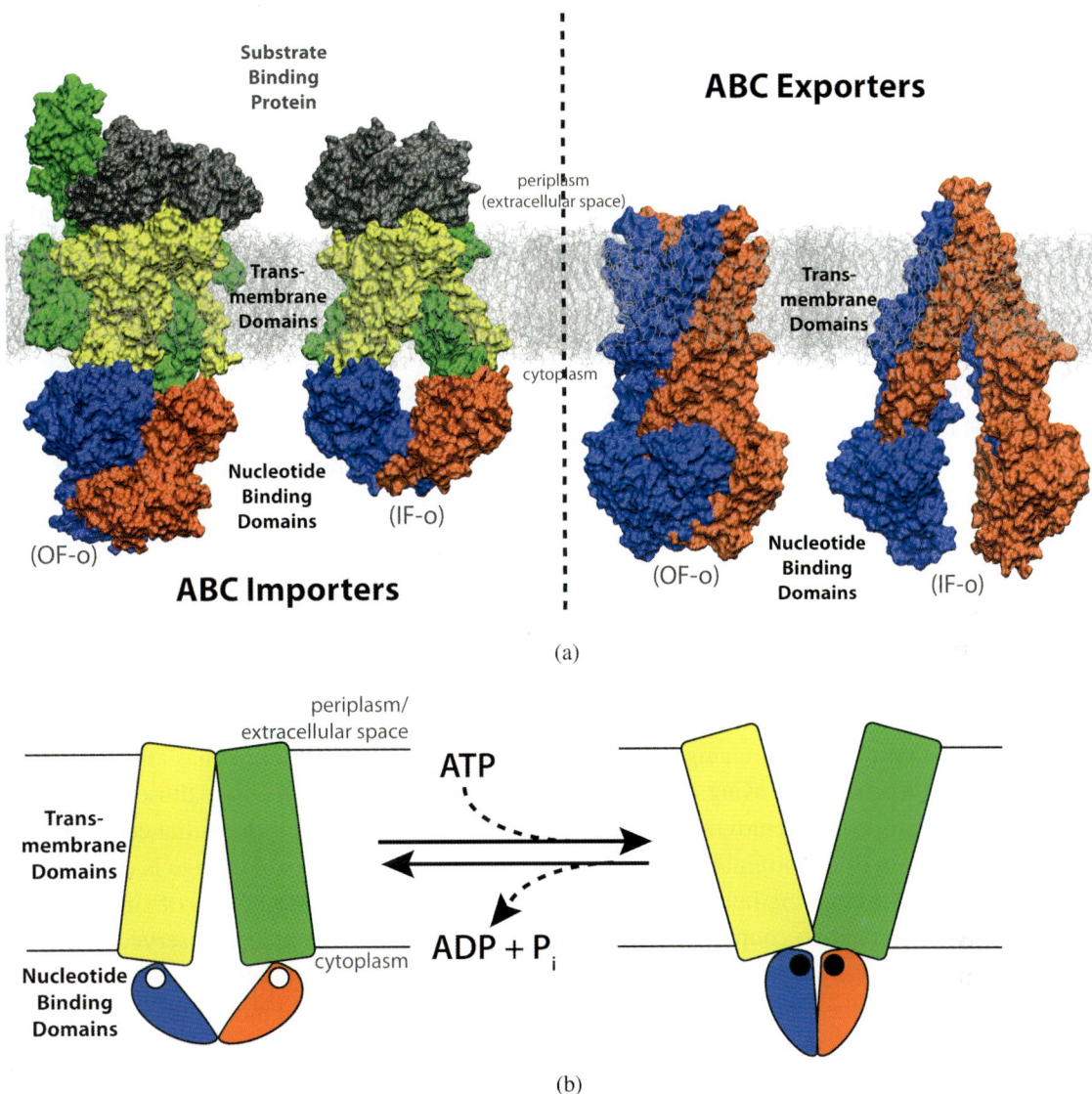

Figure 3. Structures and mechanism of ABC transporters. (a) Representative crystal structures of ABC transporters in different conformational states. From left to right: *Escherichia coli* maltose transporter in the ATP-bound state (PDB:2R6G[28]), *Archaeoglobus fulgidus* molybdate/tungstate transporter in the nucleotide-free state (PDB:2ONK[79]), *Salmonella typhimurium* lipid-A flippase in the ATP-bound state (PDB:3B60[80]), and *Mus musculus* multidrug resistance protein in the nucleotide-free state (PDB:3G5U[81]). (b) Schematic representation of the mechanism of ABC transporters. An ABC transporter is represented as two rectangles (transmembrane domains) and two half-ovals (nucleotide binding domains, NBDs). The nucleotide binding sites are shown as empty (nucleotide-free) or filled (ATP-bound) circles. The transporter exhibits either inward-facing (left) or outward-facing (right) conformations, as controlled by ATP binding and hydrolysis.

Escherichia coli maltose transporter (MalK, PDB:1Q12[21]). To simulate the effect of ATP hydrolysis, the bound ATP was converted into ADP and inorganic phosphate (P_i), either at one or at both nucleotide binding sites;[5] in addition, a control system with ATP in both sites was also simulated. Thus, four different simulations were performed to study the effect of ATP hydrolysis on the conformations of the NBD dimer, and to investigate whether or not ATP hydrolysis at both nucleotide binding sites is required for the function of the transporter.

Interestingly, the NBD dimer opening is observed in all three post-hydrolysis simulations, that is, regardless of the number (one or two) of the ATP molecules hydrolyzed. On the contrary, the ATP-bound dimer maintained its closed conformation throughout the 80 ns of simulation (Figures 4a and 4b). These results suggest that a single ATP hydrolysis event at either catalytic site is sufficient to trigger the dimer opening. It is also noteworthy that the hydrolysis products, ADP and P_i, both stay closely associated within the active sites during the entire simulation time, indicating that the dimer opening is a direct effect of the hydrolysis reaction, and not relying on the release of the hydrolysis products from the protein.[5]

The dimer opening appears to be a stochastic process, based on the observed uncertainty of the time and the location of the opening after ATP hydrolysis.[5] The process relies on the random "breathing" motions of the two monomers driven by their thermal fluctuations. As such, it is important that the simulations are performed at sufficiently long times (typically, at least 30–50 ns) in order to observe the opening, as also demonstrated in MD simulations of other ABC transporters.[22–24]

In ABC transporters, the conformational changes of the NBDs, triggered by ATP binding and hydrolysis, need to be transmitted to the transmembrane domains in order to flip their conformations from the IF-o to the OF-o state. Structurally, a pair of short helices in the permease domains, named the "coupling helices" (the labeled helices in Figure 4c), have been proposed to be responsible for the conformational coupling of the NBDs and the transporter domains.[25,26] The propagation of conformational changes from the NBDs to the coupling helices has also been captured with MD simulations,[27] using the structure of the intact *Escherichia coli* maltose transporter (PDB:2R6G,[28] Figure 4c). Removing the nucleotides from the NBDs after the initial equilibration triggers the opening of the originally closed NBDs, and this conformational change is expected to be transmitted to the coupling helices in the transmembrane domains. After 70 ns of simulation, a separation between the two coupling helices on the order of 2–3 Å was indeed observed, which is followed very closely by the separation of other tightly coupled elements in the transmembrane domains, specifically the second pair of helices within the same conserved motif (the upper shaded helices in Figure 4c). In contrast, in the control simulation where ATP is present in both binding sites, no separation was observed between any of the two helix-pairs[27] (Figure 4d).

Surprisingly, the internal coupling between the "coupling helices" and the core of the transmembrane domains was found to depend on the presence of the NBDs.[27] In an independent simulation where the NBDs were completely removed, the two coupling helices separate from the rest of the transmembrane domains and show large fluctuations on the order of ~ 10 Å (Figure 4d). The simulations, therefore, identified a more complete picture of the coupling elements involved in the communication between the NBDs and the transmembrane domains in ABC transporters.

With the availability of more crystal structures of intact ABC transporters and more powerful computational resources, it will be possible to simulate these transporters for much longer time scales (~ μs) in the near future. Such extended simulations will provide more information on the details of the mechanism of ABC transporters at an atomic resolution, especially regarding the nature of the conformational changes and how different structural elements work together to furnish the transport process, as well as the interplay between the transporter and its substrate during the conformational changes. Although the time scale for the complete transport cycle in ABC transporters is expected to be several orders longer, such simulations would be able to describe key

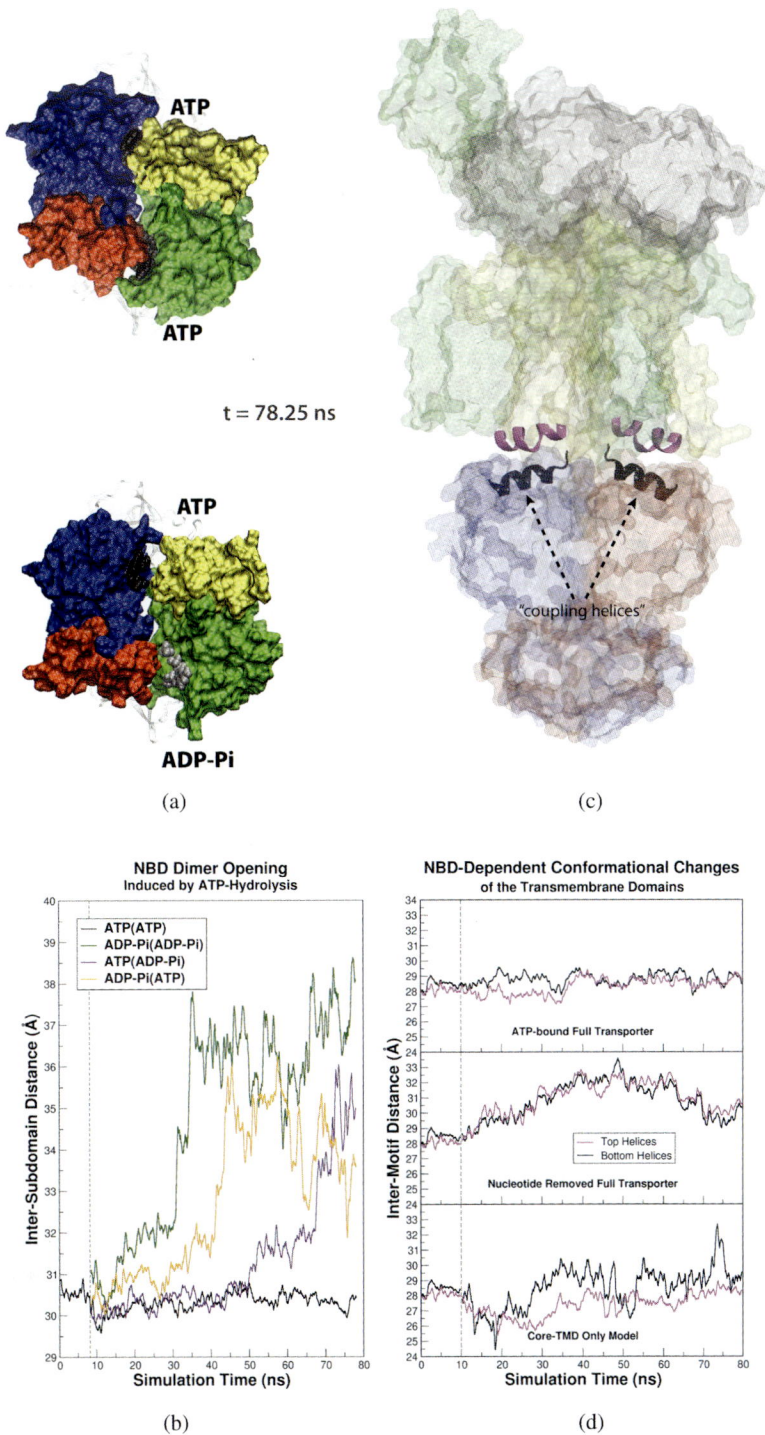

Figure 4. Conformational changes of the maltose transporter captured in MD simulations. (a) Two representative structures of the NBDs of the maltose transporter after 78.5 ns of simulations. The structures are viewed roughly perpendicular to the membrane from the extracellular side. The top structure is doubly ATP-bound and stays closed, whereas in the lower structure ATP hydrolysis in one of the binding sites results in opening at both nucleotide binding sites.

steps and transitions involved in the transport cycle, along with their detailed mechanisms, and might identify unknown conformational intermediate states, e.g. occluded states that have not yet been captured in any crystal structure of ABC transporters.

3. Ion-Coupled Neurotransmitter Uptake by the Glutamate Transporter

In order to maintain recurrent and selective signaling and to prevent reaching neurotoxic levels, neurotransmitters must be cleared from the synaptic cleft shortly after signal transmission. Glutamate is the predominant excitatory neurotransmitter in the central nervous system and plays critical role in fundamental processes such as learning and memory. Glutamate transporters (GlTs; also termed excitatory amino acid transporters, EAATs[29]) are Na^+-driven membrane transporters that drive "uphill" translocation of the neurotransmitter glutamate from the synapses across the membrane into the cell. By coupling to the co-transport of three Na^+ and one H^+, and the counter-transport of one K^+, mammalian GlT transports one negatively charged glutamate across the membrane during each transport cycle. In contrast to the mammalian GlT, the transport mechanism in an archaeal homolog (*Pyrococcus horikoshii* (Glt_{ph})) has been shown to be independent of H^+ and K^+, thus, only driven by the co-transport of three Na^+ ions along with the substrate.

The crystal structures of Glt_{ph}[30–32] have provided a structural framework for understanding the transport mechanism of all GlTs. Glt_{ph} shares about 36% amino acid identity with mammalian GlTs, with a large number of residues implicated in the binding and translocation of the substrate and co-transported ions highly conserved. Therefore, Glt_{ph} can serve as a structural model for understanding the mechanism of transport in mammalian GlTs.[33,34] The structures of Glt_{ph},[30–32] solved in both the OF and the IF states, reveal a trimeric architecture with each monomer composed of eight transmembrane helices (TM1-TM8) and two highly conserved helical hairpins (termed HP1 and HP2, respectively) forming the binding sites for the substrate and Na^+ ions. The architecture of a single monomer in the OF state[31] along with the two structurally resolved Na^+ binding sites[30–32] (Na1 and Na2 sites) are shown in Figures 5a–5c. The crystal structures, however, do not provide any information on the third Na^+ binding site (Na3). A mutagenesis study[35] on a mammalian GlT has shown that an acidic residue corresponding to D312 in Glt_{ph} is involved in the coordination of one of the Na^+ ions during the transport cycle, suggesting that this residue is likely involved in the Na3 binding site. In the crystal structure of Glt_{ph},[31] however, this acidic residue is deeply buried in the protein (Figure 5 d) and does not participate in any of the two identified Na^+ binding sites.

Figure 4. (*Continued*) (b) The degree of the subdomain separation across the nucleotide binding site that is sandwiched between the red and the green subdomains. The labels describe the nucleotide species present in the indicated binding site, whereas the nucleotide species of the opposite binding site is specified in the parentheses. The distances are measured between the two centers of masses of the opposing subdomains. (c) The structure of the intact maltose transporter used for membrane simulations. (d) The separation between the two sets of helices shown in panel c during the simulations. The separations are defined as the center-of-mass distance between the two sets of helices colored in black and magenta in panel c.

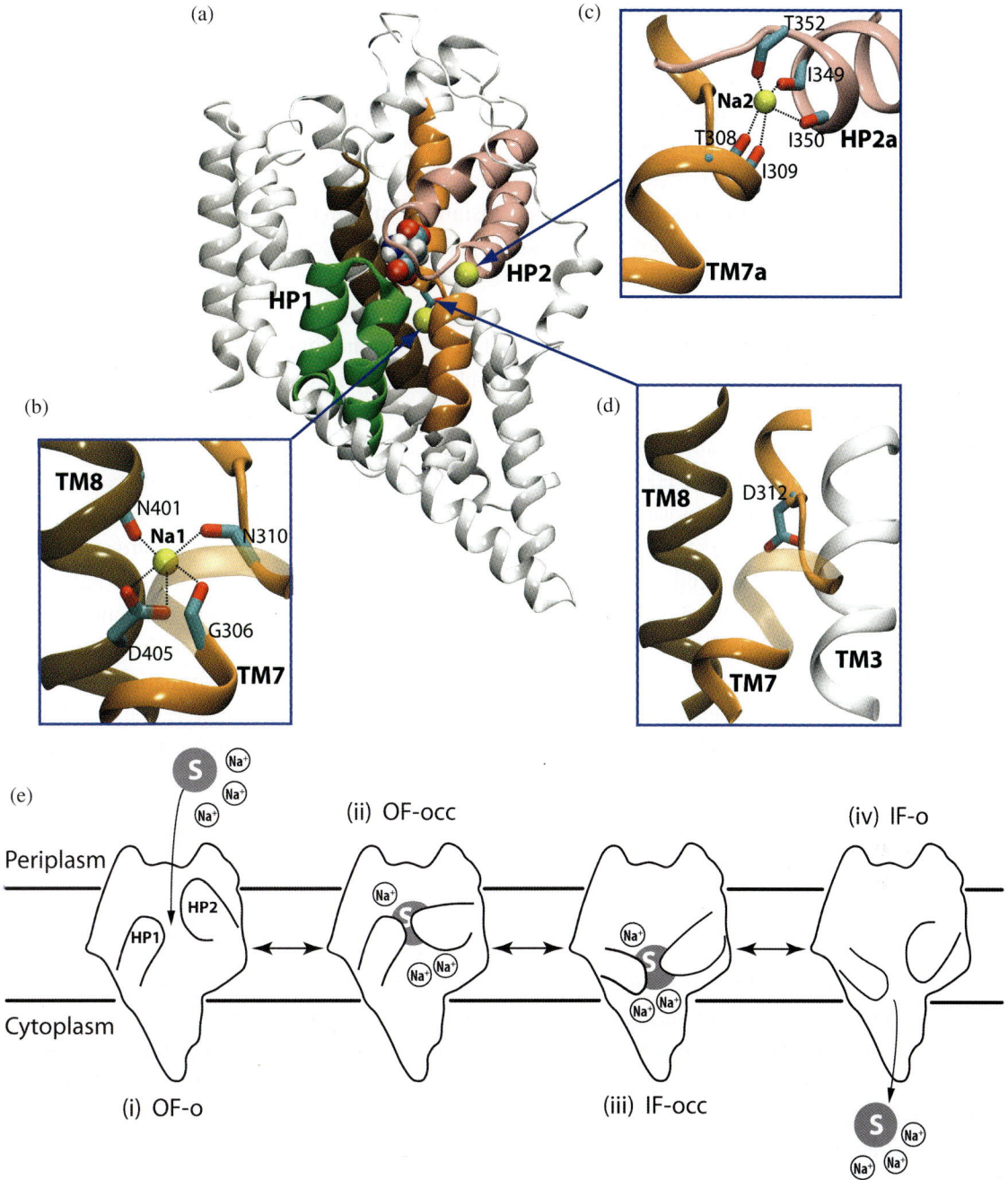

Figure 5. Structural features and hypothetical transport cycle of Glt$_{ph}$. (a) The structure of Glt$_{ph}$ monomer with bound substrate (shown in VDW) and the two structurally resolved Na$^+$ ions (at sites Na1 and Na2) (spheres). Helical hairpins HP1 and HP2, and transmembrane helices TM7 and TM8 together form the substrate and Na$^+$ binding sites. (b) The Na1 site composed of residues G306 and N310 on TM7, and N401 and D405 on TM8. (c) The Na2 site, located between two half-helices HP2a and TM7a, with the Na$^+$ coordinated by three carbonyl oxygens from HP2 and two carbonyl oxygens

Various experimental measurements have indicated that substrate binding induces conformational changes in GlT. However, the limited resolution of these studies made it difficult to draw specific conclusions about the nature and magnitude of such conformational changes. Several studies indicated that at least one of the Na^+ ions binds to the empty transporter before the substrate and that at least one Na^+ ion after the substrate.[35,36] A recent kinetic model for mammalian GlTs[37,38] suggests that the empty transporter binds two Na^+ ions before the substrate.

Although these experiments have provided insightful information on the transport cycle in GlT, the details of the mechanism of extracellular gating and the coupling between the substrate and the three Na^+ ions remained elusive (Figure 5e). In order to address these questions, taking the OF state of Glt_{ph}[30,31] as a model structure, we have performed a series of MD simulations on membrane-embedded trimeric models of Glt_{ph} with different combinations of the substrate and bound Na^+ ions.[4,7] These simulations revealed several highly relevant mechanistic details regarding the transport cycle in GlTs.

Comparison of the dynamics of the substrate-bound and the substrate-free (*apo*) simulations suggests that the helical hairpin HP2 plays the role of the extracellular gate.[4] Invariably, in all the simulations performed in the presence of the substrate, HP2 displays a very stable conformation, while removing the substrate results in its large opening motion and complete exposure of the substrate binding site to the extracellular solution (Figure 6a).[4] These results suggest that HP2 plays the role of the extracellular gate, and that, more importantly, its opening and closure are controlled by substrate binding.[4] A gating role for HP2 is supported by the structure of Glt_{ph} in the presence of an inhibitor.[31] Furthermore, recent inhibition studies in a mutant homolog of a mammalian GlT[39] suggest that HP2 serves as the extracellular gate of the transporter and that substrate induces distinct conformations of HP2. An independent MD study[12] has also provided support for this role of HP2.

Interestingly, despite its apparent structural symmetry to HP2, helical hairpin HP1 was found to exhibit a high level of conformational stability regardless of the presence of the substrate (Figure 6a).[4] This result, which might be attributed either to the shorter length of the loop of HP1 (when compared to HP2), or to its much closer contact with TM2 in the OF state, suggests that, at least during the extracellular half of the transport cycle, HP1 does not play a direct role, and its involvement might be limited to stabilization of the structure of HP2 upon substrate binding. The possibility of a gating role of HP1 on the cytoplasmic side might be determined by simulating the IF Glt_{ph}[32] using a similar methodology employed in the study described here.[4]

The second major consequence of substrate binding revealed by the simulations is the formation of a Na^+ binding site (the Na2 site)[4] at a position formed between two half-helical structures (HP2a and TM7a, see Figure 6a). In the *apo* state, the dipole moments of these half-helices were found to be completely misaligned (Figure 6a). Upon substrate binding, these two opposing half-helices align such that their dipole moments converge on the same region resulting in the formation

Figure 5. (*Continued*) from TM7a. (d) The D312 neighborhood (potential Na3 site). D312 is on the unwound part of TM7, with its side chain located behind TM7 pointing toward the interior of the transporter protein. (e) Schematic transport cycle: (i) the *apo* state with HP2 in an open conformation; (ii) binding of three Na^+ ions and the substrate induces the closure of HP2, yielding the OF-occ state; (iii) the IF-occ state is formed by large-scale conformational changes in the protein that switches the accessibility of the binding site to the cytoplasmic side of the membrane; and (iv) opening of the cytoplasmic gate(s) allows the release of the Na^+ ions and the substrate into the cytoplasm.

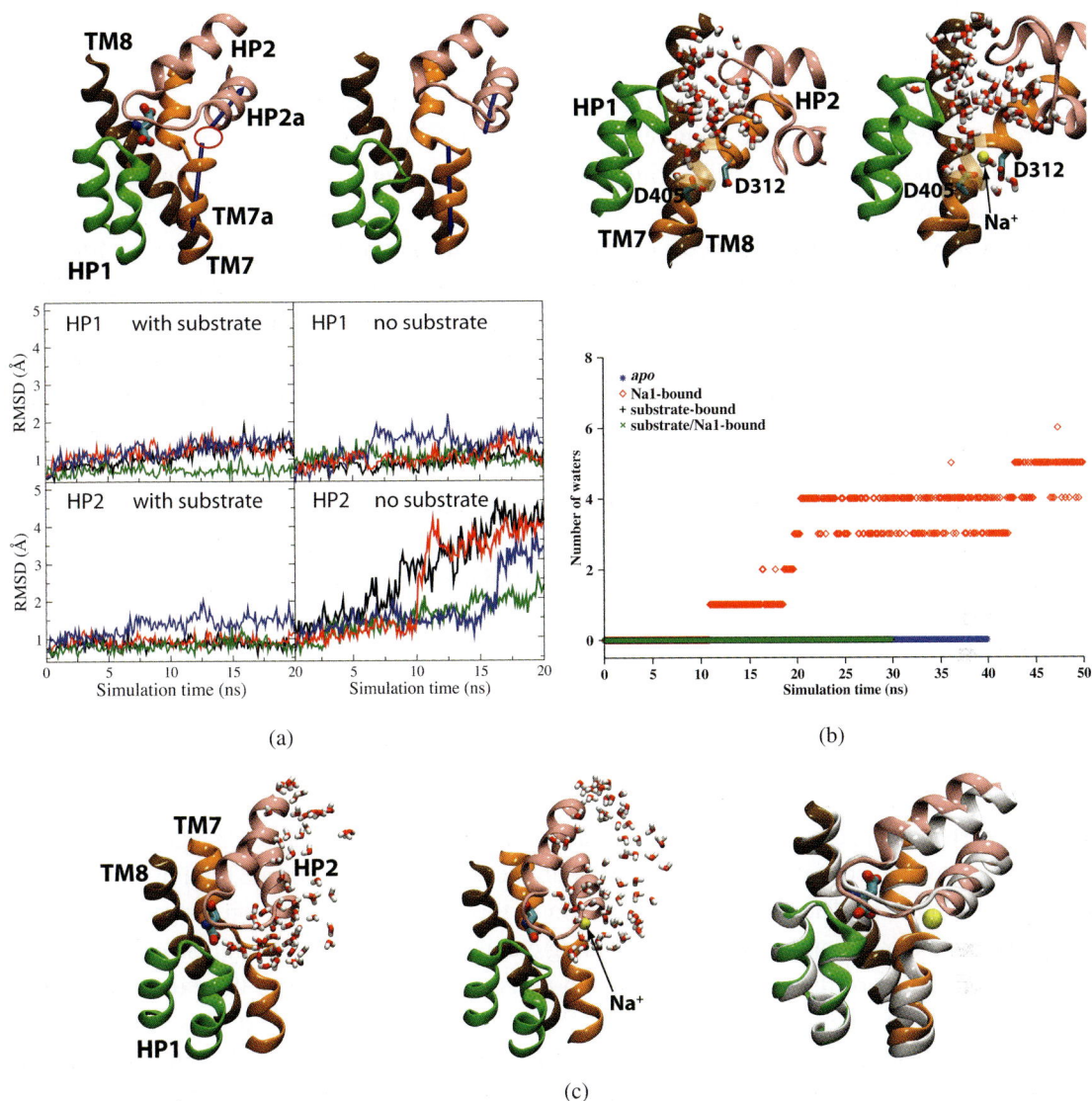

Figure 6. Dynamics of the extracellular gate and the sequence and coupling between substrate and three Na$^+$ ions in Glt$_{ph}$. (a) Dynamics of the extracellular gate in Glt$_{ph}$. Left and right panels show the results of the simulations performed in the presence and in the absence of the substrate, respectively. (Left, top panel) Substrate-bound state. Closing of HP2 (the extracellular gate) and formation of the Na2 binding site (marked with an open circle; focusing of the helical dipole moments). (Right, top panel) Substrate-free state. Opening of HP2 and exposure of the binding site. Note significant misalignment of the dipole moments of helices TM7a and HP2a (arrows). The bottom panel of (a) shows the time evolution of the RMSDs of the helical hairpins HP1 and HP2 in the presence and absence of the substrate. (b) Water accessibility of D312 in the *apo* state (left, top panel) and Na1-bound state (right, top panel), and time series of the number of water molecules within 3.5 Å of the carboxylate groups of D312 (bottom panel) in various bound states. The only system in which D312 can be accessed by extracellular water is the Na1-bound state. (c) Na2-induced formation of the occluded state. Binding of Na2 to Glt$_{ph}$ results in complete closure of the substrate binding site to the extracellular solution. In the absence of Na2 (left), the binding pocket is accessible to water, whereas upon Na2 binding (middle), the binding site is completely sealed. An overlay of the two states (white, before Na2 binding; colored, after Na2 binding) in the rightmost figure highlights the small change in the conformation of HP2 upon Na2 binding.

of the Na2 site (Figure 6a). These results strongly suggest that Na2 binding can only take place after binding of the substrate. Na2 binding further stabilizes HP2, resulting in a completely occluded form of Glt_{ph}, in which water molecules and Na^+ ions can no longer access the binding sites from the extracellular side (Figure 6c). These simulation results are supported by various experimental studies on GlT[31,40] suggesting that substrate binding enables the binding of one of the co-transported Na^+ ions.

Finally, we have also investigated a putative site for the binding of the third Na^+ ion,[7] for which no structural information was obtained from the crystal structures. The accessibility of extracellular water to the putative Na3 site, which is suggested to be around an acidic residue (D312), is monitored during the simulations of various bound states, namely, *apo*, Na1-bound, substrate-bound, and substrate/Na1-bound states (Figure 6b). These results show that neither in the *apo* state nor in the substrate-bound form is the putative Na3 site accessible by extracellular water (Figure 6b). Somewhat unexpectedly, the extracellular solution was observed to gain access to, and eventually hydrate this putative Na3 site significantly, only after the binding of a Na^+ ion in the Na1 site in the empty transporter, that is, in the Na1-bound state (Figure 6b). Therefore, it appears that the exposure of the putative Na3 site to the extracellular solution can only be achieved after Na^+ binding to the Na1 binding site, but lasts only until the binding of the substrate. We would like to note that the experimental results used to infer information on the sequence of binding events often cannot identify which of the three sites are occupied first, and only suggest that two of them are occupied before the substrate and one after it. In this regard, simulation results complement the experimental results in providing a more complete picture of the processes and events involved in the transport cycle of the transporter.

Transporter-mediated transport of solutes across the membrane involves the alternating access mechanism in which conformational changes alternately expose the substrate binding site to either side of the membrane. The simulations presented in this section shed light on the mechanisms of the opening and closure of the extracellular gating elements in Glt_{ph} and the sequence and coupling between the binding of the substrate and the co-transported Na^+ ions from the extracellular side. The mechanisms of the transition between the OF and the IF states and the details of their release from the IF-occ state into the cytoplasm (Figure 5e) are completely unknown[31,32] and call for additional simulation studies.

4. Substrate-Induced Rocker-Switch Motion in an Antiporter

Glycerol-3-phosphate transporter (GlpT) facilitates active transport of glycerol-3-phosphate (G3P) from the periplasm into the cytoplasm using the downhill gradient of inorganic phosphate (P_i) from the cytoplasm to the periplasm. As an antiporter member of the major facilitator superfamily (MFS), the crystal structure of GlpT[41] has served as a model for a wide range of homologous proteins.[42–45] MFS is the largest and the most diverse group of secondary transporters and includes a large number of medically relevant transporters.[46,47] Until now, atomic resolution structures have been obtained for only four members of this family.[41]

GlpT is composed of two six-helix halves, which are connected by a structurally unresolved loop (Figure 7a).[41,43,48] The two halves are structurally similar and have weak sequence homology. A lumen opening to the cytoplasm between the two halves implied an IF conformation. Two

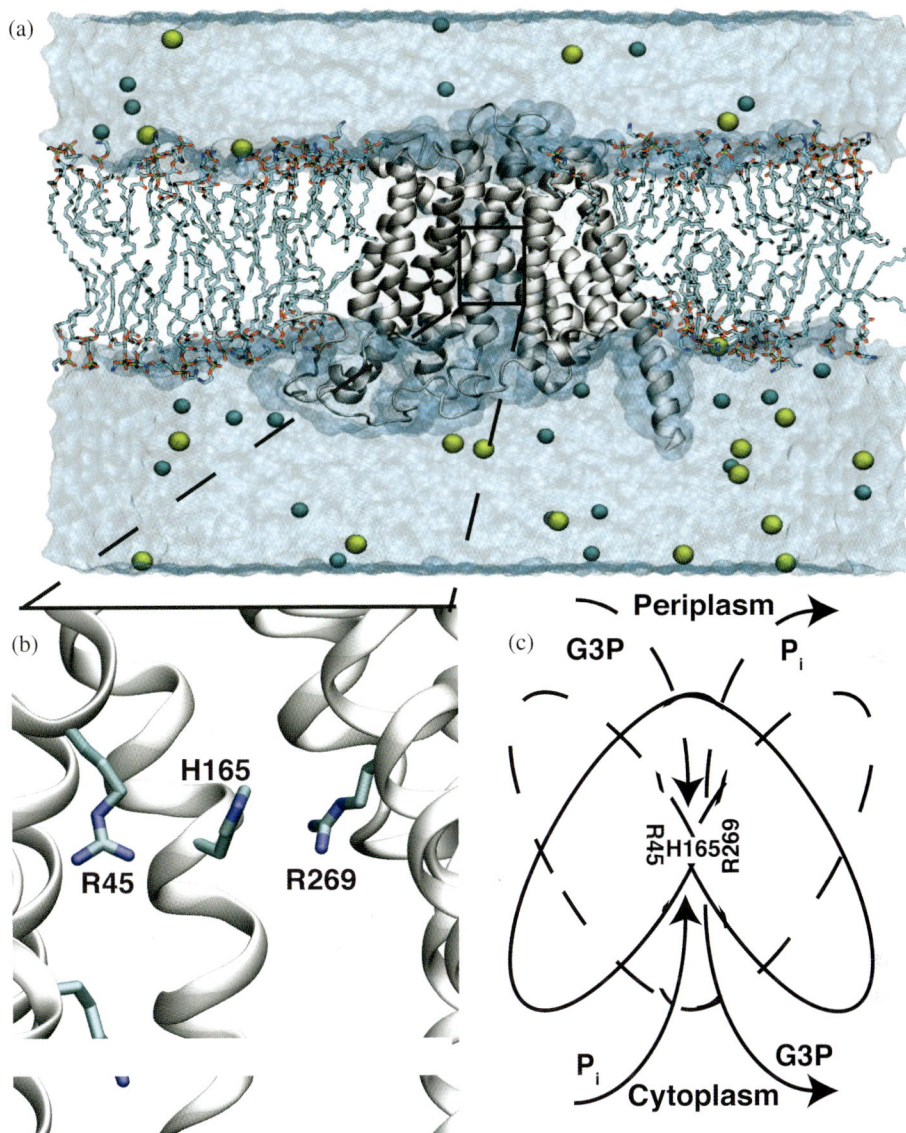

Figure 7. (a) The simulation system of GlpT showing the protein (cartoon) embedded in a lipid bilayer (licorice) with water (surface) and ions (spheres). (b) The putative binding site residues. Two symmetrically positioned arginines (R45 and R269), the histidine (H165) and the lysine (K80) identified in our simulations to act like a "fishing hook" are shown. (c) Schematic representation of the "rocker-switch/alternating access" model of GlpT. P_i replaces G3P on the cytoplasmic side in the IF structure (solid lines), resulting in the formation of the OF structure (dashed lines). G3P binds to the OF structure replacing P_i resulting in reformation of the IF-state. In this model, the binding site residues (R45, H165, R269) are only accesible from either the cytoplasmic or the perilasmic sides.

symmetrically positioned basic residues (R45 and R269), one on each half, confer a positive charge to the apex. These arginines along with a histidine in between were suggested to constitute the substrate binding site (Figure 7b). The suggested roles for these residues were supported by several mutagenesis studies on GlpT[49] and its homologs.[41,50]

The crystal structure of the IF state of GlpT implied an "alternating access mechanism" (Figure 7c).[51] The accessibility of the binding site from the two sides of the membrane was proposed to change through a "rocker-switch"-like motion of the two halves.[41,43,48,52] P_i binding to the IF state of GlpT is deemed to trigger large conformational changes that close the lumen on the cytoplasmic side and open it on the periplasmic side (formation of the OF state).[53,54] Subsequently, the release of P_i from the OF state followed by binding of G3P results in the return of the transporter to the IF state.[41,43,48,52,55]

Although the high-resolution crystal structure of GlpT[41] has provided a great deal of insight about its mechanism, characterization of the binding site, binding mode(s) and mechanism of the substrate, as well as transporter dynamics upon substrate binding are necessary to develop a molecular understanding of the function of the transporter. In order to address these questions, MD simulations of a membrane-embedded model of GlpT in the presence of its natural substrates (monovalent or divalent P_i and G3P) were performed.[2,56] The simulations not only revealed the unknown substrate binding site,[2] but also characterized initial substrate-induced conformational changes consistent with the rocker-switch model, that is, closing at the cytoplasmic side of the lumen and reorganization of the periplasmic salt bridges.[56] The simulation system was built by embedding the crystal structure of GlpT in a solvated and ionized lipid bilayer (Figure 7a). After initial equilibration of the system, individual substrates were placed at the cytoplasmic mouth of the lumen in independent simulations. The *apo*-GlpT (substrate-free) simulation was also performed and served as control. We note that no biasing forces were imposed in any of these simulations, so that they merely relied on free diffusion of the substrate inside the lumen. As described below, the simulations resulted in the identification of key residues and events along the translocation pathway of the substrate.

Similar luminal substrate translocation pathways and a common final binding site at the apex of GlpT were captured in the simulations irrespective of the type and the charge (protonation state) of the substrate (Figure 8a).[2,56] Although capturing the process of spontaneous substrate binding has been difficult in all-atom MD simulations due to the long time scales needed to sample random fluctuation and movement of the substrate towards its binding site (note that there is usually a single copy of the protein, and often a single copy of the substrate present in the simulations), we were able to capture this process in GlpT in a reproducible manner, mostly due to the presence of a strong luminal electrostatic potential attracting the negatively charged substrates and the rather small size of their substrates.[2,56] This is to our knowledge the second example of capturing complete, reproducible substrate binding to a protein with equilibrium all-atom MD simulations (The first case was reported for mitochondrial ATP/ADP carrier AAC.)[11]. The process of substrate binding was found to be accompanied/facilitated by conformational changes of side chains lining the lumen. Once bound, the substrate triggers more global conformational changes in the form of both helical and side chain reorientations, which eventually would result in the transition of the protein from one major conformational state to another. The most significant element in the recruitment of the substrate appears to be side chain motion of a highly conserved lysine (K80) which extends towards the cytoplasmic opening of the lumen. This conserved lysine acts like "a fishing hook" which rapidly captures the substrate from the mouth of the lumen and facilitates its translocation deeper into the lumen and towards the binding site. This residue remains coordinated to the substrate for the most part of its translocation through the lumen, until the latter is handed over to R45, which is one of the major components of the identified substrate binding pocket.

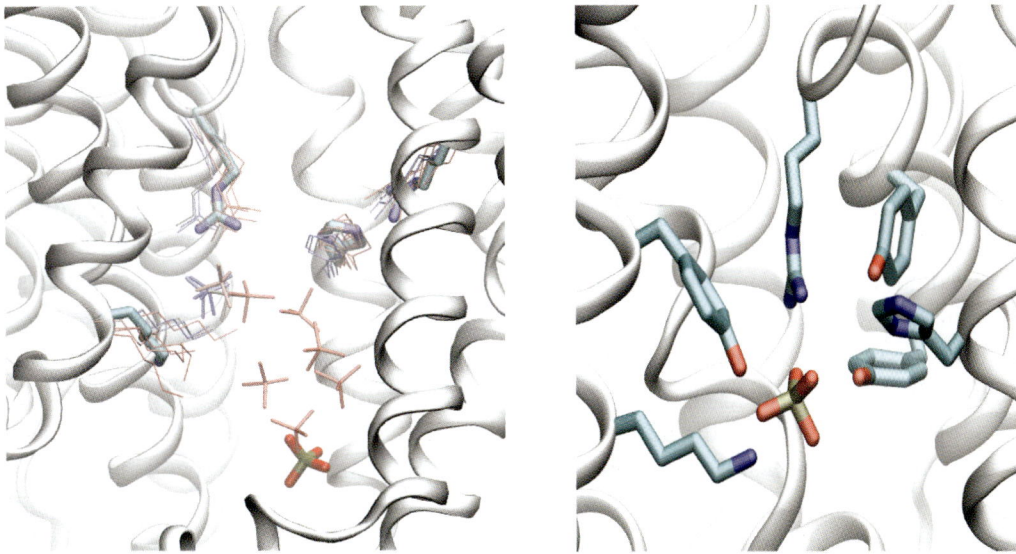

(a)

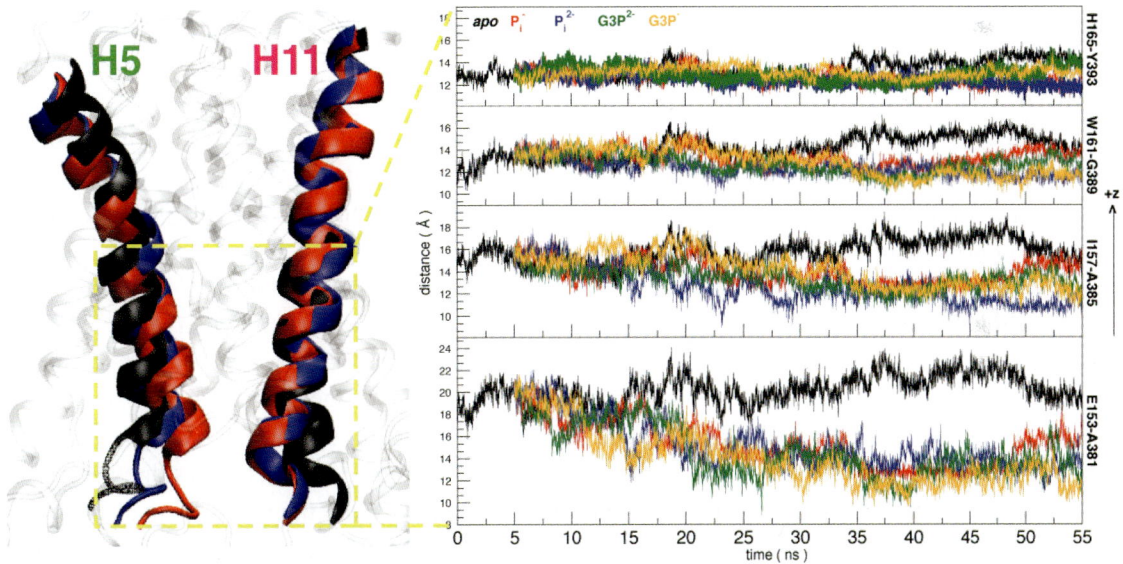

(b)

Figure 8. (a) Left: Translocation pathway of P_i^{2-} inside the lumen. Several conformations and positions of the substrate and the binding site residues are shown. The thick lines correspond to the original positons. P_i^{2-} initially interacts with K80 and finally gets coordinated stably by R45. Right: Bound P_i^{2-} in the identified binding site. The three tyrosines (Y38, Y42, and Y76) along with R45, K80, and H165 are involved in substrate binding. They form the "tyrosine-cage". (b) Left: Molecular image of helices 5 and 11 and their dynamics. Helices 5 and 11 straighten and come together at the cytoplasmic side occluding the cytoplsmic mouth of the lumen (black, blue, and red represent initial, intermediate, and final structures taken from the stimulations, respectively). Right: Distance between the C_α atoms of representative residues on helices 5 and 11. The residues are chosen to lie approximately on the same plane. While the distances between the residues on the periplasmic side (+z) are maintained (not shown), the residues on the cytoplasmic side approach each other in all simulations except the *apo*-simulation.

Fifty ns-long simulations performed on GlpT also revealed the unknown substrate binding site (Figure 8a). One of the arginines in the putative binding site (R45) keeps the substrate in a "tyrosine cage" formed by three tyrosines,[2,56] which establish hydrogen bonds with the phosphate moiety of the substrate. The histidine residue is another major side chain contributing to the substrate binding site. In contrast to what was deduced based on the crystal structure,[41] simulations revealed very different involvements of R45 and R269 (two symmetrically positioned arginines in the apex of the lumen; see Figure 7b). While one of the arginines was identified to coordinate the substrate, the other did not seem to be directly involved in substrate binding.[56] On the other hand, our simulations with the neutral form of that arginine show that substrate cannot be recruited in the absence of this residue, and thus, it is at least indirectly involved in substrate binding by maintaining the basicity of the binding site.[56] We note that R269 is still a key residue for transport, and may play a role in later stages of the transport cycle, even directly participate in binding the substrate. In fact, our preliminary simulations also showed that R269 may act as a second binding site.

Given the extended time scale of the simulations, we have also been able to observe conformational changes induced by the substrate that are beyond the level of side chain motions within the binding pocket. Substrate binding consistently triggers global conformational changes in the helices and results in a semi-closed conformation of the protein.[56] The observed conformational changes are in agreement with the proposed rocker-switch mechanism for GlpT, in that they appear to capture early events during the formation of an occluded state. Indeed, substrate binding results in partial closure of the cytoplasmic mouth of the lumen by bringing together the cytoplasmic ends of two helices on different halves (Figure 8b). This event is consistently reproduced in all substrate binding simulations but was absent in the *apo* simulation.[56] Interestingly, it was previously observed that these helices show the highest flexibility compared to other helices of GlpT, when simulated individually in the membrane.[53]

Another major effect of substrate binding manifests itself on the periplasmic salt-bridge network which we view as an important switch mechanism in the rocker-switch model.[56] This salt-bridge network is composed of basic and acidic side chains that tie the N- and C-terminal halves together.[41,47,49] Inter-domain salt bridges formed between a lysine on one half and two acidic residues on the other half are also accompanied by an intra-domain salt bridge between the binding site arginine not directly involved in substrate binding and one of the acidic residues involved in inter-domain salt bridge. Consistent with the role of this network in the transport mechanism, it has been shown that the mutation of most of these residues affects the transport rate, while substrate binding is unaffected.[49]

In summary, our simulations captured spontaneous substrate recruitment and binding as well as structural changes induced by substrate binding in GlpT. The partial closure of the cytoplasmic mouth of the lumen and substrate-induced reorganization of the periplasmic salt bridge network can be considered as some of the major steps of the "rocker-switch" mechanism, resulting in the formation of the occluded state.

5. Alternating Access Model in Leucine Transporter

Neurotransmitter transporters are specialized proteins that perform active transport of neurotransmitters across the cell membrane. The family of Na$^+$/Cl$^-$ dependent transporters derive energy for substrate transport by co-transporting ions (Na^+ and often, Cl^-) along their electrochemical gradient. These proteins form the neurotransmitter:sodium symporter (NSS, also called SLC6) family, responsible for clearing the synaptic cleft of molecules such as serotonin, dopamine, norepinephrine, and GABA.[57] Abnormality in the function of these transporters may occur due to several reasons including drug abuse or genetic factors, and lead to neurological disorders such as depression, anxiety, and hyperactivity. The NSS family thus constitutes one of the main targets for psychostimulants, anti-depressants and drugs for other neurological disorders. While these transporters have been extensively characterized functionally, structural information remains limited to that of a prokaryotic homolog, the leucine transporter (LeuT). LeuT shares significant sequence similarity with other NSS members, serving as an important model to obtain structural and mechanistic information for this family of transporters.

The "alternating access" mechanism[8] is widely accepted as the general mechanism by which several transporters function. The transporter is proposed to alternate between the OF-open (OF-o) and IF-open (IF-o) states (Figure 9b), going through several intermediate states. These would include OF-occluded (OF-occ) and IF-occluded (IF-occ) states, representing states where the overall transporter structure resembles the OF to IF states but the substrate is inaccessible to the extracellular or intracellular solution, respectively. LeuT and other NSS members are believed to follow the alternating access mechanism. The first crystal structure of LeuT was solved at a high resolution (1.65 Å) and reported as a dimer, each with bound substrate (leucine) and two Na^+ ions, termed Na1 and Na2[58] (Figure 9a). Since then, several structures of LeuT, with other bound substrates, including amino acids as well as NSS inhibitors, have been reported[59]. All these crystal structures for LeuT represent the OF states (either OF-occ or OF-o), thus providing only a partial description of the possible states involved in the transport process. Structural and dynamic characterization of the unknown alternate states is necessary to understand the transport mechanism of LeuT and other NSS members.

LeuT is composed of 12 transmembrane (TM) helices, of which helices 1–5 and 6–10 form two pseudosymmetric domains, arranged as inverted repeats. Interestingly, this typical fold of LeuT has also been observed in several secondary transporters from other families.[60] Such a common occurrence suggests a strong functional relevance of this fold, and it is possible that these structures sharing a common fold may also show similarity in mechanism. Thus, following the transport cycle in these structures may reveal some common features of transport in the LeuT-fold.

Several earlier functional studies on NSS members revealed residues involved in permeation, gating, and binding of ions and substrate or inhibitor. However, mechanistic understanding remained limited since only the OF states of LeuT were known. Recent modeling and simulation studies addressed at investigating the IF state and the putative transport mechanism[1,61–63] have provided much needed structural information to complement these experiments. Computational techniques were also employed to predict the Cl^- binding site in SERT and GABA transporter (GAT). This is important to the understanding of NSS function, since these and other eukaryotic NSS members are also known to be dependent on Cl^-, though prokaryotic NSS members including LeuT are Cl^--independent.[64] Computational simulations have also provided information about substrate/ion binding and specificity[1,65,66] and the putative transport mechanism.[1,62,63] A notable prediction from a simulation-based approach was the presence of a second substrate binding site involved in transport function.[62] The existence of this binding site and its role as an activator of conformational changes responsible for substrate transport was supported by multiple functional and structural studies.[62,67,68]

(a)

(b)

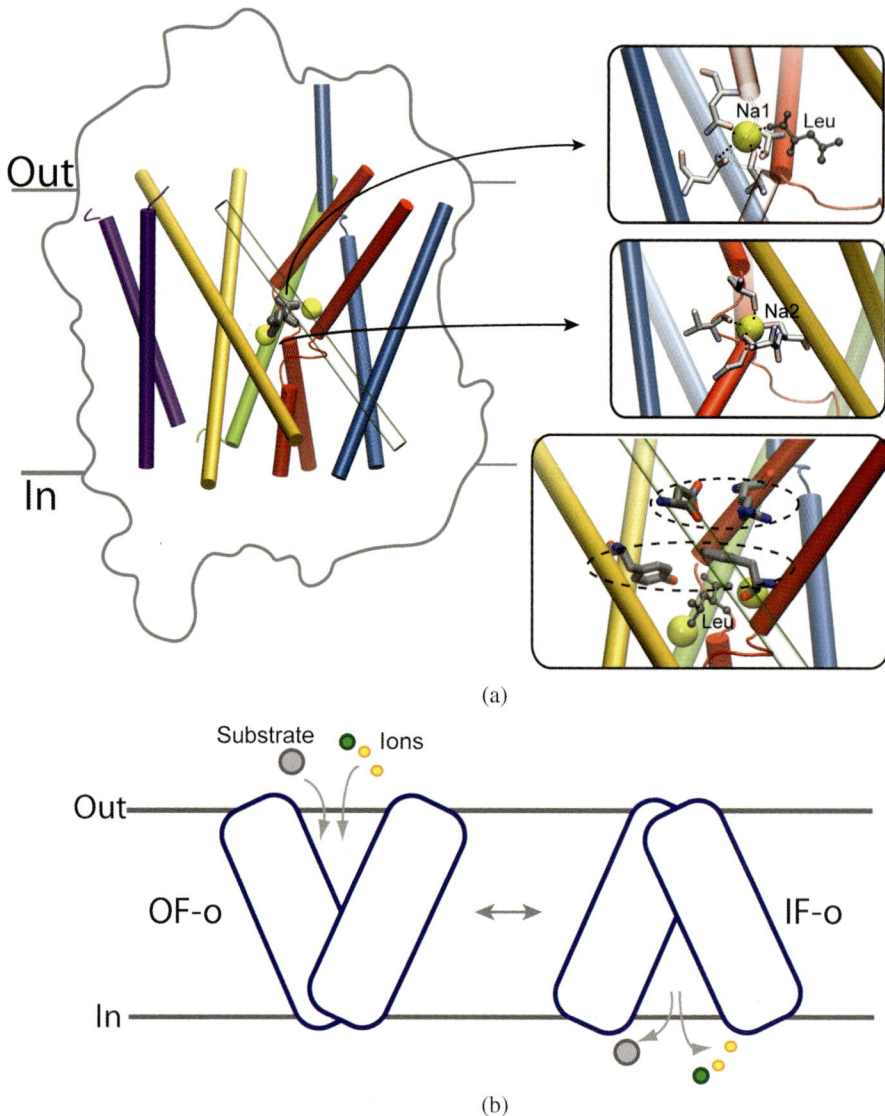

Figure 9. (a) Left: The structural arrangement of TM1-TM10, forming the "LeuT-fold" in the LeuT crystal structure (OF-occ state) is shown, with substrate (gray sticks) and bound Na^+ ions (spheres). The outline of the complete protein is also shown. Right: Enlarged views of the binding site of Na1 (top) and Na2 (center). Binding residues are shown in stick representation, Na^+ as spheres and the substrate, Leu, in ball and stick representation. Right bottom: The two putative extracellular gates (dotted ovals), including the R30:D404 salt bridge (upper oval) which acts as a secondary gate and the Y108:F253 aromatic lid (lower oval) which forms the primary gate, are shown. The binding residues and the substrate, Leu, are shown in stick and ball-and-stick representation respectively. (b) The alternating access mechanism represented schematically for LeuT and other NSS members. Substrate, Na^+ and for eukaryotic members, Cl^- ions are bound on one side of the membrane and released on the other side as the protein alternates between OF-o and IF-o states. The substrate:ion stoichiometry differs among the NSS members.

The dynamics of LeuT has been studied using equilibrium MD simulations as well as biased simulations such as steered MD (SMD) and targeted MD (TMD), which allow additional forces to be added to the simulation system,[1,62,63] in order to probe processes that cannot be described by free MD due to their long time scales. SMD simulations have been performed where additional forces were applied to enable the study of substrate unbinding in LeuT.[1,62] In the first reported SMD study on LeuT,[1] the substrate, leucine, was pulled towards the extracellular side. The substrate translocated through the extracellular vestibule, with major barriers to unbinding offered by interaction and/or physical blockage by Na1, an aromatic lid over the substrate, and a salt bridge (Figure 9a). It was seen that a mere rotation of the side chains of the aromatic lid residues allows opening of the binding site when the substrate exits. The substrate showed a common unbinding pathway in a majority of the simulations, suggesting physiological relevance of this pathway. These simulations provided one of the earliest views of the possible substrate translocation pathway in the extracellular pore, and transition from the OF-o to OF-occ state in an NSS member, obtained from computational studies on the crystal structure.

The next step to understanding the LeuT transport cycle is to study its IF states, and transitions between OF and IF states. In order to investigate the IF state, independent computational studies have been performed to model the IF-o/IF-occ states[61,63] as well as to simulate substrate unbinding in the IF state[62] and the OF-to-IF transition.[63] Based on the internal symmetry of LeuT, an IF state model was generated where the structures of the two pseudosymmetric domains of LeuT were exchanged.[61] While this model provided the first structural description of a possible IF state, it being a static model, the structural elements involved in the OF-to-IF transition could only be deduced from a comparison of the crystal OF and modeled IF states. Based on these states, the transport mechanism is described in terms of rigid body motion of a bundle of transmembrane helices 1, 2, 6, and 7 in LeuT.[61,69] In an independent simulation study, the substrate was pulled along the extracellular or cytoplasmic directions and putative substrate transport pathway and mechanism were proposed.[62] The extracellular pulling revealed a second substrate binding site (S2) which was supported by experimental studies.[62,67,68] The cytoplasmic pulling suggested that structural changes for the opening of the putative IF transport pathway were smaller compared to those proposed from the symmetry-based IF model, and involved hydrogen bond and salt-bridge dissociation and rearrangements. Computational studies thus provided vital information about the possible structural changes involved in the formation of the LeuT IF state. However, in the absence of sufficient information about IF state dynamics and that connecting the transition between the OF and IF states, the transport mechanism remained only partially understood.

The need for dynamic information about the LeuT IF state, and the opportunity presented by an increasing number of solved LeuT-fold structures resulted in a recent computational study where TMD simulations were adopted to generate IF state models, using information from a structurally related transporter.[63] The modeling methodology combined sequence- and structure-based alignment, threading and TMD.[63] The crystal structure of vSGLT, a LeuT-fold transporter crystallized in the IF-occ state,[70] was used to construct a homology model of LeuT in this state which was used as a target for TMD simulations in which the original LeuT structure (OF-occ state) was deformed into an IF-occ state while embedded in a lipid bilayer and surrounded with solvent and ions. In each simulation, the C_α atoms of a subset of the helices involved in the inverted repeats were constrained toward a target structure, while the remaining parts of the structure, including all side chains, were allowed to freely move and adapt to the new conformation. Extended TMD simulations (50 ns each) were used to induce the OF to IF transition, followed by 20 ns of free MD for relaxation, finally resulting in models of the IF state of LeuT.[63] The resulting models retained the secondary structural features of LeuT and preserved the substrate/ion binding sites, while adopting an IF state. Water permeation was monitored with the progress in transition and indicates inward-opening (Figure 10).

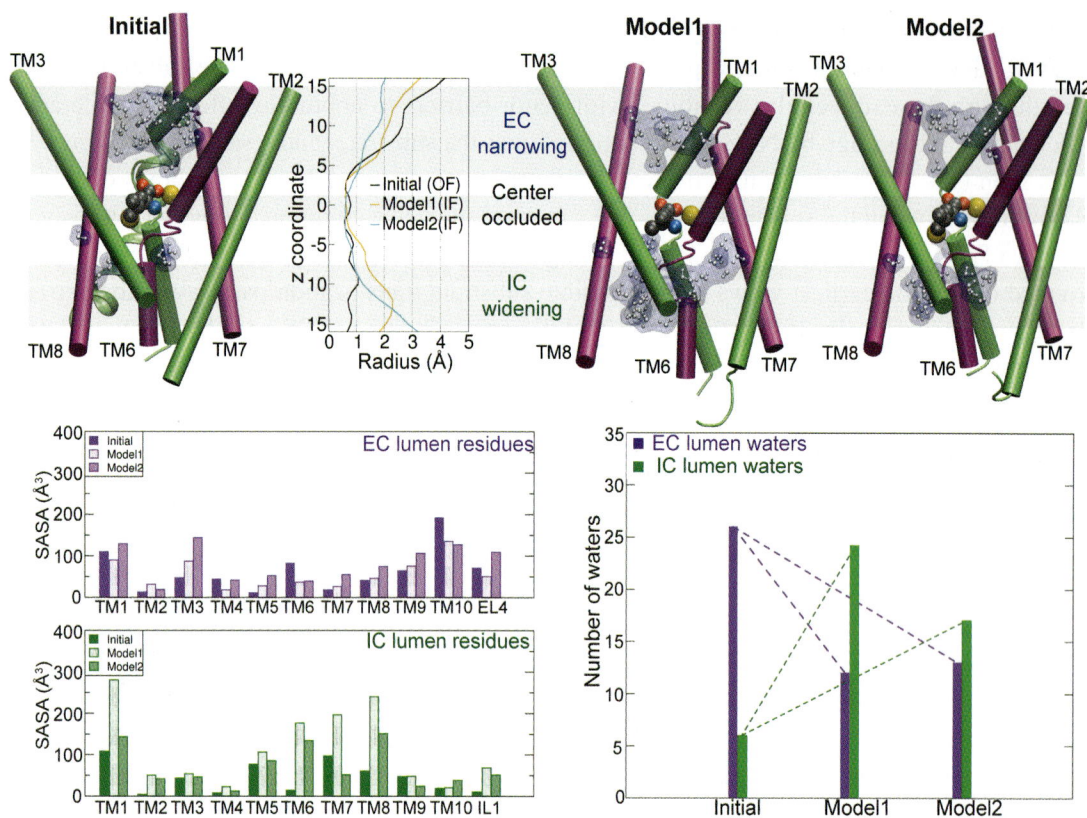

Figure 10. Top panel: Comparison of TM domains and water penetration in the initial structure input to TMD (left), and in Model1 and Model2 (right). Water molecules (white atoms, gray surface) in the lumen are shown. TM4, TM5, TM9, and TM10 are hidden for clarity. The radius profile of the lumen (center) shows EC narrowing and IC widening in both Model1 and Model2 as compared to the initial structure (black curve), and remains occluded at the substrate binding site in all structures. These regions are highlighted with grey bands. Bottom panel, Left: Solvent accessible surface area (SASA) of EC (top) and IC (bottom) lumen residues for TM1-TM10 in the initial structure, Model1, and Model2. Right: Number of water molecules in the EC and IC halves of the lumen for the initial structure, Model1 and Model2.

Release of Na2, an important step in the transport process, was observed during the transition. An interesting revelation is the possible role for pseudosymmetry in LeuT structures; modification of only one of the two pseudosymmetric domains induced an overall transition to the IF state suggesting that each of the two pseudosymmetric domains may represent a functional unit capable of inducing transition in the full protein. This study thus presented the first detailed description of the IF state dynamics of an NSS homolog and a possibly general transport mechanism in transporters adopting the LeuT fold.

Computational studies of LeuT have thus played a major role in providing missing dynamic and structural information to develop the connection between the OF and the IF states, and possible intermediate states in the alternating-access pathway. These studies, coupled with information from new crystal structures of LeuT-fold transporters, and experimental characterization of conformational states of these transporters, have enabled the emergence of a clearer picture of the transport mechanism in LeuT-fold transporters.

6. Interconversion of the Open and Occluded States in Na⁺-Coupled Galactose Transporter

Recent years have seen an increasing number of high-resolution structures of secondary transporters. Surprisingly, several of these structures, namely, leucine transporter (LeuT), Na⁺-coupled galactose transporter (vSGLT), and some other secondary transporters, bear significant architectural resemblance, despite distinct substrate specificities and almost no sequence similarity. For example, structural alignment of substrate-bound structures shows that the substrate binding site is always located at the interface of the two inverted structural repeats. Sequence and structural comparison of these transporters also suggests a conserved cation binding site which is critical for coupled substrate binding and symport.[60] These structural similarities suggest that the secondary transporters sharing the LeuT-fold topology might be a growing superfamily that might share a similar transport mechanism.

Solute sodium symporters (SSS) constitute a family of secondary transporters that couple Na⁺ symport to the transport of a wide range of solutes, including sugars, amino acids, organo-cations such as choline, nucleotides, inositols, vitamins, urea, and anions. SSS play crucial roles in human health, and malfunction of these transporters might result in various metabolic disorders.[71] As the first solved structure of an IF, substrate-bound protein in the SSS family and among transporters with the LeuT-fold topology, the structure of vSGLT,[70] took a critical step toward a better characterization of the alternating-access mechanism (Figure 1) and characterization of the sequence of events involved in the transport cycle. In the reported crystal structure,[70] the Na⁺ binding site could not be unequivocally resolved, however, a Na⁺ ion was modeled in a binding site corresponding to that of LeuT based on structural similarity. While mutagenesis studies confirmed the involvement of a number of residues from this putative site in Na⁺ binding, the occupancy of the site in the conformation captured in the crystal remained unresolved.[70] Another remaining important mechanistic question was regarding the sequence of unbinding events on the cytoplasmic side, which is poorly understood not only for vSGLT, but also for the majority of secondary transporters as they often require a dynamical description of the protein to resolve.

To address these questions, we have performed equilibrium simulations of a membrane-embedded model of vSGLT.[13] In the crystal structure of vSGLT, Na⁺ is modeled in a site located at the intersection of two transmembrane helices (Figure 11c). In this region, several residues corresponding to conserved residues in the Na⁺-binding site of LeuT, appear to be optimally positioned to coordinate a Na⁺ ion. Somewhat surprising initially, however, in all of the performed MD simulations, we observed a rapid Na⁺ unbinding from the site within a few nanoseconds. In addition, it is interesting that in all simulations, the Na⁺ ion was observed to hover around the region close to an aspartic acid (~ 4.5 Å away from the Na⁺ site in the crystal structure) after its detachment from the binding site, and before exiting the transporter on the cytoplasmic side (Figure 12a). Given that in these simulations, the protein did not show a major structural change (RMSD ≤ 2.5 Å) the observed spontaneous Na⁺ unbinding suggests that the crystal structure of vSGLT is in an *ion-releasing* state.[13]

Close examination and structural alignment of the Na⁺ binding sites in vSGLT and LeuT provide supporting evidence for this hypothesis. While at the sequence level, the residues forming the ion binding site of vSGLT are very similar to those in LeuT, there are significant geometrical

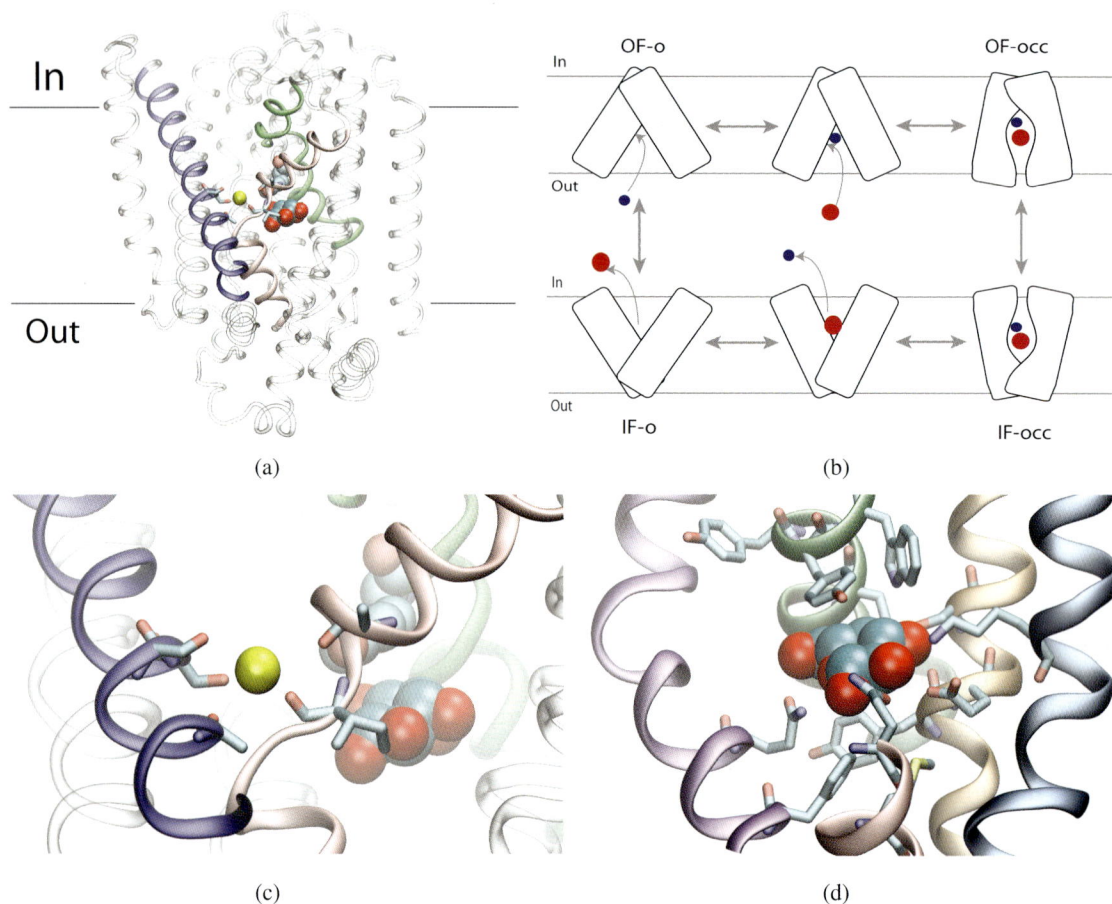

Figure 11. Structure and transport mechanism of vSGLT. (a) Overview of the crystal structure of vSGLT. The protein is shown in ribbon representation. The substrate and ion are in van der Waals (vdW). (b) The alternating access model proposed for the vSGLT function, including two major open states, outward facing-open (OF-o) and inward facing-open (IF-o), two intermediate states and two major occluded states (OF-occ and IF-occ). "Out" and "In" represent outside and inside the cell, respectively. Two dots respectively represent the substrate (big dot) and the Na+ ion (small dot). (c) The proposed Na+-binding site in crystal structure. Residues in the Na+-binding site are displayed as sticks. (d) The sub strate-binding site. Residues in the substrate binding site are displayed as sticks.

differences between the two sites. In the crystal structure of LeuT,[72] residues coordinating the Na+ ion form a typical square pyramidal arrangement with bond distances between the Na+ ion and the coordinating oxygen atoms ranging from 2.1 to 2.4 Å. In vSGLT, on the other hand, all of the corresponding distances are above 3.1 Å, resulting in an "open" site, which, consistent with the results of the simulations, is not able to hold the ion.[13]

The discovered differences between the ion binding sites of vSGLT and LeuT successfully capture the type and extent of conformational changes involved in the transition between the ion-releasing (open) and ion-binding (occluded) states (Figure 12b) (we note that the terms "open" and "occluded" are usually used with reference to the substrate binding site). The differences in the binding site geometries seem to originate mainly from the reorientation of two helices resulting in

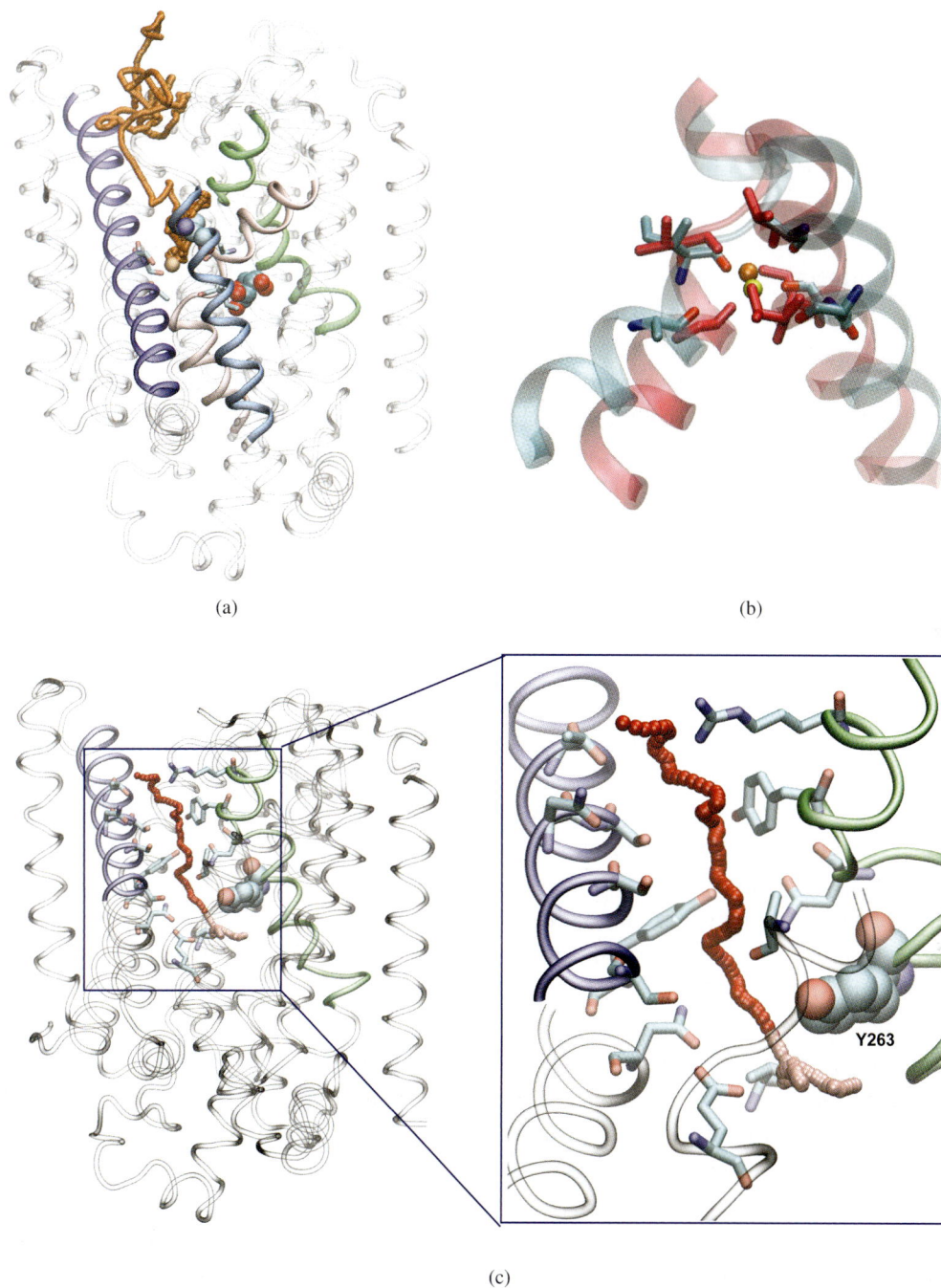

(a)

(b)

(c)

Figure 12. The release of the Na$^+$ ion and the substrate in vSGLT. (a) Overview of the release trajectory of the Na$^+$ ion. Residues in the Na$^+$-binding site are displayed as sticks, D189 and the substrate are in van der Waals (vdW). (b) Comparison of the Na$^+$-binding sites of vSGLT and LeuT. Superposition of the Na$^+$-binding sites of vSGLT and LeuT. Alignment was done using TM1 and TM8 helices from vSGLT and LeuT. (c) Overview of the release trajectory of the substrate. The release trajectory is combined by a part of equilibrium trajectory (from 80 ns to 83 ns) and the trajectory of SMD simulation. The substrate is represented by the C5 atom in the substrate. The protein is shown in ribbon representation, and the residues lined in the substrate-release pathway are displayed as sticks.

a smaller interhelical angle and, thus, a closed binding site and stable Na^+ binding in LeuT.[1] In the crystal structure of vSGLT this inter-helical angle has become larger, resulting in the opening of the Na^+ binding site, explaining the observed unbinding of the ion during the simulations.

The interaction between the Na^+ ion and the aspartic acid along its unbinding pathway, which was consistently observed in all the MD simulations, suggests that this residue, which is highly conserved within the Na^+/solute cotransporter family, might play a critical role in cation unbinding in vSGLT. The functional importance of this aspartic acid is supported by experimental studies in homologous proteins.[73,74] Based on our simulations, we propose that this residue might act as a "fishhook" above the Na^+-binding site facilitating the departure of the ion upon the opening of the binding site.

The release of the substrate is another important event on the cytoplasmic side. According to the crystal structure,[70] the substrate is bound about halfway across the membrane with the binding site flanked by hydrophobic residues on both the cytoplasmic and extracellular sides (Figure 11d). On the cytoplasmic side, a tyrosine (Y263) stacks with the pyranose ring of the substrate (galactose), a feature commonly found in sugar-binding protein structures.[75,76] This primary interaction establishes a plug that appears to prevent the exit of the substrate into the cytoplasmic opening of the lumen. Mutation of many residues in the substrate binding pocket has been shown to abolish Na^+-dependent galactose transport.[70] Despite detailed characterization of the binding site, the mechanism of substrate unbinding and its pathway have not been described. Most importantly, it is not known whether or not a local gating motion (specifically for the side chain of Y263) would be necessary for substrate unbinding. In other words, we don't know whether the binding site is occluded or open for the substrate.

In order to investigate the process and to characterize the unbinding pathway of the substrate, we have performed extended equilibrium simulations (on the order of 200 ns) complemented by pulling (SMD) simulations.[77] Through the extended equilibrium simulations, we examined whether spontaneous unbinding from the binding site could be captured without biasing the substrate with external forces. Indeed, during the simulations we were able to capture complete unbinding of the substrate from its binding pocket on several occasions, although the substrate did not completely leave the protein during the simulations.

During these unbinding events, the substrate completely loses its original contacts with the residues in the binding pocket. Most interestingly, the substrate appears to be able to go around the side chain of Y263 during its unbinding. In the unbound form, the substrate is no longer covered by this residue which was suspected to act as a cytoplasmic plug or gate. The unbinding of the substrate is rather mediated by its displacement in the membrane plane, i.e. parallel to the plane of the side-chain of Y263. Therefore, based on the dynamical behavior characterized by the simulations, we propose that Y263 does not play a gating role in the mechanism of vSGLT.

Starting from the fully unbound state identified in our extended equilibrium simulation described above, the substrate could be steered out of the protein without facing any major obstacles with SMD simulations. These simulations provided a complete description of the unbinding pathway for the substrate in vSGLT. We note that pulling simulations starting from a bound state (original location of the substrate) or from several partially unbound states, resulted in very large forces required for substrate unbinding. The large forces arising in these simulations are due to the steric clash between the substrate and the side chain of Y263 when one starts from a bound state and pulls the substrate toward the cytoplasmic side. In other words, the side chain of this residue seems to be "in the way", and that is why unbinding of the substrate in these simulations is accompanied by the rotation of this side chain. We view this mechanism an artifact of the pulling scheme, since our equilibrium simulations clearly identified a very different pathway and mechanism for substrate unbinding that does not depend on the conformational changes of this residue and can be done using much smaller forces. We also note that the entire protein, including the unbinding pathway, was found to maintain its original structure during the unbinding process, indicating that possibly no further conformational changes are needed for the substrate release in the state captured in the crystal structure.

The latest crystal structure of an IF conformation of Mhp1, another LeuT-fold secondary transporter, is highly similar to the IF structure of vSGLT.[78] Supporting our hypothesis, it is interesting that the structure of vSGLT is captured in its substrate-bound form, while that of Mhp1 is substrate-free, which suggests that the transition from the substrate-bound state to the substrate-free one in the LeuT-fold transporters might not require major conformational changes of the protein, as also observed in our simulations.

Together, the simulation studies suggest that the crystal structure of vSGLT represents a substrate-bound, but Na$^+$-free state, and during the transport cycle the cytoplasmic release of the ion precedes that of the substrate. Furthermore, cytoplasmic release of the substrate from the state captured in the crystal structure does not appear to require further protein conformational changes and can be captured in equilibrium simulations. The unbinding pathway for the substrate also goes through the Na$^+$-binding site supporting the notion that the release of the Na$^+$ ion is not only taking place before the substrate release, but is in fact necessary for it.

7. Concluding Remarks

Conformational changes are at the heart of the function of membrane transporters. These molecular machines rely on various degrees of protein conformational changes not only for the mechanisms of gating and selective transport, but also, and more importantly, for implementing energy coupling mechanisms, which effectively pump the substrate against its concentration gradient. Various modes of protein conformational changes are at work in membrane transporters. While localized gating motions are responsible for transition between the open and occluded states of the binding sites, global conformational changes are usually in charge of the change of the overall accessibility of the substrate from one side of the membrane to the other, a mechanism broadly referred to as the alternating access model in membrane transporters. Given the recent advances in structural biology resulting in an increasing number of membrane transporter structures at atomic resolutions and significant developments in algorithms and parallel computing, we have been able, over the past few years, to extend the scope of computer simulation into the realm of protein domain motions, thus, capturing more functionally relevant conformational changes of proteins in membrane transporters. In this review, we have discussed the results of application of extended large-scale molecular dynamics simulations to a number of membrane transporter proteins in our laboratory. We demonstrated that such simulations are able to capture various forms and degrees of conformational changes involved in the function of these mechanistically complex proteins. Combining atomic representations of the proteins and their surrounding (lipid, water, ions etc.)

with extended simulations, we have been able to characterize motions ranging from side chain rotations and hydrogen-bond breaking events, all the way to subdomain flipping and even domain separations in different membrane transporters. Most importantly, we show that the captured motions are all of functional significance, i.e. such motions are triggered in response to various events and elements that are involved in their function, and not random molecular events. Indeed in most cases, we have also demonstrated that the observed events are reproducible and are seen in independent simulations. The results of these studies have produced novel hypotheses regarding the function of membrane transporters. Many of these hypotheses can be and have been tested by designing new experiments that can verify the proposed mechanisms. Along with the growing computational power, we will be able to extend further the time scale of MD simulations and improve our sampling and statistics. Therefore, we should look forward to more examples of bio-molecular simulations in which key functional dynamical events will be captured. We should also expect to have a larger number of high-resolution structures for various intermediates and functional states of membrane transporters, which, combined with extended simulations and hybrid methods, will allow us to provide a more complete description of the dynamics of the entire transport cycle in membrane transporters.

Acknowledgment

This work was supported in part by the National Institutes of Health (grants RO1-GMO86749, RO1-GM67887, U54-GM087519, and P41-RR05969). All simulations have been performed using TeraGrid resources (grant number MCA06N060) and the DoD High Performance Computing Modernization Program.

Suggested Additional Reading Materials

F. Khalili-Araghi, J. Gumbart, P.-C. Wen, M. Sotomayor, E. Tajkhorshid and K. Schulten. 2009. Molecular dynamics simulations of membrane channels and transporters. *Curr Opin Struct Biol* **19**, 128–137.

H. Krishnamurthy, C. L. Piscitelli and E. Gouaux. 2009. Unlocking the molecular secrets of sodium-coupled transporters. *Nature* **459**, 347–355.

A. Nyola, N. K. Karpowich, J. Zhen, J. Marden, M. E. Reith and D.-N. Wang. 2010. Substrate and drug binding sites in LeuT. *Curr Opin Struct Biol* **20**, 415–422.

M. L. Oldham, A. L. Davidson and J. Chen. 2008. Structural insights into ABC transporter mechanism. *Curr Opin Struct Biol* **18**, 726–733.

References

1. L. Celik, B. Schiott and E. Tajkhorshid. 2008. Substrate binding and formation of an occluded state in the leucine transporter. *Biophys J* **94**, 1600–1612.
2. C. J. Law, G. Enkavi, D. N. Wang and E. Tajkhorshid. 2009. Structural basis of substrate selectivity in the glycerol-3-phosphate:phosphate antiporter GlpT. *Biophys J* **97**, 1346–1353.

3. S. A. Shaikh and E. Tajkhorshid. 2008. Potential cation and H⁺ binding sites in acid sensing ion channel-1. *Biophys J* **95**, 5153–5164.

4. Z. Huang and E. Tajkhorshid. 2008. Dynamics of the extracellular gate and ion-substrate coupling in the glutamate transporter. *Biophys J* **95**, 2292–2300.

5. P.-C. Wen and E. Tajkhorshid. 2008. Dimer opening of the nucleotide binding domains of ABC transporters after ATP hydrolysis. *Biophys J* **95**, 5100–5110.

6. J. Gumbart, M. C. Wiener and E. Tajkhorshid. 2009. Coupling of calcium and substrate binding through loop alignment in the outer membrane transporter BtuB. *J Mol Biol* **393**, 1129–1142.

7. Z. Huang and E. Tajkhorshid. 2010. Identification of the third Na⁺ site and the sequence of extracellular binding events in the glutamate transporter. *Biophys J* **99**, 1416–1425.

8. O. Jardetzky. 1966. Simple allosteric model for membrane pumps. *Nature* **211**, 2406–2414.

9. Y. Wang and E. Tajkhorshid. 2007. Molecular mechanisms of conduction and selectivity in aquaporin water channels. *J Nutr* **137**, 1509S–1515S.

10. J. Gumbart, M. C. Wiener and E. Tajkhorshid. 2007. Mechanics of force propagation in TonB-dependent outer membrane transport. *Biophys J* **93**, 496–504.

11. Y. Wang and E. Tajkhorshid. 2008. Electrostatic funneling of substrate in mitochondrial inner membrane carriers. *Proc Natl Acad Sci U S A* **105**, 9598–9603.

12. I. H. Shrivastava, J. Jiang, S. G. Amara and I. Bahar. 2008. Time-resolved mechanism of extracellular gate opening and substrate binding in a glutamate transporter. *J Biol Chem* **283**, 28680–28690.

13. J. Li and E. Tajkhorshid. 2009. Ion-releasing state of a secondary membrane transporter. *Biophys J* **97**, L29–L31.

14. F. Khalili-Araghi, J. Gumbart, P. C. Wen, M. Sotomayor, E. Tajkhorshid and K. Schulten. 2009. Molecular dynamics simulations of membrane channels and transporters. *Curr Opin Struct Biol* **19**, 128–137.

15. J. C. Phillips, R. Braun, W. Wang, J. Gumbart, E. Tajkhorshid, E. Villa, C. Chipot, R. D. Skeel, L. Kale and K. Schulten. 2005. Scalable molecular dynamics with NAMD. *J Comp Chem* **26**, 1781–1802.

16. I. B. Holland, S. P. Cole, K. Kuchler and C. F. Higgins, *ABC Proteins: From Bacteria to Man*. Academic Press, London, 2003.

17. M. L. Oldham, A. L. Davidson and J. Chen. 2008. Structural insights into ABC transporter mechanism. *Curr Opin Struct Biol* **18**, 726–733.

18. K. P. Locher. 2009. Structure and mechanism of ATP-binding cassette transporters. *Phil Trans R Soc Lond B* **364**, 239–245.

19. V. Kos and R. C. Ford. 2009. The ATP-binding cassette family: A structural perspective. *Cell Mol Life Sci* **66**, 3111–3126.

20. A. Moussatova, C. Kandt, M. L. O'Mara and D. P. Tieleman. 2008. ATP-binding cassette transporters in *Escherichia coli*. *Biochim Biophys Acta* **1778**, 1757–1771.

21. J. Chen, G. Lu, J. Lin, A. L. Davidson and F. A. Quiocho. 2003. A tweezers-like motion of the ATP-binding cassette dimer in an ABC transport cycle. *Mol Cell* **12**, 651–661.

22. P. M. Jones and A. M. George. 2007. Nucleotide-dependent allostery within the ABC transporter ATP-binding cassette. *J Biol Chem* **282**, 22793–22803.

23. P. M. Jones and A. M. George. 2009. Opening of the ADP-bound active site in the ABC transporter ATPase dimer: Evidence for a constant contact, alternating sites model for the catalytic cycle. *Proteins: Struct, Func, Bioinf* **75**, 387–396.

24. A. S. F. Oliveira, A. M. Baptista and C. M. Soares. 2010. Insights into the molecular mechanism of an ABC transporter: Conformational changes in the NBD dimer of MJ0796. *J Phys Chem B* **114**, 5486–5496.

25. K. Hollenstein, R. J. Dawson and K. P. Locher. 2007. Structure and mechanism of ABC transporter proteins. *Curr Opin Struct Biol* **17**, 412–418.

26. R. J. Dawson, K. Hollenstein and K. P. Locher. 2007. Uptake or extrusion: Crystal structures of full ABC transporters suggest a common mechanism. *Mol Microbiol* **65**, 250–257.

27. P.-C. Wen and Tajkhorshid. 2011. Conformational coupling of the nucleotide-binding and the transmembrane domains in the maltose ABC transporter. Submitted.

28. M. L. Oldham, D. Khare, F. A. Quiocho, A. L. Davidson and J. Chen. 2007. Crystal structure of a catalytic intermediate of the maltose transporter. *Nature* **450**, 515–521.

29. D. J. Slotboom, W. N. Konings and J. S. Lolkema. 1999. Structural features of the glutamate transporter family. *Microbiol Mol Biol Rev* **63**, 293–307.

30. D. Yernool, O. Boudker, Y. Jin and E. Gouaux. 2004. Structure of a glutamate transporter homologue from *Pyrococcus horikoshii*. *Nature* **431**, 811–818.

31. O. Boudker, R. M. Ryan, D. Yernool, K. Shimamoto and E. Gouaux. 2007. Coupling substrate and ion binding to extracellular gate of a sodium-dependent aspartate transporter. *Nature* **445**, 387–393.

32. N. Reyes, C. Ginter and O. Boudker. 2009. Transport mechanism of a bacterial homologue of glutamate transporters. *Nature* **462**, 880–885.

33. M. P. Kavanaugh, A. Bendahan, N. Zerangue, Y. Zhang and B. I. Kanner. 1997. Mutation of an amino acid residue influencing potassium coupling in the glutamate transporter GLT-1 induces obligate exchange. *J Biol Chem* **272**, 1703–1707.

34. A. Bendahan, A. Armon, N. Madani, M. P. Kavanaugh and B. I. Kanner. 2000. Arginine 447 plays a pivotal role in substrate interactions in a neuronal glutamate transporter. *J Biol Chem* **275**, 37436–37442.

35. Z. Tao, Z. Zhang and C. Grewer. 2006. Neutralization of the aspartic acid residue Asp-367, but not Asp-454, inhibits binding of Na^+ to the glutamate-free form and cycling of the glutamate transporter EAAC1. *J Biol Chem* **281**, 10263–10272.

36. P. H. Larsson, A. V. Tzingounis, H. P. Koch and M. P. Kavanaugh. 2004. Fluorometric measurements of conformational changes in glutamate transporters. *Proc Natl Acad Sci U S A* **101**, 3951–3956.

37. D. E. Bergles, A. V. Tzingounis and C. E. Jahr. 2002. Comparison of coupled and uncoupled currents during glutamate uptake by GLT-1 transporters. *J Neurosci* **22**, 10153–10162.

38. H. P. Koch, J. M. Hubbard and H. P. Larsson. 2007. Voltage-independent sodium-binding events reported by the 4B-4C loop in the human glutamate transporter excitatory amino acid transporter 3. *J Biol Chem* **282**, 24547–24553.

39. S. Qu and B. I. Kanner. 2008. Substrates and non-transportable analogues induce structural rearrangements at the extracellular entrance of the glial glutamate transporter GLT-1/EAAT2. *Proc Natl Acad Sci U S A* **283**, 26391–26400.

40. Z. Zhang, Z. Tao, A. Gameiro, S. Barcelona, S. Braams and T. Rauen. 2007. Transport direction determines the kinetics of substrate transport by the glutamate transporter EAAC1. *Proc Natl Acad Sci U S A* **104**, 18025–18030.

41. Y. Huang, M. J. Lemieux, J. Song, M. Auer and D. N. Wang. 2003. Structure and mechanism of the glycerol-3-phosphate transporter from *Escherichia coli*. *Science* **301**, 616–620.

42. A. Salasburgos, P. Iserovich, F. Zuniga, J. Vera and J. Fischbarg. 2004. Predicting the three-dimensional structure of the human facilitative glucose transporter Glut1 by a novel evolutionary homology strategy: Insights on the molecular mechanism of substrate migration and binding sites for glucose and inhibitory molecules. *Biophys J* **87**, 2990–2999.

43. M. J. Lemieux, Y. Huang and D. N. Wang. 2004. Glycerol-3-phosphate transporter of *Escherichia coli*: Structure, function and regulation. *Res Microbiol* **155**, 623–629.

44. J. Holyoake, V. Caulfeild, S. Baldwin and M. Sansom. 2006. Modeling, docking, and simulation of the major facilitator superfamily. *Biophys J* **91**, L84–L86.

45. M. J. Lemieux. 2007. Eukaryotic major facilitator superfamily transporter modeling based on the prokaryotic GlpT crystal structure (review). *Mol Membr Biol* **24**, 333–341.

46. S. S. Pao, I. T. Paulsen and M. H. Saier, Jr. 1998. Major facilitator superfamily. *Microbiol Mol Biol Rev* **62**, 1–34.

47. C. J. Law, P. C. Maloney and D. N. Wang. 2008. Ins and outs of major facilitator superfamily antiporters. *Annu Rev Microbiol* **62**, 289–305.

48. M. Lemieux, Y. Huang and D. Wang. 2004. The structural basis of substrate translocation by the glycerol-3-phosphate transporter: A member of the major facilitator superfamily. *Curr Opin Struct Biol* **14**, 405–412.

49. C. J. Law, J. Almqvist, A. Bernstein, R. M. Goetz, Y. Huang, C. Soudant, A. Laaksonen, S. Hovmlle and D. N. Wang. 2008. Salt-bridge dynamics control substrate-induced conformational change in the membrane transporter glpt. *J Mol Biol* **378**, 828–839.

50. M. C. Fann, A. H. Davies, A. Varadhachary, T. Kuroda, C. Sevier, T. Tsuchiya and P. C. Maloney. 1998. Identification of two essential arginine residues in UhpT, the sugar phosphate antiporter of *Escherichia coli*. *J Membr Biol* **164**, 187–195.

51. R. M. Stroud. 2007. Transmembrane transporters: An open and closed case. *Proc Natl Acad Sci U S A* **104**, 1445–1446.

52. M. J. Lemieux, Y. Huang and D. N. Wang. 2005. Crystal structure and mechanism of GlpT, the glycerol-3-phosphate transporter from *E. coli*. *J Electron Microsc* **54**, i43–i46.

53. R. S. G. D'rozario and M. S. P. Sansom. 2008. Helix dynamics in a membrane transport protein: comparative simulations of the glycerol-3-phosphate transporter and its constituent helices. *Mol Membr Biol* **25**, 571–573.

54. I. F. Tsigelny, J. Greenberg, V. Kouznetsova and S. K. Nigam. 2008. Modelling of glycerol-3-phosphate transporter suggests a potential 'tilt' mechanism involved in its function. *J Bioinformatics Comput Biol* **6**, 885–904.

55. C. J. Law, Q. Yang, C. Soudant, P. C. Maloney and D. N. Wang. 2007. Kinetic evidence is consistent with the rocker-switch mechanism of membrane transport by GlpT. *Biochemistry* **46**, 12190–12197.

56. G. Enkavi and E. Tajkhorshid. 2010. Simulation of spontaneous substrate binding revealing the binding pathway and mechanism and initial conformational response of GlpT. *Biochemistry* **49**, 1105–1114.

57. B. I. Kanner and E. Zomot. 2008. Sodium-coupled neurotransmitter transporters. *Chem Rev* **108**, 1654–1668.

58. A. Yamashita, S. K. Singh, T. Kawate, Y. Jin and E. Gouaux. 2005. Crystal structure of a bacterial homologue of Na$^+$/Cl$^-$-dependent neurotransmitter transporters. *Nature* **437**, 215–233.

59. A. Nyola, N. K. Karpowich, J. Zhen, J. Marden, M. E. Reith and D. N. Wang. 2010. Substrate and drug binding sites in LeuT. *Curr Opin Struct Biol* **20**, 415–422.

60. H. Krishnamurthy, C. L. Piscitelli and E. Gouaux. 2009. Unlocking the molecular secrets of sodium-coupled transporters. *Nature* **459**, 347–355.

61. L. R. Forrest, Y. W. Zhang, M. T. Jacobs, J. Gesmonde, L. Xie, B. H. Honig and G. Rudnick. 2008. Mechanism for alternating access in neurotransmitter transporters. *Proc Natl Acad Sci U S A* **105**, 10338–10343.

62. L. Shi, M. Quick, Y. Zhao, H. Weinstein and J. A. Javitch. 2008. The mechanism of a neurotransmitter: sodium symporter–inward release of Na⁺ and substrate is triggered by substrate in a second binding site. *Mol Cell* **30**, 667–677.

63. S. A. Shaikh and E. Tajkhorshid. 2010. Modeling and dynamics of the inward-facing state of a Na⁺/Cl⁻ dependent neurotransmitter transporter homologue. *PLoS Comput Biol* **6(8)**, e1000905.

64. Y. Zhao, M. Quick, L. Shi, E. L. Mehler, H. Weinstein and J. A. Javitch. 2010. Substrate-dependent proton antiport in neurotransmitter:sodium symporters. *Nat Chem Biol* **6**, 109–116.

65. S. Y. Noskov and B. Roux. 2008. Control of ion selectivity in LeuT: Two Na⁺ binding sites with two different mechanisms. *J Mol Biol* **377**, 804–818.

66. S. Y. Noskov. 2008. Molecular mechanism of substrate specificity in the bacterial neutral amino acid transporter LeuT. *Proteins* **73**, 851–863.

67. M. Quick, A.-M. L. Winther, L. Shi, P. Nissen, H. Weinstein and J. A. Javitch. 2009. Binding of an octyl-glucoside detergent molecule in the second substrate (S2) site of LeuT establishes an inhibitor-bound conformation. *Proc Natl Acad Sci U S A* **106**, 5563–5568.

68. Y. Zhao, D. Terry, L. Shi, H. Weinstein, S. C. Blanchard and J. A. Javitch. 2010. Single-molecule dynamics of gating in a neurotransmitter transporter homologue. *Nature* **465**, 188–193.

69. L. R. Forrest and G. Rudnick. 2009. The rocking bundle: A mechanism for ion-coupled solute flux by symmetrical transporters. *Physiology* **24**, 377–386.

70. S. Faham, A. Watanabe, G. M. Besserer, D. Cascio, A. Specht, B. A. Hirayama, E. M. Wright and J. Abramson. 2008. The crystal structure of a sodium galactose transporter reveals mechanistic insights into Na⁺/sugar symport. *Science* **321**, 810–814.

71. E. Wright and E. Turk. 2004. The sodium/glucose cotransport family slc5. *Pflgers Archiv European Journal of Physiology* **447**, 510–518.

72. S. Yamada, S. Pokutta, F. Drees, W. I. Weis and W. J. Nelson. 2005. Deconstructing the cadherin-catenin-actin complex. *Cell* **123**, 889–901.

73. M. Quick and H. Jung. 1998. A conserved aspartate residue, Asp187, is important for Na⁺-dependent proline binding and transport by the Na⁺/proline transporter of *Escherichia coli. Biochemistry* **37**, 3800–13806.

74. M. Quick, D. D. F. Loo and E. M. Wright. 2001. Neutralization of a conserved amino acid residue in the human Na⁺/glucose transporter (hSGLT1) generates a glucose-gated H1 channel. *J Biol Chem* **19**, 1728–1734.

75. M. Sujatha and P. V. Balaji. 2009. Identification of common structural features of binding sites in galactose-specific proteins. *Proteins* **55**, 44–65.

76. J. Abramson, I. Smirnova, V. Kasho, G. Verner, H. R. Kaback and S. Iwata. 2003. Structure and mechanism of the lactose permease of *Escherichia coli. Science* **301**, 610–615.

77. J. Li and E. Tajkhorshid. 2010. Spontaneous unbinding of the substrate in sodium-glucose transporter Submitted.

78. T. Shimamura, S. Weyand, O. Beckstein, N. G. Rutherford, J. M. Hadden, D. Sharples, M. S. P. Sansom, S. Iwata, P. J. F. Henderson and A. D. Cameron. 2010. Molecular Basis of Alternating Access Membrane Transport by the Sodium-Hydantoin Transporter Mhp1. *Science* **328**, 470–473.

79. K. Hollenstein, D. C. Frei and K. P. Locher. 2007. Structure of an ABC transporter in complex with its binding protein. *Nature* **446**, 213–216.

80. A. Ward, C. L. Reyes, J. Yu, C. B. Roth and G. Chang. 2007. Flexibility in the ABC transporter MsbA: Alternating access with a twist. *Proc Natl Acad Sci U S A* **104**, 19005–19010.

81. S. G. Aller, J. Yu, A. Ward, Y. Weng, S. Chittaboina, R. Zhuo, P. M. Harrell, Y. T. Trinh, Q. Zhang, I. L. Urbatsch and G. Chang. 2009. Structure of P-glycoprotein reveals a molecular basis for poly-specific drug binding. *Science* **323**, 1718–1722.

ABC Transporters

10

E. P. Coll and D. P. Tieleman

1. Introduction

In order to maintain homeostasis, the cell membrane is impermeable to nearly all biologically relevant molecules. To facilitate the flux of nutrients, toxins, secretions, and waste across the membrane, the cell expresses a number of pores, channels, and transporters.[1] Transport of a molecule against a concentration gradient requires an energy source, such as an ion gradient or the hydrolysis of an energy carrier such as ATP. ATP-binding cassette (ABC) transporters are a family of transport proteins powered by ATP hydrolysis. They are characterized by the presence of a highly conserved ATPase domain, the ABC.[2] ABC transporters are one of the largest families of transporters, and its different members transport a variety of substrates, from ions, to sugars, to amino acids and lipids.[3]

Members of the ABC family are present in all organisms, from bacteria, in which they were first discovered, to archaea and eukarya. Transport proteins containing the ABC motif comprise nearly 5% of the *E. coli* genome,[4] 48 are encoded in the human genome,[5] and over 100 are present in the *Arabidopsis* genome.[6]

The ATPase motif contained in ABC transporters is similar to that in other nucleotide-binding proteins such as the F1-ATPase and the RecA DNA-binding protein, suggesting that evolution has adapted this motif to serve a number of physiological needs. ABC transporters are integral membrane proteins composed of a membrane-spanning channel formed by two transmembrane domains (TMDs), coupled to two nucleotide-binding domains (NBDs) containing the conserved sequences of the ABC motif. While the NBD sequence remains highly conserved across all organisms, the sequence of the TMDs varies significantly.[7]

Mutations in the sequence of eukaryotic ABC transporters are responsible for a number of hereditary diseases such as cystic fibrosis, Tangier disease, and adrenoleukodystrophy.[8,9] Other eukaryotic transporters are responsible for the export of hydrophobic molecules, including drugs used in chemotherapy treatment of cancer. Expression of these transporters in cancer cells results in chemotherapy resistance. In bacteria, ABC exporters operate on toxic substances including antibiotics, leading to antibiotic resistance.[10] For these reasons, the study of these transporters is of interest with respect to human health.

2. Structure of ABC Transporters

The details of ABC transporter structure vary between different members of the family. A transporter is generally composed of two nucleotide-binding domains and two membrane-spanning domains. The NBDs exist as a dimer on the cytosolic side of the membrane and bind two molecules of ATP at the dimer interface. Depending on the structure of the TMDs, the NBDs may be flush with the membrane, or at a distance of several nanometers. The TMD dimer forms a membrane-spanning channel, composed of bundles of helices, and is coupled to the NBDs at the cytoplasmic side of the bilayer. The TMDs are able to undergo a conformational change in order to bind or release the substrate on either side of the membrane.[3]

Several hundred genes encoding ABC transporters have been identified but high-resolution structural data on ABC transporters are relatively recent. Beginning in the 1990s, X-ray crystallography and NMR have been used to solve the structures of a growing number of soluble NBD domains. In the past 8 years, a growing number of full-length structures has been solved by X-ray crystallography, providing a major impetus to functional studies that now can be placed in a detailed structural context. Figure 1 shows a number of known crystal structures.

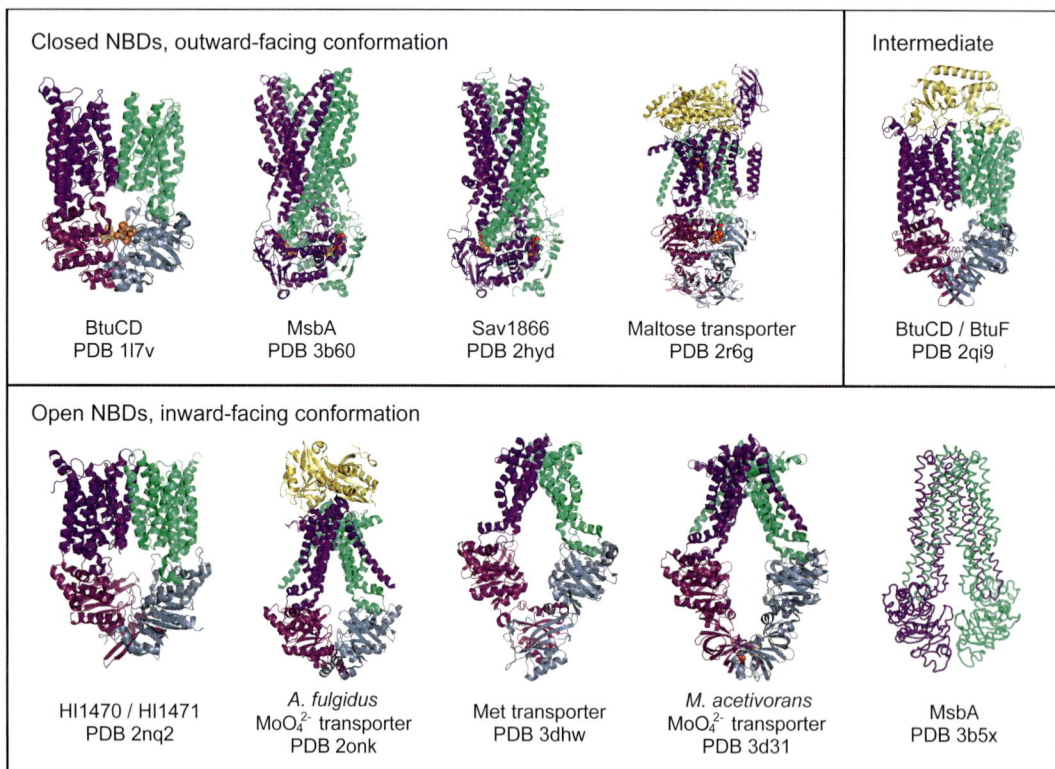

Figure 1. Gallery of crystal structures of all currently crystallized full-length transporters, except P-gp, with their Protein Data Bank identifiers. In same cases multiple crystal structures are available for the same protein, e.g. with different nucleotide analogs. Different colors indicate different polypeptide chains. Periplasmic binding proteins are shown in brick red. MsbA and Sav1866 are homodimers, with NBDs attached to the TM domain, while the others have separate TM domains. The structures are sorted by overall conformation, either outward-facing or inward-facing, with BtuCD/BtuF in what appears to be an intermediate conformation. (Figure adapted with permission from Procko *et al.*)[2]

All known ABC transporter structures can be classified into three types: Exporters, which bind substrates from the cytoplasm and transfer them across a membrane, such as the cell membrane, endoplasmic reticulum, or peroxisomal membrane, and importers. Exporters are expressed in bacteria, archaea, and eukaryotes, while importers are only found in bacteria and archaea.[11] Cells in higher eukaryotes do not require importers, whose purpose is to scavenge nutrients from the extracellular space. ABC importers can be divided into two types, Type I and Type II, based on the size of the transmembrane domains.[1] A common feature of both types of importers is a binding protein that exists in solution outside the cell membrane, in the periplasmic space between the cell membrane and outer membrane or in the form of a solute binding protein in gram-positive bateria. This periplasmic or solute binding protein (PBP) has a high affinity for the substrate of the transporter, and its role is to bind the substrate from the periplasm and return to the transporter, where the substrate can be imported. PBPs have been identified that bind sugars, small ions, and large metal-chelating molecules, and each PBP appears to be specific to one transporter.[3,12,13]

The subunits that comprise a transporter may be encoded in the genome in several different ways (Figure 2). In the case of importers, each NBD and TMD is encoded separately, and the full transporter must be assembled from four distinct proteins. Some importers are homodimeric,

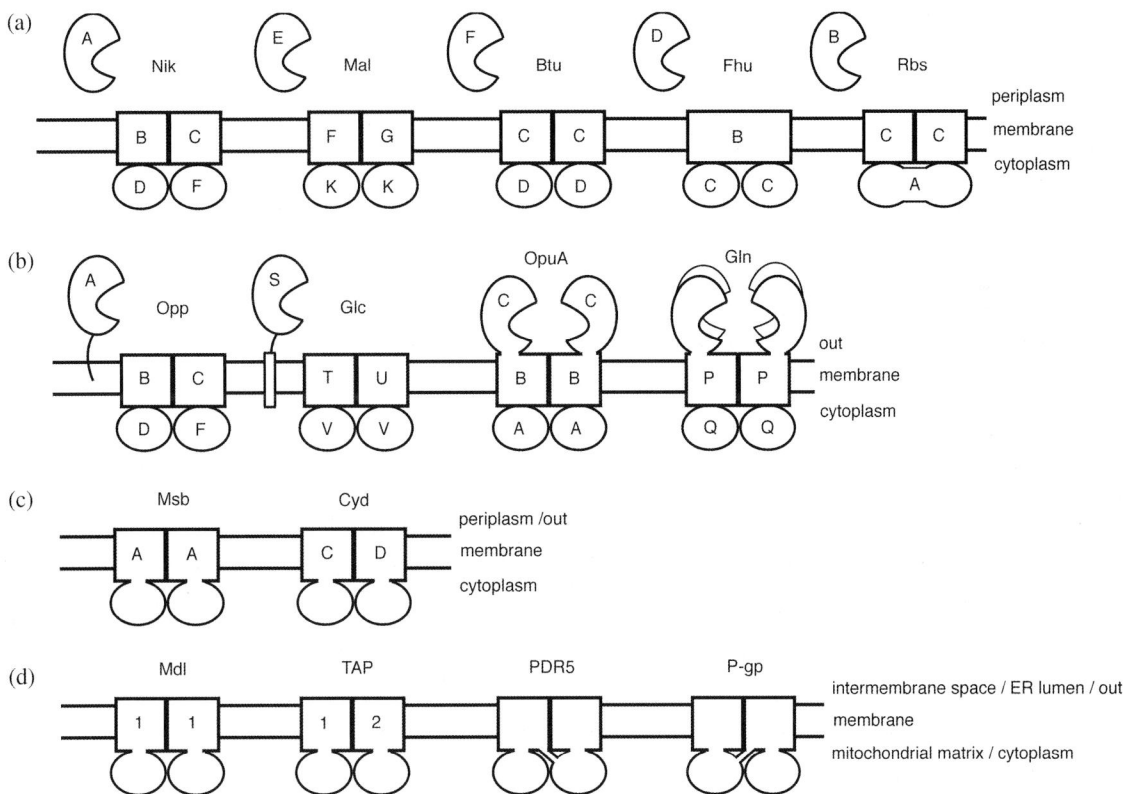

Figure 2. An overview of the diversity of domain arrangements for a number of ABC transporters. The protein names are formed by adding the character inside a shape to the overall name, e.g. OppA is the substrate binding protein of the Opp system. (a) shows importers from Gram-negative bacteria, (b) importers from Gram-positive bacteria, (c) bacterial exporters, and (d) eukaryotic ABC transporters. (Figure adapted with permission from Biemans-Oldehinkel *et al.*)[7]

formed by two identical TMDs and NBDs; other importers are asymmetric, with the TMDs, NBDs, or both being heterodimeric. In exporters, the NBD and TMD are fused into one polypeptide chain, called a half transporter, which dimerizes to form a whole transporter. Like importers, exporters may be homo- or heterodimeric. In eukaryotes, all four domains may be also be encoded in the same polypeptide chain as a full transporter,[7] as a homodimeric half-transporter or a heterodimeric transporter. The resulting arrangement often includes asymmetries between the two halves, which in some cases have significant effects on protein function.

2.1 *NBD structure*

The core sequence of the NBDs varies little between the different types of ABC transporter, and indeed between different species. Some transporters incorporate additional regulatory domains fused to the NBDs, but the ATPase section remains conserved among all members of the ABC family. Two ATP binding sites are formed at the interface between the dimer, sandwiched between the two NBDs.

The sequence of an NBD is several hundred amino acids long, with the conserved motifs spread throughout. Structurally, the NBD folds into two lobes of approximately equal length, connected by a flexible linker region (Figure 3). The first lobe is similar in structure and sequence to the ATPase domain of the DNA-binding protein RecA, as well as other nucleotide-binding proteins such as F1-ATPase, and is called the RecA-like domain.[14] This similarity points to a common heritage, both proteins deriving from the same nucleotide-binding motif, but each having been adapted to different purposes. The smaller second lobe is helical in structure and contains the sequence LSGGQ, a motif that is the signature of the ABC family.

ATP is composed of an adenine ring, ribose sugar, and three phosphate groups, named α, β, and γ from proximal to most distal to the ribose group. The mechanism of ATP hydrolysis in ABC transporters is similar to that in other members of the family of nucleotide-binding proteins. The conserved Walker A and B motifs coordinate the phosphate groups, while a negative residue following the Walker B motif binds the magnesium ion that is coordinated to the β and γ phosphates. Another residue must act as a general base at the active side to polarize an attacking water molecule to break the phosphate bond.[14] In addition to these conserved residues, certain sequences that engage in ATP binding have consensus sequences that are specific to the ABC family.[15]

In the NBD sequence, the first conserved residue is the A-loop, an aromatic residue, often tyrosine, on the RecA-like subdomain that forms a stacking interaction with the adenine ring. C-terminal to this region is the Walker A motif. This motif, with consensus sequence GXXGXGK[S/T], where X is any amino acid, contacts the β and γ phosphate groups of bound ATP. The conserved lysine residue forms hydrogen bonds with the negatively charged phosphate oxygens. As this motif contacts the phosphate groups, it is also referred to as the P-loop.[16]

The Q-loop is a loop of approximately eight residues beginning with a conserved glutamine (Q) residue, located at the linker region between the two lobes of the NBD. The conserved glutamine is hypothesized to act as a general base during ATP hydrolysis, since conformational changes in this sequence switch the glutamine residue in and out of the active site.[17] Following the Q-loop, the NBD assumes a helical structure, forming the second lobe, which contains the sequence that is

(a)

(b)

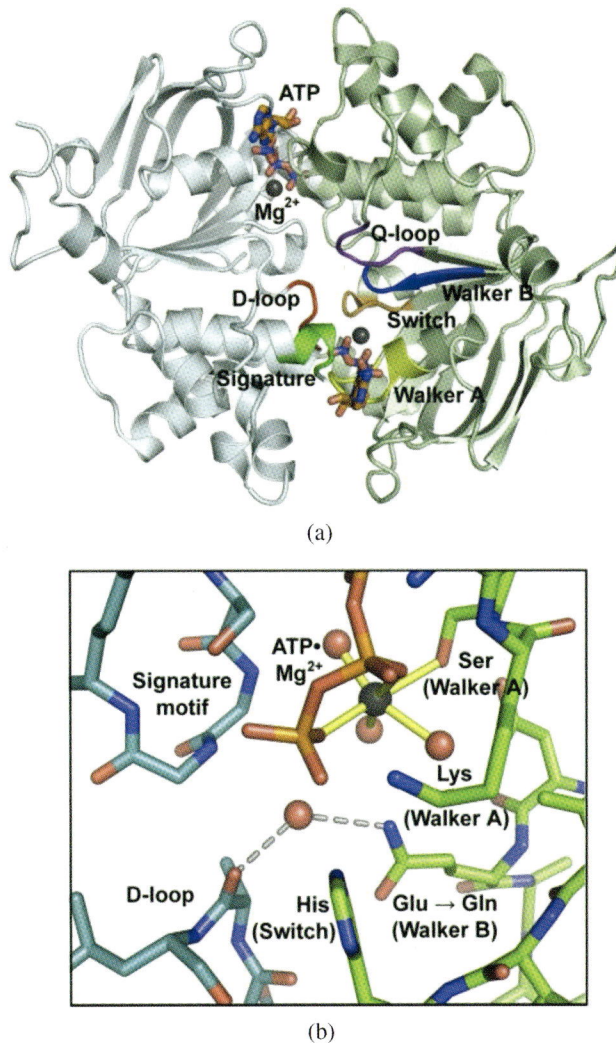

Figure 3. (a) Structure of the TAP1 homodimeric NBD with the main conserved motifs common to all NBDs high-lighted. (b) One of the two active sites with groups involved in catalysis highlighted. ATP binds across the interface of the two NBDs. Each NBD thus has two half binding sites, and a dimer is required for ATP binding. (Figure reproduced with permission from Procko *et al.*)[2]

specific to ABC transporters, LSGGQ. This is known as the signature motif. The amino groups on the backbone of the sequence interact with the phosphates of ATP, taking the place of a conserved arginine residue present in the related RecA-like family. The lack of side chains on the glycine residues allows a close approach of the backbone to ATP.[18] This sequence also acts as a "sensor" of ATP, communicating the presence or absence of bound nucleotide to the opposing lobe by conformational changes.

The sequence of the NBDs returns to the first lobe, the RecA-like subdomain, following the helical subdomain containing the signature motif, to form the Walker B motif. The consensus sequence of the Walker B motif is $\Phi\Phi\Phi\Phi DE$, where Φ is any aliphatic residue.[16] The aspartic acid residue stabilizes the positively charged magnesium ion that coordinates ATP. It is also possible that the glutamic acid residue is the one that acts as a general base to polarize the catalytic water molecule.

C-terminal to the Walker B motif is the "D-loop", a four-residue motif with consensus sequence SALD. This sequence is thought to play a role in communication between the two ATP-binding sites of the ATP-bound dimer.[2] The last conserved residue is a histidine near the C-terminus, in the "H-loop". This residue interacts with the γ phosphate of ATP, and also with the aspartic acid residue of the D-loop. The H-loop is thought to regulate ATPase activity by "switching" in and out of the active site.

In a whole transporter, the NBDs are arranged in a head-to-tail orientation, such that each ATP binding site is formed from the RecA-like subdomain of one NBD and the helical subdomain of the opposite NBD. This conformation sandwiches two ATP molecules between the NBDs, and ensures tight dimerization when ATP is bound.[19] This orientation allows the conserved residues on the RecA-like subdomain of one NBD and the LSGGQ motif of the second NBD to sandwich one ATP molecule between them, and for the complete dimer to bind two ATP molecules. The association of the NBDs around ATP, and subsequent dissociation upon ATP hydrolysis, provides the "power stroke" driving the rearrangement of the protein during transport. The movement of the NBDs during these events is transmitted to the TMDs by a helix in the TMD structure that inserts into a cavity in the NBDs. This coupling helix is a common element of all ABC transporters, and provides a mechanism for the binding and release of ATP to drive the conformational changes that are required for substrate translocation.[3]

2.2 *TMD structure*

The transmembrane section of ABC transporters is responsible for the formation of a translocation pathway to allow movement of substrates across the bilayer, governed by the state of the NBDs and the substrate binding protein (for importers). The three different subtypes of transporter achieve this function in several different ways, and while the sequences vary significantly, all TMDs are composed of bundles of helices.

2.2.1 *Importers*

Importers are only expressed in bacteria or archaea with an outer membrane, as they scavenge required nutrients such as ions, vitamins, and sugars from the periplasmic space.[3] The structures of several ABC importers have been solved, and can be classified into two groups. As in the case of exporters, a transporter is composed of two NBDs and two TMDs. However, these subunits are all encoded on separate polypeptide chains, requiring the association of four proteins to form a whole transporter. Importers are also dependent on a soluble protein to bind the substrate and deliver it to the transporter. These periplasmic binding proteins (PBP) or more generally solute binding proteins (SBP) are formed of two lobes connected by one, two, or three flexible helices that allow the proteins to open and close around their substrates. The PBPs have high affinity for their substrates in solution. Upon binding to the periplasmic side of the transporter, the PBP spreads its lobes, losing its affinity for the substrate and delivering it to the TMDs of the transporter. It was initially believed that PBPs operated via a "Venus fly-trap" mechanism, remaining open until the substrate associated with the PBP, resulting in the PBP snapping shut around the substrate.[20] However, it seems likely that these proteins exist in an equilibrium between open and closed states both in the absence and presence of substrate. One indication for this are computer simulations of the vitamin B12 binding protein BtuF that suggest PBPs exist in equilibrium between open and closed states, until the substrate is bound.[21]

2.2.2 *Type I importers*

The molybdate transporter ModB was the first protein in the ABC family to have its 3D crystal structure solved.[22] Several other proteins were later found to have a similar architecture, possessing a transmembrane channel composed of bundles of helices. The helices of importers are significantly shorter than those of exporters, so that, unlike in exporters, the NBDs are located at the edge of the cell membrane. The TMDs of importers may be homodimers, as in the case of ModB, or heterodimers, for example the maltose transporter, where the transmembrane channel is formed from the proteins MalG and MalF. The exact number of helices comprising each subunit varies between members of this family. The ModB TMD consists of six transmembrane helices, the methionine transporter TMD MetI has only five helices per subunit, whereas the proteins that compose the maltose transporter, MalF and MalG, have eight and six helices respectively. Some Type I importers exhibit a swapping of helices, where helices encoded in one TMD associate structurally with the opposite TMD subunit, resulting in a closer entanglement of TMDs.[12]

Structural and biochemical studies of these transporters suggest that they exist in an outward-facing state when ATP is bound to the NBDs, causing tight dimerization of the NBDs and draws the cytoplasmic side of the TMDs together (see also Figure 1). This exposes the transmembrane channel to the periplasm, allowing the PBP to release its substrate into the channel. The outward-facing state forms a binding pocket with high affinity for the substrate, inducing the release of the substrate from the PBP.[23] Structures of ATP-free proteins show an inward-facing conformation, from which the substrate would be released into the cytoplasm. This inward-facing state of the transmembrane channel may have a lower affinity for the substrate than does the binding pocket formed by the outward-facing state, resulting in the ejection of the substrate into the cytoplasm. The structure of the TMDs allows the translocation pathway to be closed to the periplasm or to the cytoplasm, preventing leakage through the channel.[23]

2.2.3 *Type II importers*

The resolution of the structure of BtuCD, the *E. coli* vitamin B12 transporter, identified a second type of ABC importer. Thus far, members of this group have been identified as transporters of chelated metal ions, such as cobalamin and heme.[24] The TMD arrangement of this type consists of 20 helices, 10 from each subunit. Transmembrane helices (TM) 3 to 5a of each subunit have been observed to take one of two conformations, either outward-facing or inward facing, while there is less evidence for large scale motions of the TMDs observed in the Type 1 importers. The resolution of a structure of BtuCD-F, the BtuCD transporter in complex with the PBP BtuF, showed that it is possible for each TMD subunit to take different conformations, in an asymmetric intermediate state following substrate translocation.[24] This structure, although lacking the transporter substrate vitamin B12, showed the PBP in its open state, suggesting that binding of the PBP to the TMDs forces the lobes of the PBP apart.[13] In contrast to Type I importers, no high-affinity binding site has been identified in the translocation pathway of Type II importers. It is therefore believed that the pathway simply provides a pore for diffusion of the substrate through the channel once released into the channel by the PBP, rather than forming a high-affinity binding pocket as in Type I.

The helices adjacent to the translocation pathway, namely the fifth helix, TM5, and an adjoined short helix, TM5a, have significantly different conformations in each of the known protein structures. The BtuCD structure reveals an outward-facing conformation, while the structure of another Type II importer, HI1470/71, is inward facing. The complete BtuCD-F structure is an intermediate structure, with one subunit in an outward-facing state and one in an inward-facing state.[13] However, none of these structures were solved in the presence of ATP, so it is unclear which conformation corresponds to the ATP-bound state. Spectroscopic data suggests that ATP binding results in the opening of the transporter to the cytoplasmic side. This would be directly opposite to the mechanism proposed for Type I transporters, where ATP-bound states are all outward-facing.

2.2.4 *Exporters*

The structure of ABC exporters is of particular interest because they are found in all organisms, from bacteria to humans, and homology exists between some transporters in all organisms, as opposed to ABC importers, which are expressed only in prokaryotes, and whose transmembrane domains bear little similarity to the exporters present in eukaryotes. All known exporters have a transmembrane structure with a common architecture, usually consisting of two TMDs of six helices each. These helices cross back and forth across the membrane, and possess intracellular loops that extend beyond the edge of the cell membrane and into the cytoplasm, with the result that the NBDs in exporters are spaced approximately 25 Å from the membrane.[25] Structurally, exporters are composed of two NBDs and two TMDs, although additional auxiliary domains may be present. These domains may be encoded on a single polypeptide chain, as in the human transporter P-glycoprotein, or two chains, each comprising one TMD and one NBD. Those that are composed of two separate chains may be homo- or heterodimers. The bacterial ABC transporter Sav1866 is a homodimer, composed of two identical chains, each encoding one TMD fused to an NBD. The human TAP transporter, responsible for antigen processing, is composed of two separate proteins, TAP1 and TAP2, which are significantly different in the sequence of the conserved motifs, although the overall structure remains similar to the exporter template.[2]

In contrast to importers, the TMDs are not simply present beside one another, but twisted together tightly. Two structural subdomains are formed, composed of six helices each in a wing-like arrangement. Unusually, a domain-switching exchange of helices occurs, wherein two helices from each TMD form a part of the opposing wing, with the result that the wings are composed of four helices from one TMD, and two from the opposite TMD. To accommodate this arrangement, the TMD monomers are rotationally symmetrical about an axis normal to the membrane, similar to the head-to-tail orientation of the NBDs.[25]

Crystallographic data have shown that the TMDs of exporters are capable of significant motion. The "wings" of the TMDs have been observed in two different conformations, one open towards the inside of the cell, exposing a large cavity to receive substrates from the cytoplasm. The large size of the binding cavity, and its deepness into the membrane, also allows the possibility of binding substrates, such as lipids, from the inner leaflet of the membrane. Conversely, the opposite conformation, with the NBDs dimerized around ATP and the wings tightly associated at the cytoplasmic side, reveals an outward-facing pocket with low hydrophobicity, which is thought to represent an extrusion pocket, intended to release substrate.[25]

Studies of polymorphisms in exporter structure have revealed that a variety of amino acid mutations at different locations in the TMDs have a significant effect on substrate specificity. This suggests that there exist a number of binding sites for different substrates at different sites in the binding cavity.[26] Overall, although the direction of transport is reversed, ABC exporters are thought to share a similar mechanism with ABC importers, switching between an outward and inward-facing state depending on ATP binding.

3. Transport Cycle

Although the exact sequence of steps in the transport cycle are not known, Figure 4 shows hypothetical mechanisms for importers and exporters that are consistent with most currently available data. Figure 5 zooms in on the interface between the NBDs and TMDs. In the absence of bound ATP, the NBDs exist in a dissociated state. A gap exists between the two halves of each ATP-binding site, which become accessible to water.[2] The state of the NBDs is communicated to the

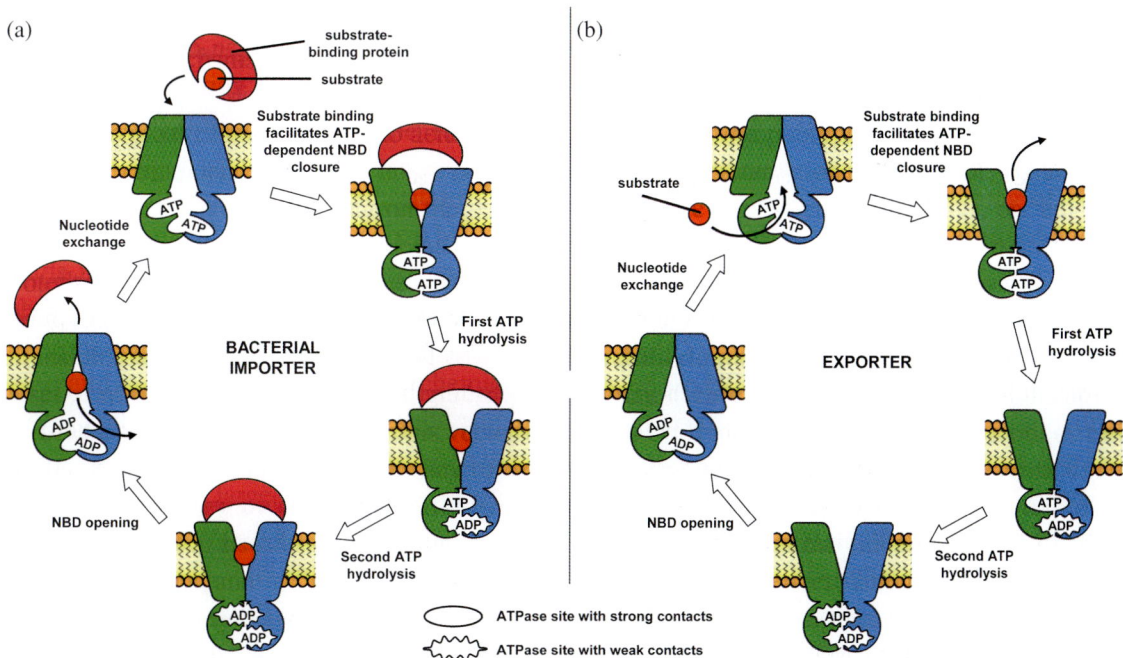

Figure 4. (a) General mechanism for bacterial importers, with two symmetric ATP binding sites and a role for the periplasmic binding protein in the transport cycle. (b) General mechanism for exporters, which do not require a periplasmic binding protein and in bacteria thus far have symmetric ATP-binding sites. (Figure adapted with permission from Procko *et al.*)[2]

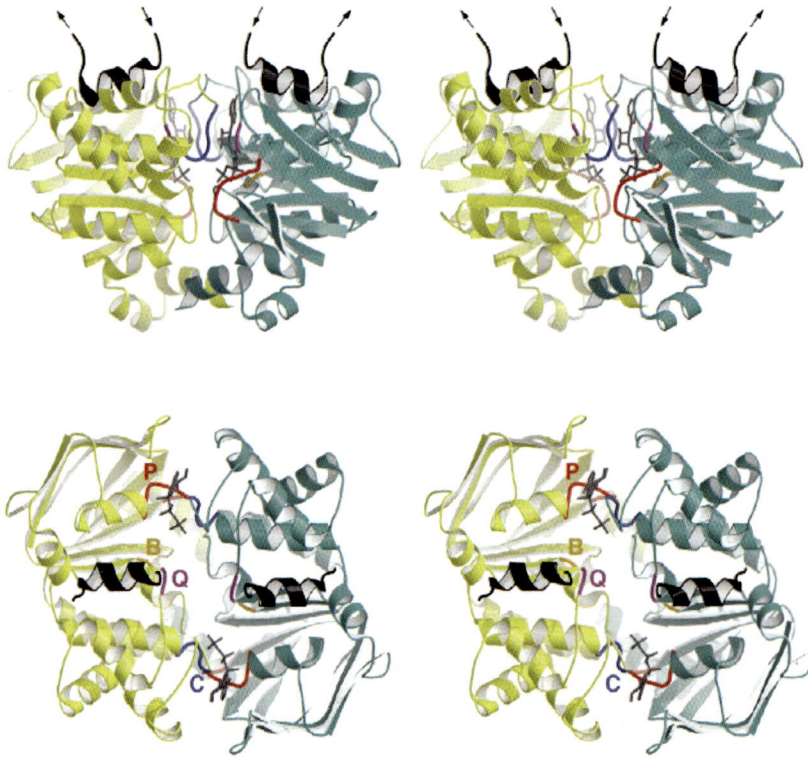

Figure 5.   The NBDs of the multidrug ABC transporter Sav1866, crystallized with bound AMP-PNP, are shown in stereo in a view from the side (top panel) or from the membrane (showing the surface facing the TMDs, bottom panel). The two NBDs are in green and yellow, with mechanistically important sequence motifs colored and labeled with single letters: P (P-loop), B (Walker-B motif), Q (Q-loop), C (LSGGQ motif or C-loop) (see also Fig. 3). Bound AMP-PNP is shown as black sticks. The short, black helices are the coupling helices from the TMDs, with arrows indicating the direction of the polypeptide chain from the N-terminus to the C-terminus. (Figure reproduced with permission from Hollenstein *et al.*)[3]

TMDs by the shared interface between domains such as the Q-loop, and in particular the coupling helices. The criss-crossing of helices back and forth across the membrane results in a number of intracellular loops formed between each membrane-spanning helix, and at least one of these connecting loops takes the form of a short helix that associates with the NBDs. This coupling helix is observed in importers as well as exporters, however the sequence of the helix is not conserved. In exporters, two loops act as coupling helices, one of which is domain-switched, in that the coupling helix of the first subunit interacts with the NBD of the second subunit, whereas in importers there is no such interaction between opposite halves of the transporter. In all transporters, the helix inserts into a groove between the two subdomains of the NBD, and as a result of these contacts between domains, the TMDs open at the cytoplasmic side when the NBDs are dissociated.[3] In the case of exporters, this exposes the substrate binding pocket to the cytoplasm, whereas in importers, this state presents a closed translocation pathway to the outside of the membrane.

In exporters, the binding of substrate to the binding pocket is thought to allow ATP binding at the NBDs, resulting in their dimerization. This motion triggers a rearrangement of the TMDs, switching to an outward-facing conformation, and the release of the substrate from the low-affinity extrusion pocket at the extracytoplasmic side of the membrane. Crystallographic results tentatively support this model, as crystal structures of an ABC exporter have been solved in an outward-facing state, as well as an ATP-free, inward-facing state. The outward-facing "closed" state also lacks bound ATP, and as a result the NBDs do not form complete ATP-binding pockets, rather the domains have "slid" along one another, such that the Walker A motifs of both NBDs are aligned.[19] The inward-facing structure shows a large separation between the NBDs, and consequently one of the coupling helices from each TMD is not in contact with its opposing NBD. In addition to linear separation of the NBDs, the TMDs are twisted by approximately 30 degrees. If accurate, these data suggest that the TMDs must undergo a large rotation about the extracellular "hinge" located at the point of the domain-swapped helix in order for the NBDs to bind ATP.[27]

Importers depend on PBPs to deliver substrate to the extracellular face of the transporter. One model proposes that the transporter exists in an ATP-free state until the loaded PBP binds to the NBDs. The NBDs then communicate to the TMDs that substrate is present, and ATP is bound, accompanied by NBD dimerization.[28] This triggers the closing of the TMDs at the cytoplasmic side, and the formation of a binding pocket at the extracytoplasmic side, allowing the binding protein to release its substrate into the TMDs. A binding pocket has been observed in type I transporters with affinity specific to the substrate, as in the maltose transporter. Type II importers, however, do not possess any sequences resembling binding motifs for their substrates, which has led to the hypothesis that these transporters use a "Teflon" model where the translocation pathway is inert, only providing a pathway through the membrane, with low binding affinity for the substrate. This model requires that the binding protein lose its affinity for its substrate in order for the substrate to enter this low-affinity translocation pathway. Biochemical data suggest that this is indeed the case, and that association of the PBP with the TMDs results in a loss of substrate affinity in the PBP.[29]

Following the release of the substrate into the translocation pathway, ATP hydrolysis occurs, allowing NBD dissociation, and a transition of the TMDs from their outward-facing state to an inward-facing pathway, resulting in the release of substrate into the cytoplasm and a resetting of the transporter to its initial state, ready to engage in another round of transport. In the case of exporters, ATP hydrolysis flips the TMDs back to the inward-facing state, ready to bind more substrate from the cytoplasm or from the membrane.[28]

This model depends on precise communication between domains to regulate when ATP binding and hydrolysis is to occur. This process may not be perfect, as *in vitro* systems show an inconsequential level of ATP hydrolysis in the absence of substrate. This suggests that while the binding of substrate stimulates ATP binding and hydrolysis, some ATPase activity occurs regardless of the state of the TMDs.[29]

4. Methods of Study

ABC transporters and other membrane proteins can be studied by a wide range of methods that often give complementary information. Here we briefly review a number of structural approaches only because they promise to give the most direct information on the global mechanism of ABC transporters, but there is a large and important body of literature on other aspects of ABC transporter structure-function relationships, including substrate binding studies, spectroscopic measurements, and extensive mutagenesis to identify key residues in specific transporters.

X-ray crystallography has provided a number of high-resolution structures of ABC transporters and their components. In some cases, crystal structures have been resolved of the same transporter at different points in the putative transport cycle, however even with these images, reaction mechanisms must still be pieced together using data from multiple sources, including structures of different transporters, non-crystallographic experiments, and simulations. These structures have provided a wealth of information about the mechanism of ABC transporters, but the number of structures successfully crystallized is only a tiny fraction of the total catalogue of ABC transporters. In addition, due to the transient nature of some of the stages of the catalytic process, it is difficult to capture every stage in the transport cycle by crystallography alone. Spectroscopic methods such as electron paramagnetic resonance (EPR) allow the dynamics of proteins to be observed *in vivo*, in real time,[30] and benefit greatly from existing structures to interpret the results.

EPR spectroscopy exploits the properties of paramagnetic chemical species, nuclei with an unpaired electron, to generate a detectable resonance signal analogous to that obtained in NMR experiments. Since few biologically occurring molecules possess an unpaired electron, a suitable marker molecule must be introduced into a protein for it to be probed using EPR. The chemistry of cysteine residues is used to introduce such a marker, for example a nitroxide radical, into the protein. Mutagenesis can be used to delete all cysteine residues from a protein, and create cysteine residues at one or more points of interest in order to selectively label the protein.[31] Spectroscopic analysis can yield structural information about the mobility and solvent accessibility of a label, as well as the proximity of labeled sites to one another.[32] This technique, known as site-directed spin labeling, is now well established as a method of probing the structure and dynamics of membrane proteins.[13,33] Comprehensive cysteine mutagenesis, and recording of EPR spectra, at many sites throughout a protein can yield detailed information about the structure of the protein.[30] Spin labeling may be used, for example, to mark the cytoplasmic or periplasmic gates of an ABC transporter. Since the TMDs of ABC transporters are dimeric, this would result in two marked residues, one per subunit. The proximity of the marked residues to one another could then be detected, in addition to the mobility of individual residues. The dynamics of the gate throughout the transport cycle can be determined from the change in mobility of the markers, and the distance between them upon the addition of ATP, or an ATP analog used to trap the protein in an ATP-bound state. Low mobility implies a closed, tightly associated gate, whereas high mobility is associated with an open gate.[33]

Computer simulations of biological systems may also be used to study dynamic proteins, such as ABC transporters. Simulations can, in principle, be used to study transient states that are unlikely to be captured experimentally as well as transitions between states, which would be generally useful in studies of complex membrane proteins. Molecular dynamics (MD) simulations of proteins allow a protein to be studied at a high temporal and spatial resolution, from femtoseconds to microseconds at single-atom resolution. The increasing speed of computers continues to expand the capabilities of computer simulations, and produce results that can be more easily compared to experimental data.[34]

MD simulations rely on the availability of a high-resolution structure of a protein, from which a model biological system may be created, at an atomistic level of detail. A simulation

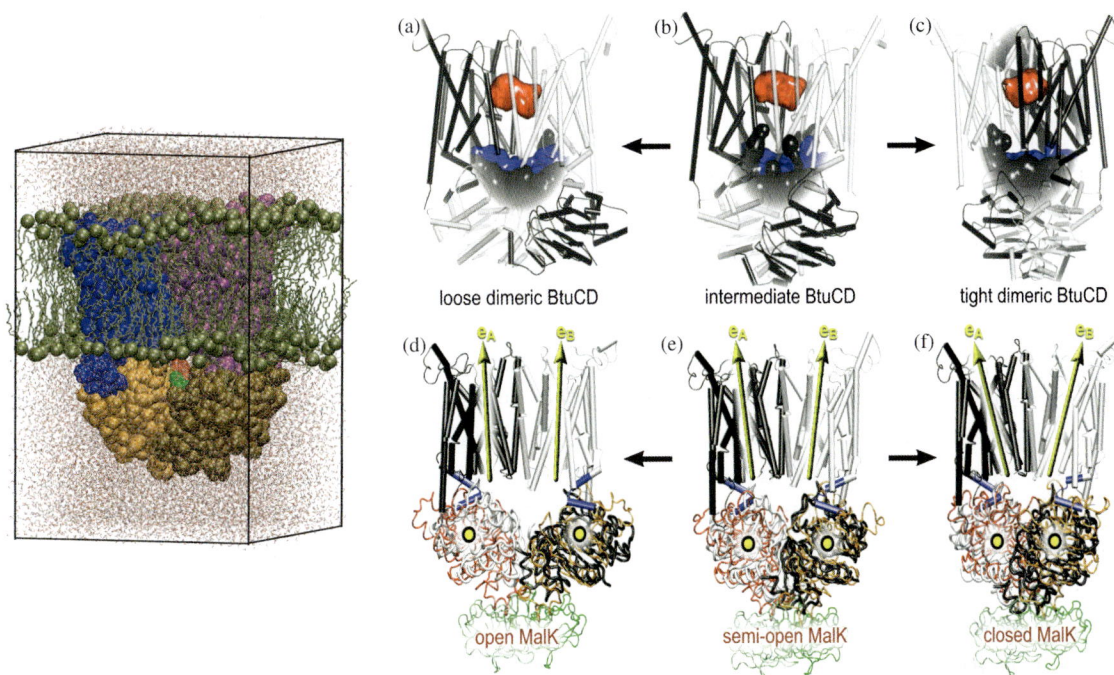

Figure 6. Left: A view of a computer model for MD simulations of the ATP-bound BtuCD protein, showing water (red and white), lipid with phosphorus atoms enlarged, and the protein. The transporter consists of two transmembrane domains (blue and purple) and two nucleotide-binding domains (orange and ochre). The two docked MgATP molecules are partially visible (green and red). (Figure reproduced with permission from Oloo *et al.*)[36] Right: A–F: simulation results that combine structural information from the NBD MalK with the crystal structure of BtuCD to investigate the response of the transmembrane domains to conformational changes in the NBD (e.g. induced by ATP binding). (Figure reproduced with permission from Sonne *et al.*)[38]

of an ABC transporter, for example, might consist of the protein complex embedded in a lipid bilayer, surrounded by water and ions, and other relevant molecules such as ATP or substrates. MD models the interactions between all atoms and integrates the classical equations of motions to generate a trajectory of all atoms over time. From this trajectory, thermodynamic and dynamic properties can be calculated.[35] Figure 6 shows an example of an MD simulation of BtuCD, embedded in a lipid bilayer with water, ions, and ATP. A modern high-performing computing system could simulate approximately 1 microsecond of the real-time motions of such a system in a few weeks, putting a limit on the length of dynamics that can be studied using atomistic MD simulations. Molecular dynamics may be especially useful in the study of ABC transporters, as these are a class of proteins that are known to undergo dramatic conformational changes that appear to depend entirely on which ligands are bound to the protein. Simulations of the components of ABC transporters in the presence and absence of bound ATP have reproduced some of the steps that occur during ATP binding and ADP release, at a level of detail that would be impossible to obtain through any experimental means. Various simulations have shown that the binding of two ATP molecules is necessary to form the tight NBD dimer, and that the hydrolysis of one or both ATP molecules will result in the dissociation of the dimer.[36,37]

Figures 6a–6f show results from a simulation that attempts to combine structural information on two distinct proteins — the NBD MalK from the maltose importer and the B12 importer BtuCD — into one mechanism by investigating the response of BtuCD to structural changes inferred from different MalK crystal structures.[38] As the library of crystal structure grows (Figure 1), this general approach may be useful to connect the individual snapshots in more complete mechanisms.

Despite the existence of good quality crystal structures for several ABC transporters, the details of the structure of the majority of ABC transporters remains unknown. However, due to the high degree of sequence similarity between members of the same subtype of ABC transporter, homology modeling can be used to generate a structure for a protein with no solved crystal structure. Homology modeling of a protein with no known structure, or a structure with missing parts, depends on the existence of a high-resolution structure of a homologous protein. If sufficient sequence similarity exists, the sequence of the unknown protein can be mapped onto the homologous structure to produce a likely three-dimensional model. While this method does not produce structures of the same degree of quality as an X-ray crystal structure, the resulting models can be very useful in the interpretation and design of experimental studies of ABC transporters.[39,40]

5. Conclusions

ABC transporters are molecular pumps capable of binding substrates from the cytoplasm or the lipid bilayer and exporting them from the cell, or working in the reverse direction as modular transport systems, receiving substrates from the periplasm by specialized carrier proteins. Irrespective of the direction of transport, ATP binding causes the cytoplasmic NBDs to dimerize tightly, translating to a conformational change in the TMDs to allow transport to occur. The hydrolysis of ATP and release of ADP and phosphate allows the TMD dimer to dissociate for another transport cycle. Some of the specific details of this model remain unclear, but with the present rapidly increasing volume of ABC transporter structural information, increasingly sophisticated spectroscopic studies, and the accelerating power of computational tools these proteins may give up their secrets before long.

Acknowledgments

This work is supported by the Canadian Institutes for Health Research. DPT is an Alberta Heritage Foundation for Medical Research Scientist.

References

1. D. C. Rees, E. Johnson and O. Lewinson. 2009. ABC transporters: The power to change. *Nat Rev Mol Cell Biol* **10**, 218–27.
2. E. Procko, M. L. O'Mara, W. F. D. Bennett, D. P. Tieleman and R. Gaudet. 2009. The mechanism of ABC transporters: General lessons from structural and functional studies of an antigenic peptide transporter. *FASEB J* **23**, 1287–1302.

3. K. Hollenstein, R. J. P. Dawson and K. P. Locher. 2007. Structure and mechanism of ABC transporter proteins. *Curr Opin Struct Biol* **17**, 412–418.

4. K. J. Linton and C. F. Higgins. 1998. The *Escherichia coli* ATP-binding cassette (ABC) proteins. *Mol Microbiol* **28**, 5–13.

5. M. Dean, A. Rzhetsky and R. Allikmets. 2001. The human ATP-binding cassette (ABC) transporter superfamily. *Genome Res* **11**, 1156–1166.

6. R. Sánchez-Fernández, T. G. Davies, J. O. Coleman and P. A. Rea. 2001. The *Arabidopsis thaliana* ABC protein superfamily, a complete inventory. *J Biol Chem* **276**, 30231–30244.

7. E. Biemans-Oldehinkel, M. K. Doeven and B. Poolman. 2006. ABC transporter architecture and regulatory roles of accessory domains. *FEBS letters* **580**, 1023–1035.

8. D. C. Gadsby, P. Vergani and L. Csanády. 2006. The ABC protein turned chloride channel whose failure causes cystic fibrosis. *Nature* **440**, 477–483.

9. M. M. Gottesman and S. V. Ambudkar. 2001. Overview: ABC transporters and human disease. *J Bioenerg Biomembr* **33**, 453–458.

10. H. Nikaido. 1998. Multiple antibiotic resistance and efflux. *Curr Opin Microbiol* **1**, 516–523.

11. J. Lubelski, W. N. Konings and A. J. M. Driessen. 2007. Distribution and physiology of ABC-type transporters contributing to multidrug resistance in bacteria. *Microbiol Mol Biol Rev* **71**, 463–476.

12. M. L. Oldham, D. Khare, F. A. Quiocho, A. L. Davidson and J. Chen. 2007. Crystal structure of a catalytic intermediate of the maltose transporter. *Nature* **450**, 515–521.

13. R. N. Hvorup, B. A. Goetz, M. Niederer, K. Hollenstein, E. Perozo and K. P. Locher. 2007. Asymmetry in the structure of the ABC transporter-binding protein complex BtuCD-BtuF. *Science* **317**, 1387–1390.

14. I. R. Vetter and A. Wittinghofer. 1999. Nucleoside triphosphate-binding proteins: Different scaffolds to achieve phosphoryl transfer. *Q Rev Biophys* **32**, 1–56.

15. P. M. Jones, M. L. O'Mara and A. M. George. 2009. ABC transporters: A riddle wrapped in a mystery inside an enigma. *Trends Biochem Sci* **34**, 520–531.

16. J. E. Walker, M. Saraste, M. J. Runswick and N. J. Gay. 1982. Distantly related sequences in the alpha- and beta-subunits of ATP synthase, myosin, kinases and other ATP-requiring enzymes and a common nucleotide binding fold. *EMBO J* **1**, 945–951.

17. J. E. Moody, L. Millen, D. Binns, J. F. Hunt and P. J. Thomas. 2002. Cooperative, ATP-dependent association of the nucleotide binding cassettes during the catalytic cycle of ATP-binding cassette transporters. *J Biol Chem* **277**, 21111–21114.

18. J. Ye, A. R. Osborne, M. Groll and T. A. Rapoport. 2004. RecA-like motor ATPases — lessons from structures. *Biochimica et biophysica acta* **1659**, 1–18.

19. D. Khare, M. L. Oldham, C. Orelle, A. L. Davidson and J. Chen. 2009. Alternating access in maltose transporter mediated by rigid-body rotations. *Mol Cell* **33**, 528–536.

20. J. S. Sack, M. A. Saper and F. A. Quiocho. 1989. Periplasmic binding protein structure and function. Refined X-ray structures of the leucine/isoleucine/valine-binding protein and its complex with leucine. *J Mol Biol* **206**, 171–191.

21. C. Kandt, Z. Xu and D. P. Tieleman. 2006. Opening and closing motions in the periplasmic vitamin B12 binding protein BtuF. *Biochemistry* **45**, 13284–13292.

22. K. Hollenstein, D. C. Frei and K. P. Locher. 2007. Structure of an ABC transporter in complex with its binding protein. *Nature* **446**, 213–216.

23. M. L. Oldham, A. L. Davidson and J. Chen. 2008. Structural insights into ABC transporter mechanism. *Curr Opin Struct Biol* **18**, 726–733.

24. H. W. Pinkett, A. T. Lee, P. Lum, K. P. Locher and D. C. Rees. 2007. An inward-facing conformation of a putative metal-chelate-type ABC transporter. *Science* **315**, 373–377.

25. R. J. P. Dawson and K. P. Locher. 2006. Structure of a bacterial multidrug ABC transporter. *Nature* **443**, 180–185.

26. S. Hoffmeyer, O. Burk, O. von Richter, H. P. Arnold, J. Brockmöller, A. Johne, I. Cascorbi, T. Gerloff, I. Roots, M. Eichelbaum and U. Brinkmann. 2000. Functional polymorphisms of the human multidrug-resistance gene: Multiple sequence variations and correlation of one allele with P-glycoprotein expression and activity *in vivo*. *Proc Natl Acad Sci U S A* **97**, 3473–3478.

27. V. Kos and R. C. Ford. 2009. The ATP-binding cassette family: A structural perspective. *Cell Mol Life Sci* **66**, 3111–3126.

28. C. F. Higgins and K. J. Linton. 2004. The ATP switch model for ABC transporters. *Nat Struct Mol Biol* **11**, 918–926.

29. K. P. Locher. 2009. Review. Structure and mechanism of ATP-binding cassette transporters. *Philos Trans R Soc Lond, B, Biol Sci* **364**, 239–245.

30. E. Perozo, D. Cortes and L. Cuello. 1998. Three-dimensional architecture and gating mechanism of a K+ channel studied by EPR spectroscopy. *Nat Struct Biol* **5**, 459–469.

31. W. L. Hubbell, H. S. Mchaourab, C. Altenbach and M. A. Lietzow. 1996. Watching proteins move using site-directed spin labeling. *Structure* **4**, 779–783.

32. L. Columbus and W. L. Hubbell. 2002. A new spin on protein dynamics. *Trends Biochem Sci* **27**, 288–295.

33. B. A. Goetz, E. Perozo and K. P. Locher. 2009. Distinct gate conformations of the ABC transporter BtuCD revealed by electron spin resonance spectroscopy and chemical cross-linking. *FEBS letters* **583**, 266–270.

34. E. O. Oloo, C. Kandt, M. L. O'Mara and D. P. Tieleman. 2006. Computer simulations of ABC transporter components. *Biochem Cell Biol* **84**, 900–911.

35. D. van Der Spoel, E. Lindahl, B. Hess, G. Groenhof, A. E. Mark and H. J. C. Berendsen. 2005. GROMACS: Fast, flexible, and free. *J Comput Chem* **26**, 1701–1718.

36. E. O. Oloo and D. P. Tieleman. 2004. Conformational transitions induced by the binding of MgATP to the vitamin B12 ATP-binding cassette (ABC) transporter BtuCD. *J Biol Chem* **279**, 45013–45019.

37. P. C. Wen and E. Tajkhorshid. 2008. Dimer opening of the nucleotide binding domains of ABC transporters after ATP hydrolysis. *Biophys J* **95**, 5100–5110.

38. J. Sonne, C. Kandt, G. H. Peters, F. Y. Hansen, M. Ø. Jensen and D. P. Tieleman. 2007. Simulation of the coupling between nucleotide binding and transmembrane domains in the ABC transporter BtuCD. *Biophys J* **92**, 2727–2734.

39. M. L. O'Mara and D. P. Tieleman. 2007. P-glycoprotein models of the apo and ATP-bound states based on homology with Sav1866 and MalK. *FEBS Lett* **581**, 4217–4222.

40. G. Oancea, M. L. O'Mara, W. F. D. Bennett, D. P. Tieleman, R. Abele and R. Tampe. 2009. Structural arrangement of the transmission interface in the antigen ABC transport complex TAP. *Proc Natl Acad Sci U S A* **106**, 5551–5556.

Sodium-coupled Secondary Transporters

11

Insights from Structure-based Computations

Elia Zomot, Ahmet Bakan, Indira H. Shrivastava, Jason DeChancie,
Timothy R. Lezon and Ivet Bahar

1. Introduction: Biological Function and Classification

The biological membrane bilayer is impermeable to almost all polar or charged molecules. In order for the various solutes to cross this barrier, integral membrane proteins have evolved to provide a hydrophilic environment within the membrane that can bind and translocate these solutes into or out of the cell, often against their electrochemical gradient. These transporters are conventionally classified into three classes on the basis of the energy source used for transport: (1) primary active transporters rely on light, hydrolysis of ATP or redox reactions, (2) secondary active transporters require the electrochemical gradient of ions across the membrane to power the "uphill" translocation of the substrate, and (3) precursor/product antiporters exchange one molecule with its metabolic product independent of another source of energy.[1,2]

A major family of secondary transporters involves sodium- (and less often proton-) dependent symporters that couple the energy-costly translocation of the solute into the cell to that of sodium down its electrochemical gradient. These sodium-coupled transporters are found in all species and participate in a myriad biological functions, e.g. maintenance of efficient neurotransmission,[3,4,5] absorption of nutrients in the intestine,[6] regulation of pH and cytoplasmic $[Na^+]$[7,8] and osmoregulation[9,10] are only some of their physiological functions.

Of particular interest among sodium-coupled symporters are two families: the dicarboxylate/amino-acid:cation symporters (DAACS) and the neurotransmitter sodium symporters (NSS). The DAACS and NSS keep the extracellular (EC) neurotransmitter concentrations sufficiently low at the synaptic cleft, which enables postsynaptic receptors to detect signaling by the presynaptic nerve cell in the form of exocytotically released transmitters (Figure 1). The DAACS and NSS are key elements in the termination of the synaptic action of neurotransmitters, which is accomplished by diffusion and re-uptake of neurotransmitters into neuronal or glial cells; the single exception in this regard involves acetylcholine-mediated signal transmission.

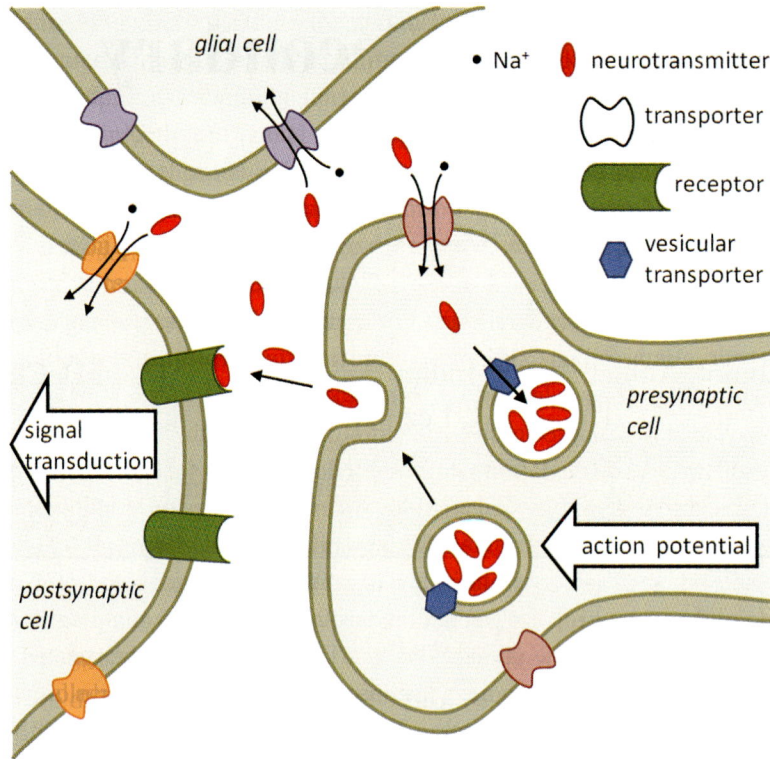

Figure 1. Simplified depiction of the role of neurotransmitter transporters. An arriving action potential in the presynaptic cell causes the fusion of neurotransmitter-loaded vesicles with the cell membrane and the release of their content into the synaptic cleft. Therein, the released neurotransmitter binds receptors on the postsynaptic cell membrane initiating a signal transduction pathway by inducing an action potential and/or a metabolic cascade. In order to induce a subsequent postsynaptic signal, the neurotransmitter has to be cleared from the synaptic cleft and this function is carried out by neurotransmitter transporters which couple the intake of the neurotransmitter to the transport of sodium ions into the pre-synaptic, postsynaptic or glial cells. The vesicular transporters — which are proton-dependent — load the pre-synaptic vesicles with neurotransmitter from the cytoplasm.

The DAACS family includes the glutamate (Glu^-) transporters (GluTs), also called excitatory amino acid transporters (EAATs) in humans. GluTs are located on neurons and glia (astrocytes) where they clear the excess Glu^- released at the synapses (Figure 1). The concentration of Glu^- in the EC space may increase by 10^3–10^4 fold during periods of synaptic activation. Accumulation of Glu^- above physiological (mM) levels may cause neurotoxic effects. Glu^- uptake and re-uptake by GluTs is essential for controlling EC levels of Glu^-,[11,12] thereby regulating glutamatergic signal transmission, i.e. preventing sustained activation and desensitization of ionotropic receptors, modulating the activation of metabotropic receptors,[13] and assisting in the glutamate-glutamine cycle.[14] The transport of Glu^- by GluTs is facilitated by the sodium electrochemical gradient.[15,16,17] The aspartate (Asp^-) transporter from *Pyrococcus horikoshii*, Glt_{Ph}, is the only member of this family that has been structurally resolved to date,[18,19,20] and has significantly assisted in improving our understanding of the potential molecular mechanisms that control the activity of GluTs.

The NSS family members, on the other hand, translocate small molecules that may be neurotransmitters, such as γ-aminobutyric acid (GABA), dopamine, serotonin or norepinephrine, or even amino acids such as leucine or tyrosine. The leucine transporter from *Aquifex aeolicus* (LeuT), later found to be a more efficient transporter of alanine, has served as a structural paradigm for this family after the resolution of its structure with bound substrate and sodium ions.[21]

Since transporter function is to regulate neurotransmitter activity by removing the transmitters from the synaptic cleft, specific transporter inhibitors (or enhancers/agonists) can potentially be used as novel drugs for neurological diseases. Inhibitors that block the biogenic amine transporters include antidepressant drugs (e.g. Prozac) and stimulants (e.g. amphetamines and cocaine).[22,23] GluTs, in particular GLT-1, play a central role in preventing both hyperexcitability and excitotoxicity,[11] which are implicated in epilepsy, stroke and Huntington's disease. Therefore, Na^+-coupled neurotransmitter transporters are of considerable medical interest.

Notably, many Na^+-coupled transporters belonging to other families have been found — upon resolution of their crystal structure — to share the LeuT structural fold despite their lack of significant sequence similarity, inviting attention to the functional versatility of this particular fold for transporting solutes across the membrane (Table 1). These include the galactose transporter from *Vibrio parahaemolyticus* (vSGLT) of the sodium solute symporters (SSS) family, the benzyl-hydantoin transporter (Mhp1) of the nucleobase cation symporters (NCS1) family, and the betaine transporter (BetP) of the betaine/choline/carnitine transporters (BCCT) family. Of these, the SSS family members transport sugars, inorganic ions, vitamins and choline.[24] SGLT1-6, for instance, are found in various tissues, such as the heart, kidney and intestine, and are responsible for the transport and absorption of glucose, galactose and mannose.[24,25] Genetic disorders of SGLTs include glucose-galactose malabsorption and familial renal glucosuria.[25,26,27] SGLT1 plays a central role in oral rehydration therapy to treat secretory diarrhea (e.g. cholera) and increased attention is being focused on SGLTs as drug targets for diabetic therapy.

In addition, the LeuT fold has been recently found to be shared by transporters that are not sodium-coupled. These are the arginine/agmatine antiporter (AdiC) of the APC family and the carnintine/γ-butyrobetaine antiporter (CaiT) of the BCCT family. It is worth noting that whereas prokaryotic and eukaryotic members of the NSS family share about 20–25% sequence identity, there is practically no sequence similarity between LeuT and the above listed structurally similar members of the SSS, NCS1 or BCCT families. As to GluTs, the five isoforms, EAAT1-5, share 50–60% sequence identity among themselves, and exhibit about 20-30% sequence identity with the bacterial transporters in the DAACS family, but share no significant homology with the NSS family members.

A full summary of the sodium-coupled secondary transporters with resolved structures (along with the transporters that have the same fold but are not sodium-coupled) is presented in Table 1. We note that in addition to the LeuT and Glt_{Ph} folds, a third fold of a representative member (NhaA) of the Na^+/H^+ antiporter family has been recently identified. In the present chapter, we focus on the functional dynamics of the DAACS and NSS families, using the respective Glt_{Ph} and LeuT folds as structural prototypes. Section 2 provides a macroscopic description of their Na^+-coupled mechanism, based on thermodynamics arguments. Sections 3 and 4 summarizes our current understanding of their substrate translocation mechanisms at the microscopic/molecular level, based on the resolved structures and on recent structure-based computations.

Table 1. Properties of Na⁺-coupled transporters and their structural homologs resolved to date by X-ray crystallography.

	LeuT							GltPh	NhaA
Fold	LeuT							GltPh	NhaA
Transporter	LeuT	Mhp1	BetP	CaiT	vSGLT	AdiC1	ApcT	GltPh	NhaA
Family	Neurotransmitter/Sodium Symporters (NSS)	Nucleobase/Cation Symporters (NCS1)	Betaine/Choline/Carnitine Transporters (BCCT)		Sodium/Solute Symporters (SSS)	Amino Acid/Polyamine/Organocation Transporters (APC)		Dicarboxylic Amino-Acid: Cation Symporters (DAACS)	Sodium/Proton Antiporters
Oligomeric state	Dimer	Dimer	Trimer		Dimer	Dimer	N/A	Trimer	Dimer
Repeating helices	'1–5' + '6–10'	'1–5' + '6–10'	'3–7' + '8–12'		'2–6' + '7–11'	'1–5' + '6–10'		'1–3, 7, HP1' + '4–6, 8, HP2'	'3–5' + '10–12'
Crystallized conformation	1) Outward-facing occluded 2) Outward-facing open	1) Outward-facing occluded 2) Outward-facing open 3) Inward-facing open	Intermediate-occluded	Inward-facing	Inward-facing occluded*	1) Outward-facing occluded 2) Outward-facing open	1) Inward-facing open	1) Outward-facing open 2) Outward-facing-occluded 3) Inward-facing occluded	Inward-facing
Substrate(s)**	Leucine, glycine, alanine,.... + Na*	Benzyl-hydantoin + Na*	Glycine-betaine + Na*	Carnitine/γ-butyrobetaine	Galactose + Na*	Arginine/agmatine	Alanine, glyine, glutamine,.... + H*	Aspartate + Na*	2H⁺/Na+
PDB accession code	2A65, 3F4G, 3F3A, 3F4J, 2Q6H, 2QE1	2JLN, 2JLO, 2X79	2WIT	3HFX	3DH4	3LRC, 3L1L	3GIA	1XFH, 2NWL, 2NWW, 2NWX, 3KBC	1ZCD

* Only the galactose, but not the sodium ion, is occluded from the IC in this conformation.[28,29,30]

** "+ Na*" indicates co-transport with one or more Na⁺ ions whereas "/" indicates antiport.

2. Na⁺-Coupled Mechanism of Transport: A Macroscopic View

2.1 *Transport cycle and alternate access mechanism*

In order for transporters to carry out their function, the substrate needs to be alternately exposed to either side of the membrane: the binding of the substrate and co-transported ions takes place on one side, and their release, on the other. The notion that a transporter can exist in either an outward- or an inward-facing state has been subject to debate for decades.[31] Recently obtained crystal structures of several transporters clearly demonstrate the occurrence of an outward- or inward-facing state consistent with the alternate access mechanism.[19,20,32–34]

A typical transport cycle for a Na⁺-coupled symporter is initiated by binding the substrate and co-transported sodium ions (or protons) from the EC medium. For this to occur, the transporter has to be in a conformation where the external aqueous cavity provides access to the binding site (Figure 2a). Following binding, the external cavity may remain open, but the substrate may be shielded from the external medium by one or more hydrophobic side-chains of the protein, as in the cases of LeuT and BetP,[10,21] or by a small loop as in Glt$_{Ph}$. Such a state is often referred to as the outward-facing occluded form (Figure 2b). The transporter then undergoes a large-scale conformational change to a conformation where neither the external nor internal cavity becomes accessible to the environment: the occluded-intermediate state (Figure 2c). Subsequently, the transition to the inward-facing state is completed, which predisposes the transporter to substrate and Na⁺ release (Figure 2d). Following the release of the substrate and sodium ion(s) to the cytoplasm, the empty inward-facing transporter (Figure 2e) is presumed to reconfigure back to its outward-facing form, thus completing one transport cycle.

Alternatively, in the outward-facing state, a competitive blocker may bind to the primary site (where the substrate binds) (Figure 2f) or a non-competitive blocker may bind to a secondary site (usually located within the cavity "above" the primary site) locking the transporter in that state and preventing progress through the transport cycle. Inhibitors of transport may also bind to the intracellular (IC) cavity that is exposed in the inward-facing state.

The scheme in Figure 2 is a simplified description of the transport cycle since in many cases the cycle does not merely include the substrate and Na⁺ but other cations and/or anions alike. For instance, in the serotonin and γ-aminobutyric acid transporters, SERT and GAT-1, respectively — which belong to the NSS family — a chloride ion is co-transported with the substrate and Na⁺ ion;[35,36] but in the neuronal Glu⁻ transporter EAAT3, a potassium ion is counter-transported.[12]

In the following, for clarity, we will distinguish between the local and global structural changes. Generally, the transporter can assume an inward- or an outward-facing state, and in each state, it may assume alternative conformations accessed by structural changes on a smaller scale. In the outward-facing conformation, the aqueous cavity leading to the binding site is exposed to the extracellular medium whereas in the inward-facing one, the binding pocket becomes accessible to the intracellular medium. The passage between the inward- and outward-facing states involves cooperative changes in the overall structure — hence termed "global" — whereas in each state, changes on a smaller scale may occur which cause the binding pocket to become accessible (*open*) or inaccessible (*occluded*) to the inner or outer media — hence termed "local changes". The diagrams (a), (b) and (f) in Figure 2 represent the outward-facing state, which may assume an

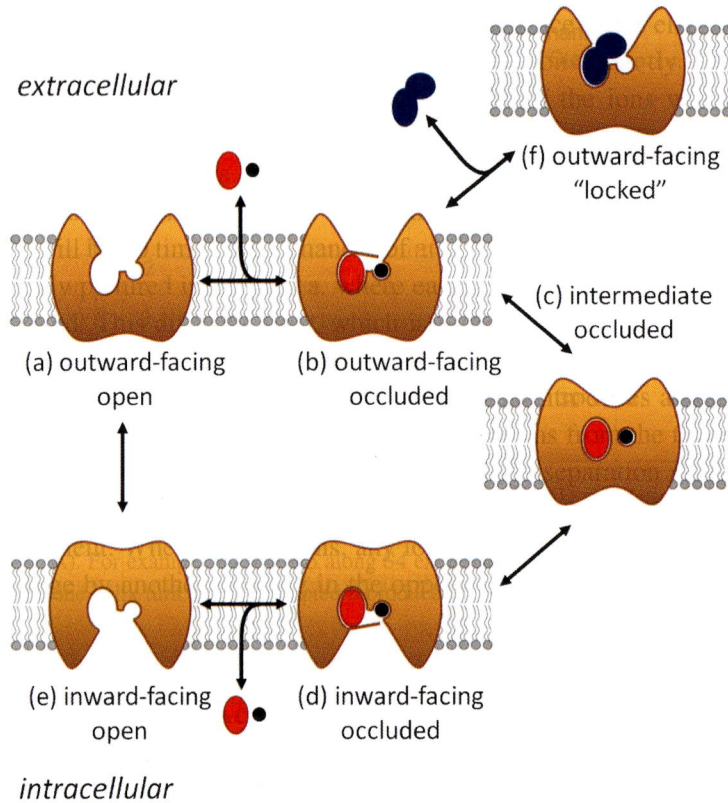

Figure 2. Schematic depicting the general transport cycle in a sodium-coupled transporter. The transporter with the binding site accessible to the EC medium (a) binds the substrate (*red ellipse*) and one or more sodium ions (*black circle*) (b). In the latter conformer, referred to as the outward-facing occluded conformation, the substrate and sodium ion(s) are mostly shielded from the EC medium by a small loop or one or more hydrophobic side chains. Following binding, the transporter shifts to an intermediate occluded state where the substrate and sodium ion(s) are inaccessible from either side of the membrane (c) before undergoing further conformational changes that cause the cavity leading to the binding site to become accessible to the IC medium (d). The substrate and sodium ion(s) are then released into the cytoplasm and the empty inward-facing transporter (e) reorients to face the EC medium again before a new cycle can begin. Alternatively, the transporter may bind an EC competitive inhibitor (*dark blue*) and become "locked" in an open conformation (f) inhibiting the completion of the transport cycle.

open (a), occluded (or closed) (b) or an inhibitor-bound (locked) (f) conformation. Likewise, (d) and (e) refer to the occluded and open forms of the inward-facing state, respectively. (c) represents an intermediate state.

2.2. Balance of forces/potentials across the membrane favors Na^+ inward flow

Two factors govern the movement of ions across selectively permeable membranes: the electric potential, or voltage difference across the membrane, and the chemical potential, or ion concentration gradient across the membrane. The sum of these potentials constitutes the electrochemical potential. When a Na^+ ion moves from outside to inside the cell, the associated free-energy change due to Na^+ concentration gradient is given by

$$\Delta G_c = RT \ln([Na^+_{in}]/[Na^+_{out}]), \tag{1}$$

where R is the gas constant and T is the absolute temperature. Under physiological conditions typical of many mammalian cells, $[Na^+_{in}] = 12$ mM and $[Na^+_{out}] = 145$ mM, which leads to $\Delta G_c = -1.5$ kcal/mol at $T = 300$K, i.e. there is a free energy release of 1.5 kcal/mol accompanying the transport of each Na^+ to the cell interior, assuming that there is no membrane electric potential. On the other hand, the free energy change due to the membrane electric potential (using a resting membrane potential $E = -70$ mV in a typical neuron) is

$$\Delta G_m = FE = -1.6 \text{ kcal/mol}, \tag{2}$$

for the transport of each monovalent cation (e.g. Na^+) from outside to inside the cell. Here F is the Faraday constant. Since both Eq. (1) and Eq. (2) apply to Na^+ ions, the total free energy change ΔG associated with the translocation of one Na^+ is the sum

$$\Delta G = \Delta G_c + \Delta G_m = -3.1 \text{ kcal/mol}. \tag{3}$$

Thus, the inward movement of Na^+ in a typical neuron is energetically favored by 3.1 kcal/mol.

2.3 *Sodium ion co-transport as a driving potential for the transmembrane translocation of solutes by secondary transporters*

Polar or charged molecules are unable to diffuse across the hydrophobic lipid bilayer due to unfavorable interactions. Instead, they cross the membrane by permeating through membrane proteins such as transporters or channels. The translocation of a solute molecule is even more difficult if its diffusion is an "uphill" process, i.e. against its concentration gradient or the membrane electric potential. A mechanism to accomplish the uphill translocation is to couple the solute transport to the flow of the ion(s) that is favored by the electrochemical gradient. Secondary transporters utilize such electrochemical gradients. The counter- or co-transport of ions (by antiporters or symporters, respectively) drives the translocation of the solute (substrate). For example, as described above, the free energy change generated from the movement of a Na^+ ion from outside to inside the cell is approximately -3 kcal/mol [Eq. (3)]. The transport of three Na^+ ions, for example, would permit the cell to accumulate a higher concentration of solute (e.g. a neurotransmitter such as dopamine or Glu^-) relative to the exterior concentration to counterbalance this potential: under equilibrium conditions, in the absence of other effects and assuming the solute is neutral,

$$RT \ln[S_{in}]/[S_{out}] = 3(-3.1) = -9.3 \text{ kcal/mol} \tag{4}$$

which leads to $[S_{in}]/[S_{out}] \approx 3 \times 10^6$ at 310 K. Thus, the free energy release associated with the transport of three sodium ions allows for the transport of an uncharged neurotransmitter against a

10^6-fold concentration increase. In the case of human EAATs, the transport of each Glu^- is accompanied by the co-transport of $3Na^+$ and a proton (H^+), followed by the counter-transport of a potassium ion (K^+), such that there is a net influx of two cations.

3. Structural Information

Up until a few years ago, no high-resolution structural data were available for Na^+-coupled secondary transporters, mainly due to their hydrophobic and dynamic nature, which makes them difficult to crystallize. Recently, however, the crystal structures of several distantly related transporters with bound substrate and/or sodium ions have been resolved (Table 1). Two prime examples, Glt_{Ph} and LeuT, both in the outward-facing state, are illustrated in Figure 3. Notably, the

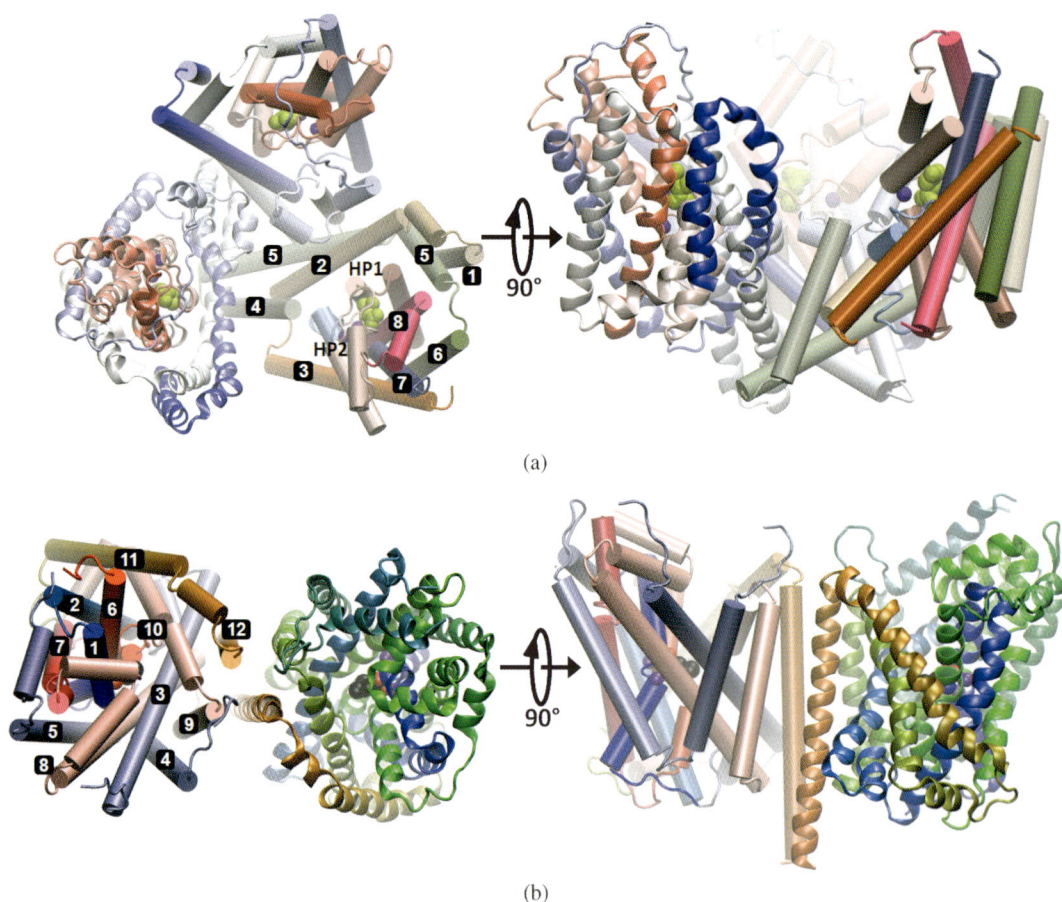

(a)

(b)

Figure 3. Crystal structures of two widely studied homologs of neurotransmitter transporters, Glt_{Ph} trimer and LeuT dimer. The outward-facing states of Glt_{Ph} (a) and LeuT (b) are shown here. The structures are viewed from the EC side (left) and through the plane of the membrane (right) together with the bound sodium ions (purple spheres), Asp- (a) (yellow space-filling) or Leu (b) (black space-filling). The structures are generated using the respective PDB files 1XFH and 2A65 and the visualization software *VMD*.[41] For clarity, one of the monomers is shown in ribbon representation in each case, and the remaining, in cylinders.

structures resolved to date capture these proteins in different conformations providing us with full atomic coordinates data on alternative forms visited along the transport cycle depicted in Figure 2. Not only have these structures confirmed previous experimental work, but they have also revealed two main distinct features of this class of transporters: an internal two-fold structural symmetry and a discontinuity of transmembrane (TM) helices at the substrate/ion binding sites.[10,18–21,28,37–40] Most importantly, they provide important insights into the molecular mechanisms that mediate substrate translocation and permit us to perform structure-based computations toward elucidating time-dependent events.

Symporters that share the LeuT fold exhibit significant commonalities in their core regions' structure and interactions, despite their lack of sequence similarity, which suggests that the LeuT fold encodes a common mechanism of substrate transport. The emerging molecular details about the structure and mechanism of these transporters have consolidated them into one main structural family. Below we present more details on the Glt_{Ph} and LeuT structures.

3.1 *Structure and topology*

Glt_{Ph} is a homotrimer. The three subunits form a bowl-shaped concave basin towards the EC side (Figure 3a). Each subunit is comprised of eight TM helices, TM1-TM8, and two helical hairpins, HP1 and HP2. These structural elements are organized into an N-terminal, outer cylinder region made up of TM helices TM1-TM6, and a C-terminal core comprised of TM7, HP1, HP2 and TM8 (Figure 4a). Within each monomer, the core region encompasses the elements of substrate binding and the transport machinery whereas the outer region (known as the scaffold) provides stability and interacts with the two other subunits.[20]

In the case of NSS family and the families with a similar structural fold, the transporters form homo-dimers or -trimers in the membrane upon crystallization. These oligomeric forms are, at least for many of them, the physiological states while the monomer is the functional unit. The monomer is composed of 11-14 TM helices and the common core is formed by 10 TM helices, the numbering of which can vary: in LeuT and Mhp1 they are helices 1 to 10 (Figure 4b), in vSGLT, 2 to 11, and in BetP, 3 to 12. The core, in the case of LeuT for example, can be divided into two structurally symmetric "halves", or repeats: TM1-TM5 and TM6-TM10. The first helix of each repeat contains a break (unwound portion) about halfway across the membrane. This break exposes backbone polar groups that interact with the substrate and the co-transported sodium ion(s) (Figure 4b).

Only recently, after the elucidation of a number of high-resolution structures did we gain deeper insights into the mechanisms of substrate transport by Na+-coupled secondary transporters.[42,43] These structures reveal that transporters may indeed share similar local motifs, even if they have different folds and topologies. LeuT (and symporters with a similar fold) and Glt_{Ph} as well as the sodium/proton antiporter, NhaA, share the internal pseudo-symmetry and the breaks in the TM helices.

3.2 *Substrate and sodium binding*

In Glt_{Ph}, the substrate binding site in each subunit is covered by the loops of HP1 and HP2. There are several conserved residues within 5–7 Å of the substrate binding site. These residues include

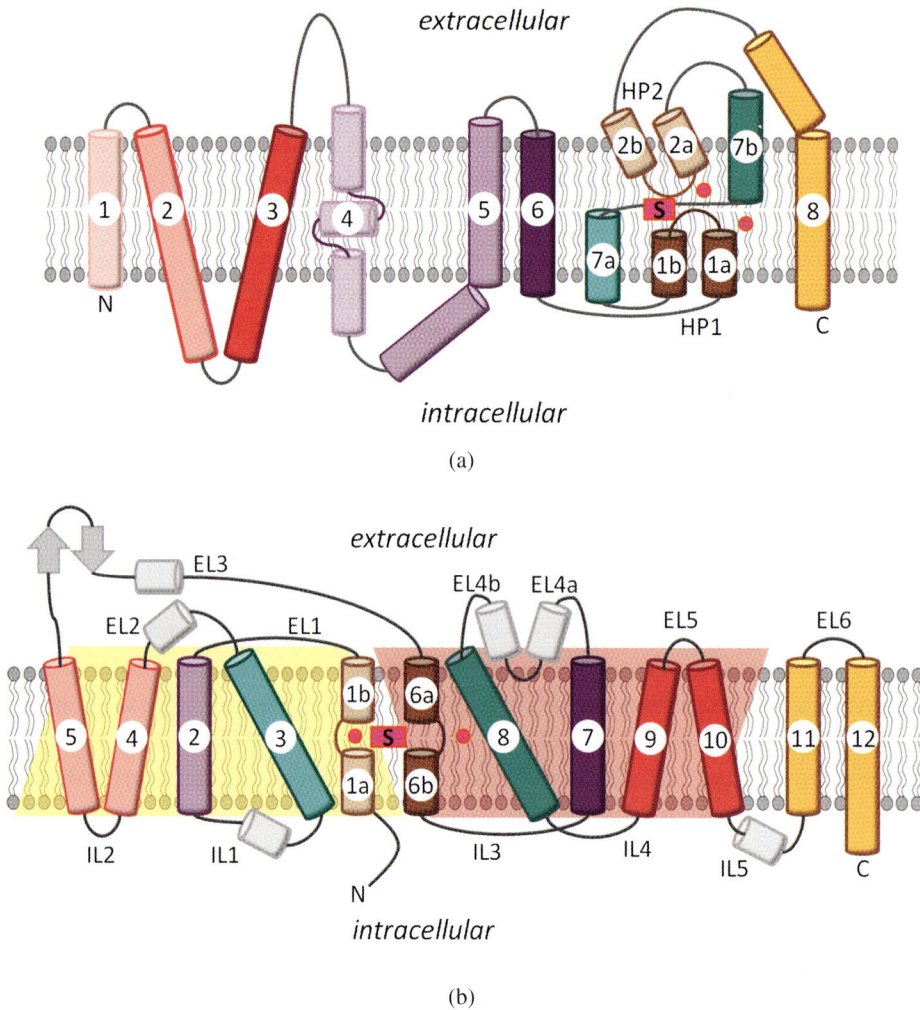

Figure 4. Schematic description of the topology of Glt$_{Ph}$ and LeuT. Helices are shown as cylinders and TM helices are numbered. The substrate and sodium ions (yellow rectangle and circles, respectively) are shown at the binding site of Glt$_{Ph}$ (a) and LeuT (b). In (b), the two pseudo-repeats, TM helices 1-5 and 6-10, are shaded (pink and orange, respectively). Extracellular and intracellular loops are labeled as EL and IL, respectively, and numbered according to their sequence position.

the highly conserved serine motif (Ser277-Ser278-Ser279) on the HP1 loop, two glycines (Gly354 and Gly357) on the HP2 loop, the NMDGT motif between Asn310 and Thr314 on TM7, Asp394, Arg397, Thr398 and Asn401 on TM8. Of these residues, Gly354 and Gly357 are considered to coordinate one of the two bound Na$^+$ ions (Figure 5a). In the higher resolution Asp-bound outward-facing structure,[18] a Na$^+$ ion (referred to as Na2) is bound close to this position, interacting with Thr352, Ser349 and Ile350. Additionally, there is a second Na$^+$ ion binding site (Na1),

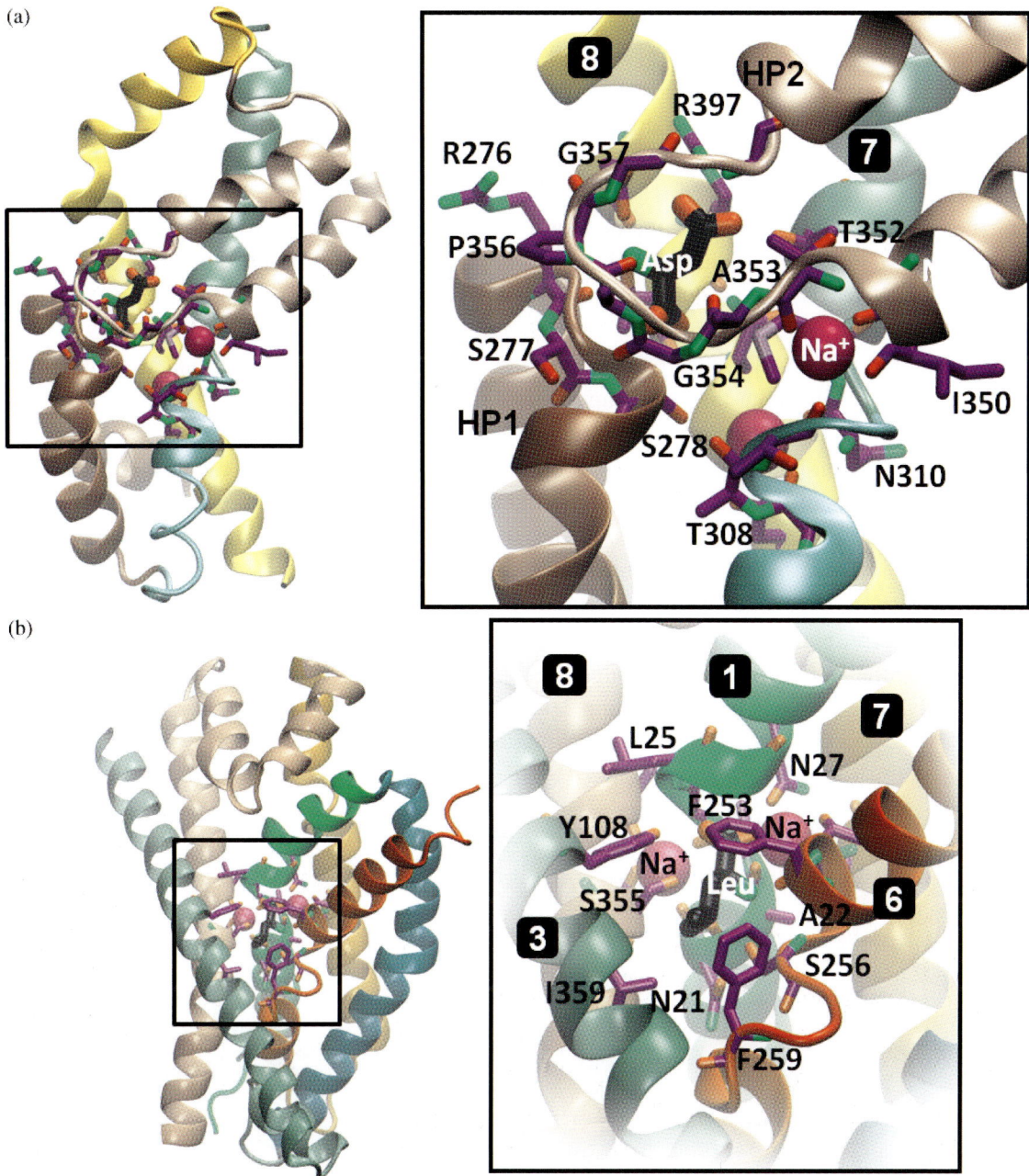

Figure 5. The substrate binding site in Glt_{Ph} and LeuT. A close-up on the core domain of Glt_{Ph} composed of HP1, HP2, TM7 and TM8 (a), and the substrate- and Na^+-binding site of LeuT between TM1, TM3, TM6, TM7 and TM8 (b). The structures are viewed through the plane of the membrane. The substrate is shown in black in each case, and the bound Na^+ ions in yellow. The residues interacting with the substrate are shown in green stick representation. The ribbon diagrams were generated using the same PDB structures and software as those used in Figure 3.

located deeper in the core near Gly306 and Asn310 on TM7 and Asn401 and Asp405 on TM8 (Figure 5a).

In the case of secondary transporters with the LeuT fold, as the identities of the substrate vary among and within the families, so do the residues that coordinate them. The substrate may be a biogenic amine in the case of NSS family members, a sugar in the case of an SGLT (SSS family) or a nucleobase in the case of Mhp1 (NCS1 family). The substrate is usually located at the center of each monomer, halfway across the membrane and at the interface between the two structural repeats (Figures 3b and 5b). The TM helices taking part in binding the substrates, and their spatial organization, however, are similar in most cases: TM1, 3, 6 and 8 in LeuT and Mhp1, TM2, 3, 7, 8 and 11 in vSGLT (the counterparts of TM1, 2, 6, 7 and 10 in LeuT and Mhp1); and TM1, 3, 6, 8 and 10 in the structurally related but not sodium-coupled antiporters AdiC1 and CaiT.[39] Figure 6 panels (a) and (b) illustrate the close superposition of these equivalent helices for LeuT, Mhp1 and vSGLT.

Two sodium ion-binding sites have been identified for the transporters with the LeuT fold, which, on the basis of those initially identified in the LeuT structure, are referred to as: (1) Na1, observed only in the crystal structure of LeuT so far, where Na^+ is coordinated by residues on TM1, TM6 and TM7, and by the carboxyl group of the substrate (Figure 5b), and (2) Na2, which appears to be a more general site, observed in LeuT and proposed in Mhp1 and vSGLT. In the latter two structures, Na2 is likely to be coordinated by residues on TM1 and TM8, as illustrated in Figure 6c. In the case of BetP, even though the residues observed at the Na2 site are on TM1 and TM5, the binding cavity is lined by residues on TM8.

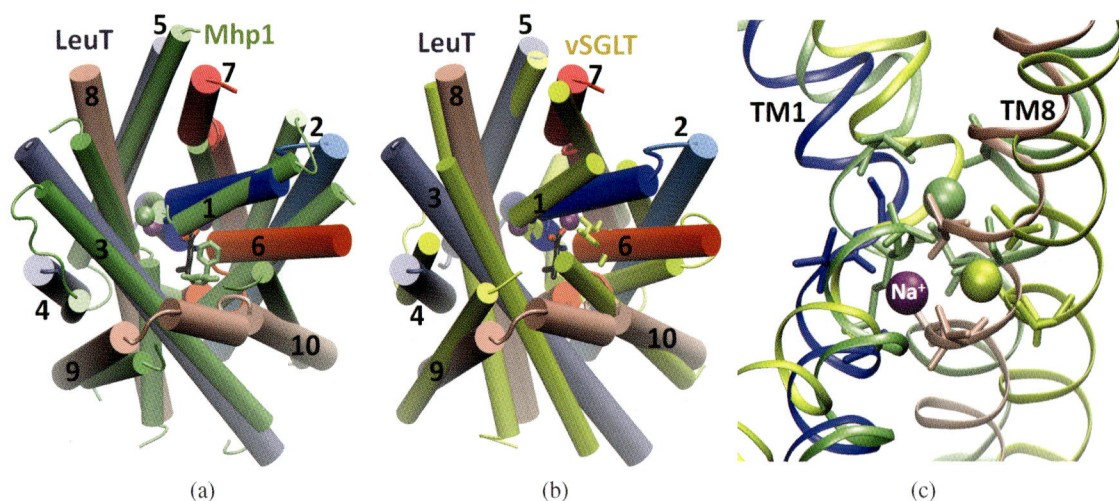

(a)   (b)   (c)

Figure 6. Superposition of three members of NSS family of transporters. Crystal structures of LeuT in the outward-facing state (blue and red) superposed on that of (a) Mhp1 (lime green) in the outward-facing state, and (b) vSGLT (yellow) in the inward-facing state. The molecules are viewed from the EC side (a and b), and through the membrane zooming in on the sodium-binding site equivalent to that of Na2 in LeuT (c). The substrates (licorice) and sodium ions (spheres) are colored with respect to the transporters. TM helices are numbered according to LeuT. Only the 10 core helices are shown here.

The fact that sodium designated as Na1 is directly coordinated by the negatively charged carboxyl group of the substrate (regardless of whether it is leucine, glycine, alanine, or otherwise)[38] suggests a direct coupling and possibly a cooperativity between the substrate and sodium in binding and/or transport. In contrast, the second sodium ion (Na2) seems to be required for the structural stability of the binding pocket and the increased selectivity for Na^+ at Na1.[44] Computational approaches have shown that in the substrate-occluded conformation of LeuT, there is no preference for Na^+ over the larger K^+ at Na1, where Na^+ is coordinated by six ligands including the carboxyl group of the substrate, and the same is true for Na2 where Na^+ is coordinated by five neutral ligands. In contrast, only modest preference for Na^+ over the smaller Li^+ was observed at both sites using the same approach based on free energy calculations.[45]

It is interesting to note that TM1 and TM6 in LeuT, Mhp1 and AdiC1, and their counterparts in vSGLT and CaiT, are partially unwound (or broken) (see the topology diagram for LeuT in Figure 4b). These particular regions contain several "frustrated" backbone carbonyl and amino groups that lack hydrogen bond-forming partners, and as such, serve as avid binding sites for substrate and ion binding. Notably, the TM7 and TM8 helices in the core domain of Glt_{Ph} also have such partially unwound segments containing highly conserved residues (e.g. the NMDGT motif) that play a critical role in binding the substrate or cations, consistent with the same "design principle". However, this mechanism of substrate binding may not hold for all conformational states of the transporters. In the crystallized intermediate occluded conformation of BetP (see Table 1), for instance, the substrate is liganded by residues on the intact helices TM4 and TM8, which are equivalent to TM2 and TM6 in LeuT.[10]

4. Molecular Mechanisms Revealed by Structure-Based Modeling and Computations

With the elucidation of LeuT, Glt_{Ph} and other Na^+-coupled transporters' structures (Table 1), several structure-based computational studies have been launched in different laboratories, which are now nicely complementing experimental findings in helping us gain insights into the transport mechanisms and visualize time-dependent events at the atomic scale. It is now clear that the description of the transport mechanism in terms of only two macrostates, inward-facing and outward-facing, is an oversimplification. In principle, transporters — like all other proteins — are subject to a multitude of conformational motions that enable their activity. A protein of N interaction sites enjoys $3N-6$ internal degrees of freedom; thus, its equilibrium motions are described in terms of $3N-6$ collective mode directions that form an orthonormal basis set. Each mode has its own "time constant" and "mechanism" described by the respective eigenvalue and eigenvector of the covariance matrix for the fluctuations of interaction sites. It is customary to describe this complex space of conformational changes in terms of a few kinetic parameters, e.g. by adopting single-, bi- or tri-exponential time-dependent functions for describing the time evolution of experimentally detected events. However, in a strict sense, the function involves an ensemble of motions.

Of the broad spectrum of motions accessible to transporters, state-of-the-art molecular dynamics (MD) simulations can explore those occurring on the order of nanoseconds; perhaps those up to tenths of microseconds with adequate computing resources. Such motions typically involve local changes in structure. Examination of longer time, larger-scale movements, or so-called global motions, on the other hand, requires the adoption of simplified models. Simplified models such as *elastic network models* (ENMs) permit us to learn about potential cooperative movements (e.g. domain/subunit rearrangements) that involve the entire molecule, at the cost of losing atomic level accuracy.[46] In many cases, the first step before launching the simulations is to model the structure, in the absence of available experimental structure. The most feasible way to achieve this goal is *comparative/homology modeling*, provided that a sequence homolog of the investigated protein has been structurally resolved.

Therefore, structure-based computations consist of three major groups: (1) MD simulations for exploring local events in the nanoseconds regime, e.g. gating, substrate/ion binding or release, or fluctuations between open and occluded forms in either the outward-facing state or the inward-facing state, i.e. the horizontal steps in the transport cycle depicted in Figure 2, but not the passage between the two states (vertical steps), (2) homology modeling for predicting the structures of family members not resolved to date, or for predicting the alternative (functional) structures of a known transporter, using as template the known structure of a family member, and (3) coarse-grained approaches, usually based on simplified models such as the ENMs and normal mode analysis (NMA), toward exploring the passage between outward-facing and inward-facing states (the vertical steps in Figure 2). Results from these three respective groups of computations will be summarized in the Sections 4, 5.1 and 5.2, respectively. Results in Section 4 will be presented in two subsections: Section 4.1 for Glu⁻/Asp⁻ transporters, and Section 4.2 for NSS family members or their structural homologs.

4.1 Local Motions and Substrate/Sodium Interactions in Glutamate Transporters

4.1.1 *Gate opening in the outward-facing state of Glt$_{Ph}$*

The resolution of the X-ray structure of the archaeal Asp⁻ transporter Glt$_{Ph}$ provided a major breakthrough in understanding the structural underpinnings of the transport mechanism at the molecular level.[20] Extensive MD simulations performed with this outward-facing structure in the presence of explicit lipid and solvent molecules revealed various aspects of the substrate recognition and binding events of the transport cycle.[47,48]

First, the simulations performed in the absence of substrate exhibited striking motions in the HP2 loop of the protein. This loop has a tendency to open up within a couple of nanoseconds in one subunit, followed by the opening up of the HP2 loop in one of the two other subunits within approximately 15 ns. The opening of the HP2 loop exposes several highly conserved charged/polar residues (e.g., Arg276, Ser277-Ser278-Ser279, on HP1 and the N310-MDG-T314 motif on TM7) at the substrate binding site to the EC region. The exposure of these residues with high-affinity to attract the substrate from the aqueous environment seems to be a pre-requisite for substrate binding. Simulations suggest that HP2 acts not only as an EC gate but also as an attractor driving the diffusion of the substrate from the aqueous basin towards the binding site, as evidenced by the involvement of HP2 glycines in the initial recognition events.

The opening of the HP2 loop in the fast regime observed by MD simulations supports the view that the transporter core domain possesses an intrinsic ability to open an EC gate for substrate entry. This propensity is supported by the X-ray structure of Glt_{Ph} in a complex with the non-transportable blocker DL-*threo-β*-benzyloxyaspartate, where HP2 is blocked in an open conformation.[18]

Second, a series of simulations have been performed, with and without substrate, in the absence and presence of sodium ions, to examine the effects of the substrate on the binding pocket.[47] Figure 7a displays a cumulative histogram derived from the analysis of the HP2 motions in these MD runs, MD0-MD6. The distance between two residues, Ser278 and Gly354 (Figure 5a), at the respective tips of HP1 and HP2 was monitored therein as basis for probing the state of the EC gate. In all runs, the HP1 loop was observed to be highly stable and almost rigid while the HP2 loop (EC gate) would always open up in at least one of the subunits, irrespective of whether the substrate was bound or not. The two peaks in the histogram indeed correspond to the open and closed conformations sampled by the EC gate. The open conformation exhibits a broader distribution consistent with the high conformational variability of the HP2 loop in this conformation, while the closed conformation exhibits a sharper peak centered around 3 Å. These simulations thus confirmed that the HP2 hairpin serves as an EC gate in the outward-facing state, in either substrate-bound or -unbound forms.

4.1.2 *Role of Na^+ ions in stabilizing the closed conformation of the EC gate in the substrate-bound outward-facing Glt_{Ph}*

The data compiled in Figure 7a refer to multiple runs conducted under different conditions: In the absence of the substrate (empty conformer, outward-facing state), the Glt_{Ph} subunits exhibit a high tendency to sample the open state (MD0 and MD6), and this tendency is maintained, albeit to a lesser extent, in the presence of bound substrate (occluded form, outward-facing state) (MD1 and MD2). Binding of sodium ions, on the other hand, strongly favors the closed conformer. The stabilizing role of Na^+ ions at the binding site is clearly evident in the runs (MD3 and MD4) conducted in the presence of bound Na^+ ions. Panel (b) in Figure 7 compares the time evolution of the distance between the HP1-HP2 tips in different runs (left) and displays snapshots from the runs conducted without (top) and with (bottom) sodium ions (right). The results clearly demonstrate the tight interaction of HP1 and HP2 tip residues in the presence of Na^+ ions. This interaction keeps the EC gate closed, in the occluded (outward-facing) conformation, suggesting the role of at least one sodium ion at the binding site as "gate-keeper" in addition to its role in assisting substrate transport. Notably, the substrate molecule located within the EC basin is observed to readily recognize the binding site, provided that the HP2 loop of a given subunit opens up, and diffuses toward that particular subunit to enter the binding site and interact tightly therein with the conserved polar residues on HP1, TM7 and TM8.[47] Simulations showed that once the substrate enters the binding site, it remains tightly bound for the entire duration of the simulation (40–50 ns), without exiting the transporter. The translocation indeed requires a global structural change to the inward-facing structure (see Section 5), which is a several orders of magnitude slower process currently beyond the time scale of MD simulations.

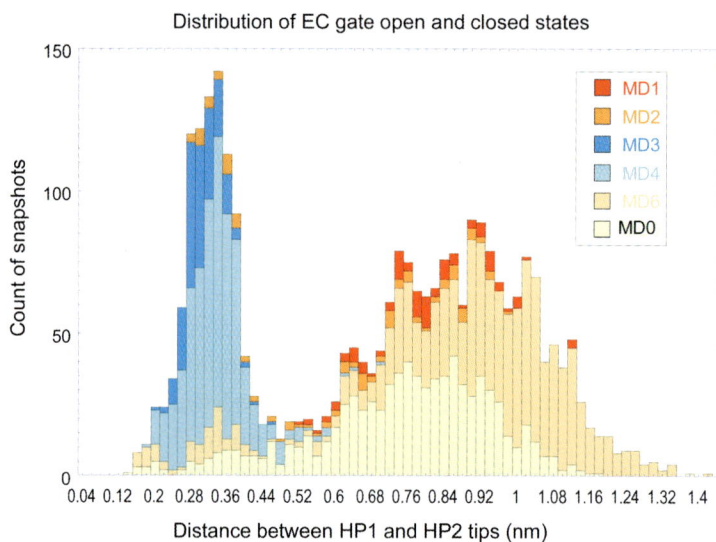

(a)

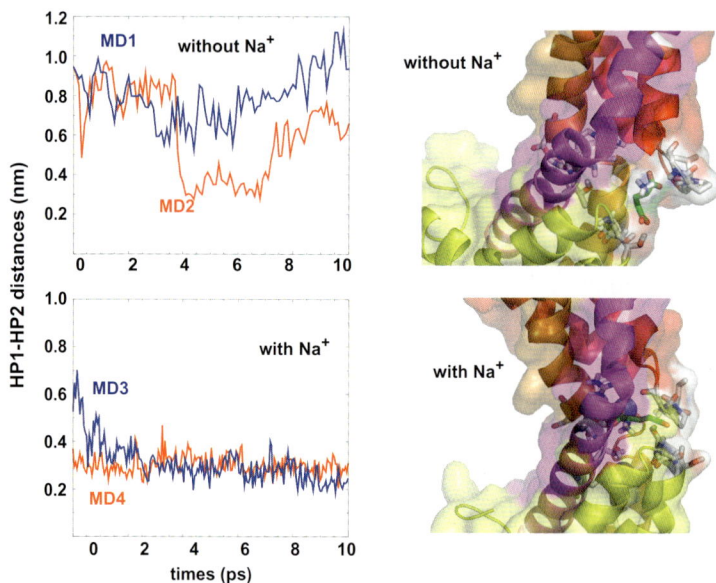

(b)

Figure 7. Local motions observed in MD simulations of Glt$_{Ph}$ outward-facing state. Panel (a) shows the occurrence of two conformers, open and closed, for the EC gate (HP2 loop). The distance between the residues Ser278 and Gly354 at the HP1 and HP2 tips (Figure 5a), respectively, exhibited a bimodal distribution in multiple simulations (MD0-MD6), indicative of the two distinctive states of the EC gate. For details of these simulations see Ref. 47. The narrow peak at shorter separation corresponds to the closed conformations, and the broader peak around 0.9 nm, to open conformations. The runs MD3 and MD4 (cyan and blue) were conducted in the presence of bound Na$^+$ ions; and all other runs, without Na$^+$ ions. Panel (b) shows the role of Na$^+$ ions in stabilizing the closed conformations of the EC gate. The curves on the left show the time evolution of HP1-HP2 tip distances in MD1 (blue) and MD2 (red) top panel, and MD3 (blue) and MD4 (red), bottom panel. The diagrams on the right display the substrate-protein interactions at the binding site in an open (top) and closed (bottom) conformation of the EC gate. In both panels, the substrate is shown in stick representation, with the backbone colored green; the Na$^+$ ions are shown as purple spheres; HP1 and HP2 are colored yellow and red, respectively; TM7 and TM8 are orange and magenta; and the residues interacting with the substrate are shown in stick-representation.

4.1.3 *Na⁺ and substrate release by Glt$_{Ph}$ in the inward-facing state*

The mechanism of release of the substrate into the IC region remained unclear until the recent determination of the inward-facing structure of Glt$_{Ph}$.[19] The popular hypothesis has been to view the two helical hairpins, HP2 and HP1, in the core domain of each subunit, as the respective EC and IC gates for substrate entry and release. The former has indeed been confirmed by both MD simulations[47,48] and experiments[18,20] to serve as an EC gate (Figure 7). The latter (HP1), on the other hand, could not be computationally confirmed up until now, due to the absence of an atomic structure for the inward-facing structure, which could serve as a starting point for MD simulations. With the recent elucidation of this structure by Boudker lab, we have been able to gain insights into the mechanism of substrate release. The emerging sequence of events that control Asp⁻ release to the cytoplasm is the following. First, in parallel with the behavior in the outward-facing form, the HP2 loop is by far more flexible than the HP1 loop at early stages of simulations. Figure 8 displays snapshots from an unbiased MD run in the presence of explicit water and lipid molecules, where the opening of the HP2 loop, the accompanying dissociation of Na2, and ensuing solvation of the binding site may be clearly seen. Notably, the release of Na2 further destabilizes the HP2 loop, and the resultant enhanced fluctuations facilitate a more massive influx of water molecules, which effectively compete with the polar and charged residues that ligate the bound substrate, and lead to the dislodging of Asp⁻ from its binding site (Figure 8). Yet, the release from the binding pocket also requires the opening of the HP1 loop. In fact, as shown in Figure 9, persistent hydrogen bonds with the conserved amino acids S277 and S278 at the HP1 loop prevent the substrate dissociation (at 44 ns) and even drive the substrate back near the binding pocket (at 45 ns). This momentary setback in substrate release is however overcome at 49 ns where the disruption of a hydrogen bond with S277 remains as the last step to complete dissociation. Notably, the inner sodium ion (Na1) is observed to remain bound and almost fixed during all these events.

Thus, substrate release necessitates both the fluctuations in HP2 and the opening of HP1. Although both HP1 and HP2 share a helix-turn-helix motif, the identities of the conserved amino acids that comprise the loops are very different.[20] The HP1 loop is composed of four polar/charged residues (Arg276, Ser277, Ser278, and Ser279), whereas the HP2 loop is composed of hydrophobic side-chains of (Ala353, Ala358, and Val355), four glycines (351, 354, 357, and 359) and Pro356. Furthermore, the chemical identity of the HP1 loop residues is consistent with their strong interactions with the substrate: the side-chains of Ser277 and Ser278 remain hydrogen-bonded to the substrate and Arg276 is locked in a salt-bridge with the side-chain of Asp238 of TM8 during a large portion of the simulations. The floppier, hydrophobic chemical identity of the HP2 loop, therefore, makes this loop intrinsically predisposed to undergo larger magnitude motions, but final substrate release is not achieved until there is a conformational change (opening) in HP1 loop as well.

The intrinsic ability of Glt$_{Ph}$ to sample the various conformational states to facilitate both substrate binding and release, which is enabled by the global change in structure from outward- to inward-facing, highlights the adroit dynamic functionality of the HP1 and HP2 loops in assisting the transport of amino acids during synaptic transmission.

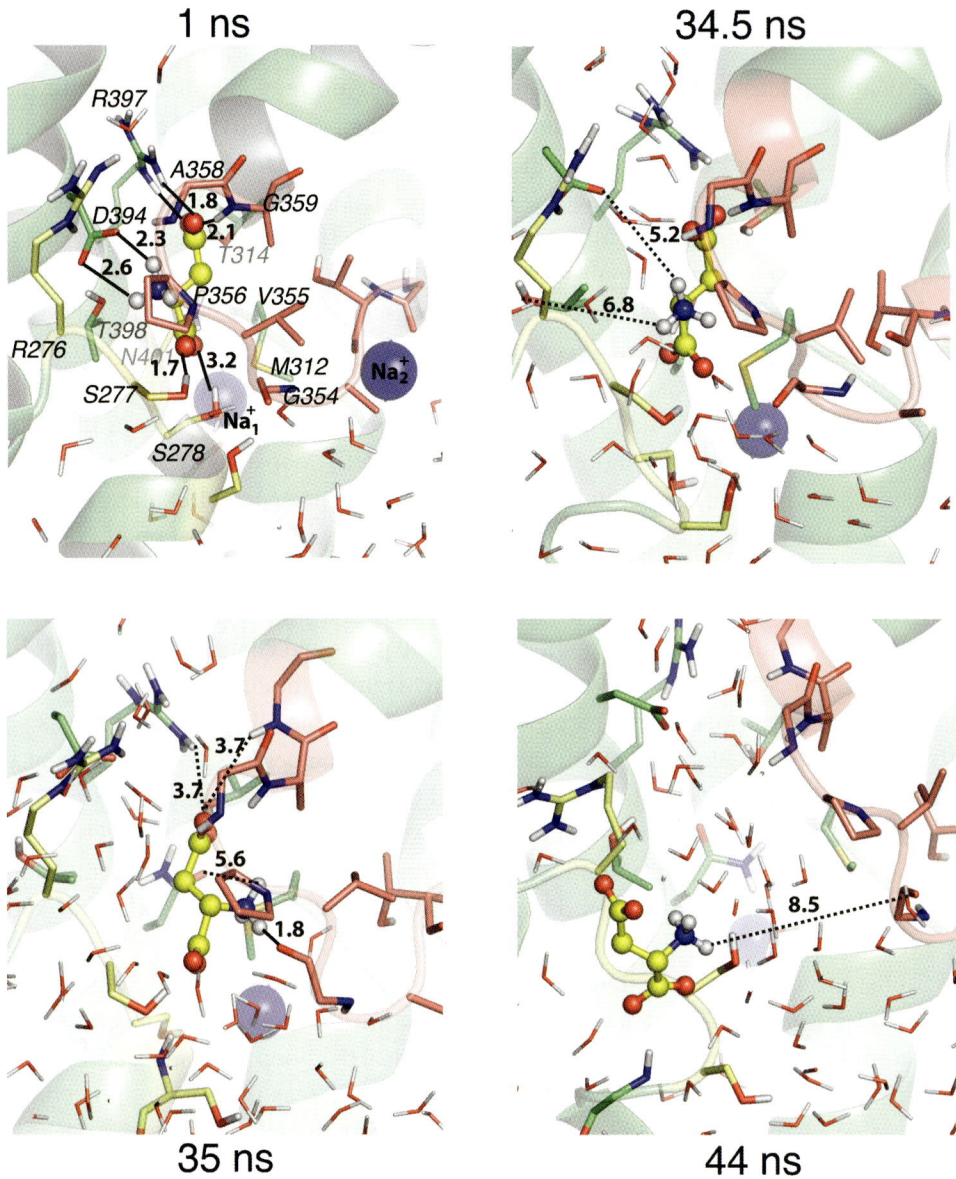

Figure 8. Succession of events leading to the dislocation of substrate in the inward-facing state of Glt$_{Ph}$. Distances between substrate atoms and the closest atoms on the labeled amino acids are reported in Ångstroms. At $t = 1$ ns, the bound substrate is sequestered from the IC solvent by the closed loop conformation of HP2 and HP1. At $t = 15$ ns (not shown), Na$^+_{(2)}$ begins to dislodge from its binding site. Snapshots at 34.5 and 35 ns demonstrate the significant change in the conformation of the substrate within 0.5 ns, induced by the severed hydrogen bonds to Asp394 and Arg397, which arise from binding site solvation following the opening of the HP2 loop. At $t = 44$ ns, the substrate is dislodged from the binding cavity. Note that the substrate is completely dissociated from HP2, as evidenced by the increase in the distance between the Asp- NH group and HP2 Gly354 backbone hydroxyl from 1.8 Å at $t = 35$ ns to 8.5 Å at $t = 44$ ns, while there is a substrate-protein hydrogen bond (1.8 Å) between the Asp- β-carboxylate oxygen and the hydroxyl group of Ser277 on the 3-Ser motif at the HP1 loop.[49]

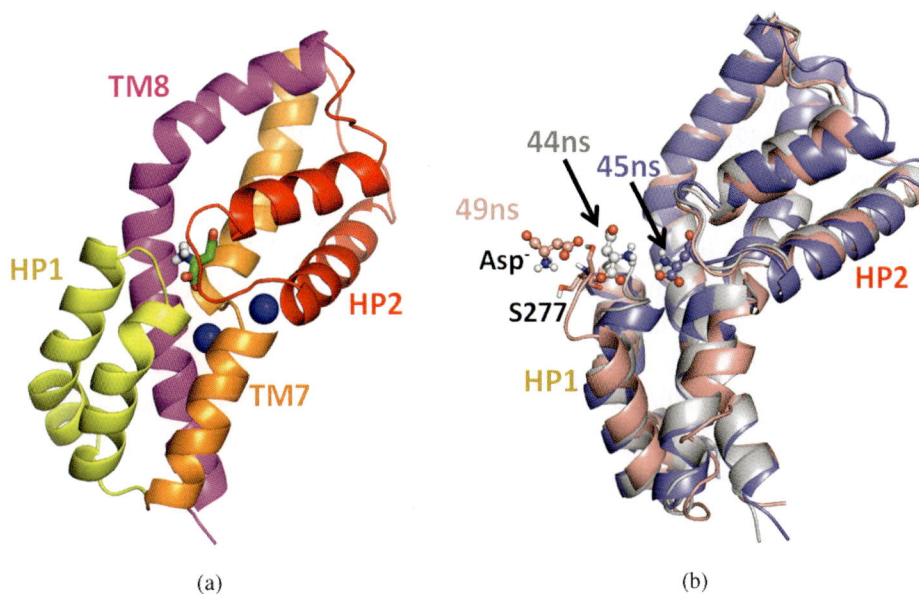

(a)   (b)

Figure 9.   Last step of aspartate transport cycle by Glt$_{Ph}$: involvement of HP1 in the final release of substrate in the inward-facing state. Panel (a) displays the original structure of the core domain in one of the subunits, color coded by the secondary structural elements, including both the bound substrate and two sodium ions. Panel (b) displays three snapshots of the same subunit taken at 44, 45 and 49 ns, succeeding the influx of water molecules and dislocation of substrate (see Figure 8). The substrate which was partially dissociated at $t = 44$ ns is attracted back toward the binding pocket at 45 ns, but moves back towards the cytoplasm this time enabled by the substantial "opening" of the HP1 loop while the hydrogen bond between Asp- β-carboxylate oxygen and Ser277 hydroxyl group is maintained. The structure in panel B is slightly rotated along the vertical axis to allow for better visualization of the opening of HP1 loop.

4.2 *Conformational dynamics of NSS family members and their structural homologs*

The elucidation of the crystal structure of either the substrate-bound or empty conformers of transporters such as LeuT, Mhp1, vSGLT and AdiC1 has provided insights into local motions that take place in this particular (LeuT fold) family of transporters upon binding the substrate (and/or Na$^+$). Despite some commonalities in their local motions, a single unifying mechanism cannot yet be inferred, suggesting that on a local scale the particular types of motions depend on amino acid identities and specificities. This is in contrast to global motions that are predominantly defined by the fold/architecture (and therefore amenable to coarse-grained analyses), irrespective of the amino acid sequence. In this section (4.2), we focus on these local motions (mainly involved in substrate/Na$^+$ binding or release in a given state), and in Section 5 we will examine the global motions (transitions between inward- and outward-facing states) undergone by secondary transporters.

4.2.1 *Molecular dynamics of transporters that share the LeuT fold*

With regard to local changes near a given state, a comparison between the occluded and open conformers of the outward-facing state of LeuT (based on the crystal structures of the substrate or competitive inhibitor-bound states, respectively) shows that more significant movements occur in TM helices 1b, 2a and 6b which rotate by ~9 degrees to partially close the external cavity.[38] In Mhp1, the major change seen is bending of the N-terminal part of TM10 [40] and in AdiC1, which is not sodium-coupled, TM2, TM6a and TM10 are repositioned following substrate binding, with the most prominent motion occurring in TM6a that rotates by about 40 degrees.[50] Even though these motions are not mutually exclusive when the overall changes in structure from the outward-facing empty state to the inward-facing one are taken into consideration, they do necessitate caution before assuming a general mechanism of conformational change induced upon substrate binding/unbinding in the LeuT fold family.

Several studies have investigated the free energy of binding and the specificity of interactions at the sodium and substrate-binding sites in NSS family members. These include studies performed on the LeuT structure itself [51,52] as well as LeuT-based homology models of the dopamine and the γ-aminobutyric acid transporters (DAT and GAT-1, respectively).[53-55] Such studies proved to yield good agreement with experimental data in general.[53,54] It should be noted that with regard to the energetics of binding, these studies have used as a structural template an occluded conformation of the transporter (in the outward-facing state), which may slightly differ from the empty conformation that is recognized by the substrate and Na$^+$, as indicated by the different substrate- and inhibitor-bound structures of LeuT.[38] Even though the changes between these two conformers may be small in terms of backbone structural deviations, they may still affect the calculated free energies of binding and specific interactions which depend on the detailed geometry of the local environment.

Another common approach has been to conduct steered MD to analyze translocation events that are beyond the time scale of conventional MD.[30,56-58] Again, caution should be exercised in interpreting the binding/unbinding pathways observed upon application of such external forces. A foreseeable "red flag" would be the application of energies (or pulling forces) much higher than that required for the physiologically relevant conformational changes, which would result in unrealistic events or deformations. To alleviate such effects, it is important to adopt small pulling forces and/or rates of deformation. Yet, such simulations are resorted to in the interest of observing local events relevant to the transport of substrate and suggesting new experiments to test or validate the observed mechanisms. A successful application to LeuT revealed, for example, the existence of a second binding site, termed S2 and located ~11 Å "above" the primary site, where the substrate settled while being pulled away from its original binding site towards the EC region.[56,58] Notably, the S2 site has been observed by experiments to serve as an antidepressant (and detergent) binding site.[37,59,60] These studies suggest that binding of a second Leu to S2 might promote Na$^+$-coupled symport, whereas inhibitor (or detergent) binding to the same site leads to a functionally blocked form.[58,60]

4.2.2 *Spontaneous release of Na$^+$ and dislodging of substrate observed in MD simulations of galactose transporter vSGLT*

We recently performed a series of molecular dynamics (MD) simulations using the crystal structure of the sodium/galactose symporter (vSGLT).[30] This particular structure[28] has been reported as the inward-facing, occluded form (Figure 2). Whereas a bound galactose molecule was clearly discernible in the substrate-binding cavity, no sodium ion was observed in this particular structure. A putative Na$^+$-binding site was proposed based on the structural homology between vSGLT and LeuT, the sequence similarity among members of the SSS family and mutational studies.[61] Accordingly, the putative Na$^+$ site would be accessible to the IC medium, but the substrate would be occluded from the IC medium.[28] Investigating the conformational dynamics of this structure via MD simulations showed that a Na$^+$ ion placed at this putative location would form competing interactions with residues on TM2 and TM9 (the counterparts of TM1 and TM8 in LeuT) and escape from the binding site within nanoseconds to migrate to the IC medium with the help of Asp189, located on TM6 (equivalent to TM5 in LeuT (Figures 10a–10c)).[29,30]

These MD simulations further showed that even though the substrate was mostly shielded from the IC by an aromatic side-chain (Tyr263), it was accessible to water molecules that entered the binding site. The substrate exhibited significant rotational mobility within the binding site and formed versatile interactions with the protein (Figures 10d–10e). This is in contrast to the tightly-bound Leu and two Na$^+$ ions in the outward-facing occluded state of LeuT (Figure 5b).[44,56]

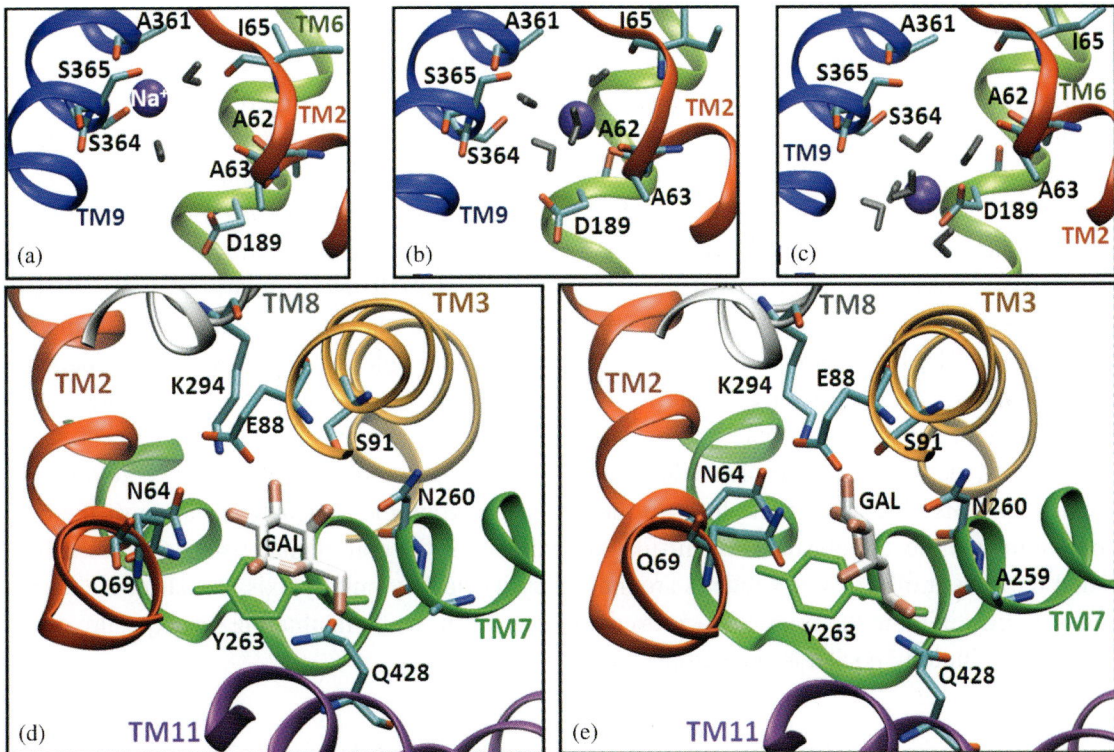

Figure 10. Mobility and interactions of sodium ion and galactose at the binding site of vSGLT. (a–c) Snapshots illustrating the time evolution of the coordination of the sodium ion Na2 (purple sphere) in vSGLT taken at 1, 5 and 13 ns (a–c, respectively). The Na2 site, viewed through the plane of the membrane, is shown together with the interacting residues on TM9 (blue), TM2 (red) and TM6 (green) and water molecules (gray). (d–e) Snapshots of the galactose binding site, viewed from the external side, at 10 and 25 ns, respectively. Tyr263 lying between galactose and the IC medium is shown in green. Water molecules interacting with the galactose have been omitted for clarity. See Ref. 30 for more details.

5. Transitions between Inward- and Outward-Facing States Explored by Computational Analyses

5.1 *Structure prediction via homology modeling based on structural symmetry*

The inverted structural repeats in Glt_{Ph} and LeuT have been used to develop homology models for the inward-facing states of Glt_{Ph}[62] and LeuT[63] based on the outward-facing crystal structures. We present the results for the respective cases in Sections 5.1.1 and 5.1.2.

5.1.1 *Modeling of Glt_{Ph} in the inward-facing state based on the outward-facing structure*

As described above, each Glt_{Ph} subunit is composed of two domains, N-terminal (TM1-TM6) (scaffold) and C-terminal (HP1-TM7-HP2-TM8) (core). Each of these domains is in turn composed of two structural components, also called topological repeat units, which have been noted to be symmetrically arranged with respect to each other in the 3D structure of the particular domain. Thus, each subunit is composed of four topological units overall, as described in Figure 11a. In the N-terminal domain, the helices TM1-3 exhibit an inverted symmetry with respect to TM4-6, and in the C-terminal domain, HP1-TM7 is symmetrically related to HP2-TM8. The two substructures of the scaffold can be closely superimposed onto each other, after two rigid-body rotations of one of them by ~180 degrees as shown in Figure 11b. Likewise, a similar type of superposition is obtained for the core domain. A homology model for the inward-facing state of Glt_{Ph} has been generated[62] by joining the inverted topological repeats, that is, by merging inverted cores 1 and 2 and scaffolds 1 and 2 (note that Figure 11b shows the inversion of core 2 and scaffold 1, only, and similar operations were performed for core 1 and scaffold 2). This resulted in the core-domain moving ~25 Å toward the cytoplasmic region, relative to the outward-facing conformation (Figure 11c). The homology model thus predicted proved to agree very well with the independently resolved structure of Glt_{Ph} in the inward-facing state,[19] in support of the alternate access mechanism for substrate transport.

Notably, the Asp^- molecules embedded in the binding pocket of each of the three core domains are readily "lifted" to a position closer to the cytoplasm upon the *en bloc* translation of the core domain toward the cytoplasm. The "lift-like" movement of the core domain is accompanied by a radial movement of the three subunits around the cylindrical symmetry axis to restrict the access from the EC region to the originally exposed aqueous basin, while facilitating the exposure of the "translated" binding pocket to the cytoplasm.

5.1.2 *Modeling of LeuT in the inward-facing state based on known outward-facing structure*

Unlike the Glt_{Ph} fold, where high-resolution crystal structures have been determined for both the outward- and inward-facing conformations, the LeuT structure has been resolved in the outward-facing form only. Among other Na^+-coupled secondary transporters found by X-ray crystallography to share the LeuT fold (Table 1), Mhp1 is the only one captured in both the outward-facing (occluded and open)[40] and inward-facing[64] forms.

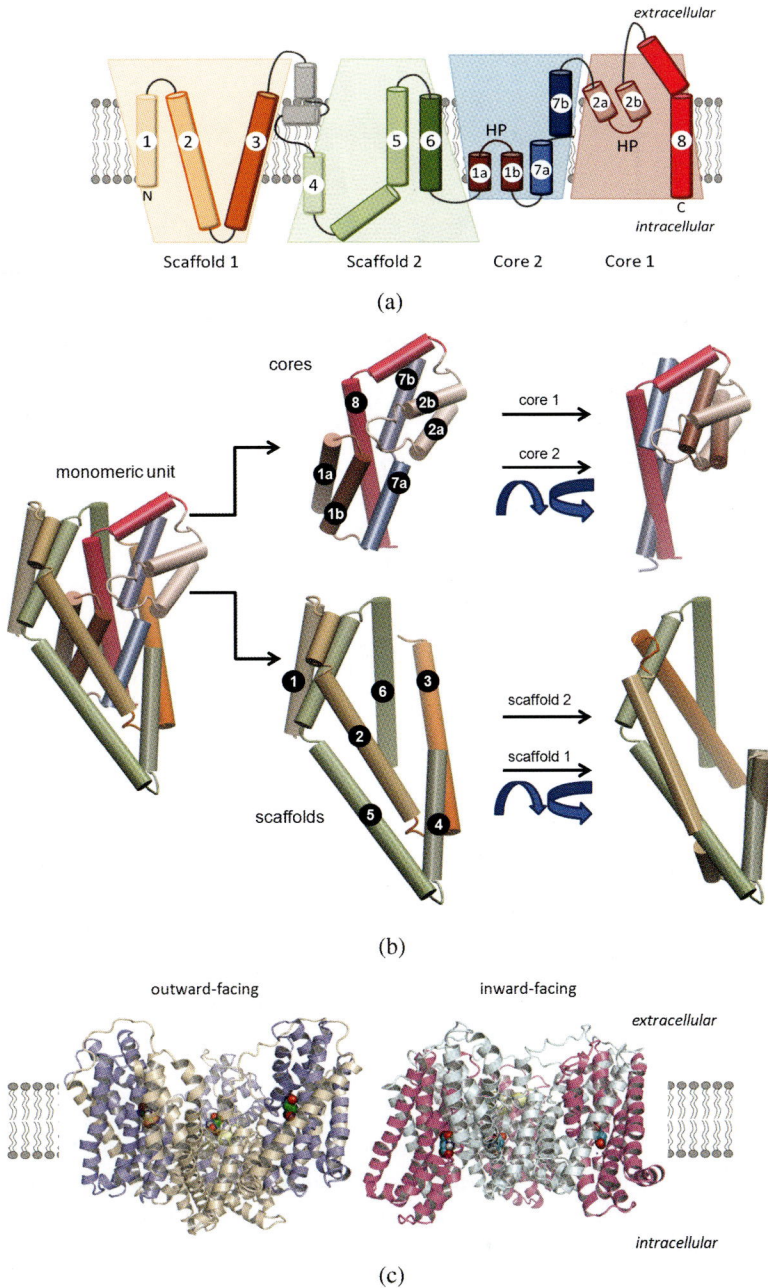

(a)

(b)

(c)

Figure 11. Generation of Glt$_{Ph}$ structure in the inward-facing state, by homology modeling based on outward-facing structure and symmetry considerations. (a) Glt$_{Ph}$ subunits consist each of two domains: scaffold and core, which, in turn, contain two symmetry-related topological repeats. (b) The two repeats in either the scaffold or the core can be closely superimposed upon rotation of one of the repeats by 180 degrees twice. (c) Crystal structures of the outward- and inward-facing states of the Glt$_{Ph}$ trimer. Note the vertical shift in the position of the core domains (purple/magenta ribbons) with respect to the scaffold (gray/cyan) in each subunit. The Asp$^-$ in each subunit (CPK spheres) shifts from an EC-exposed state to an IC-exposed state upon the translation of the binding pocket rigidly embedded in the core domain.

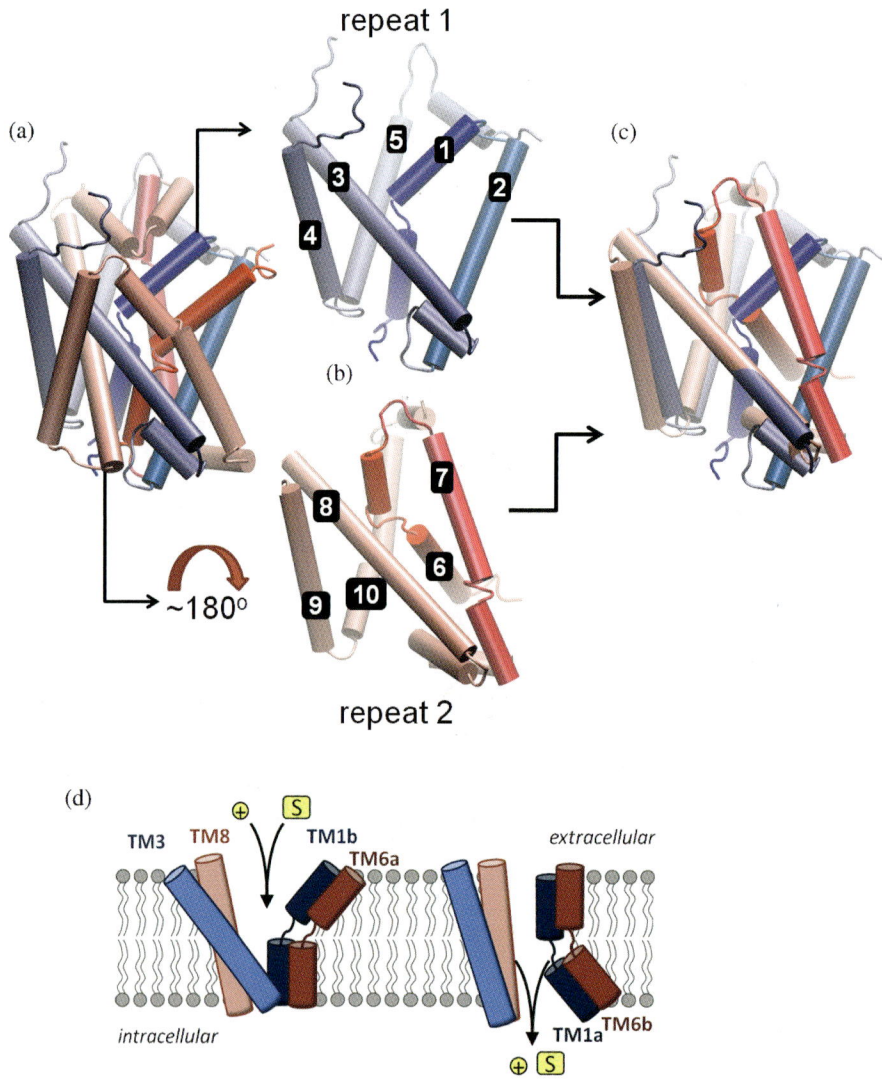

Figure 12. Homology modeling of the structure of NSS family members in the inward-facing state. (a–c) Pseudosymmetric organization of the first (blueish) and second (reddish) LeuT repeats. In this case, the first repeat is taken as is whereas the second is rotated by ~180 degrees (b) and aligned onto the first by superposition of TM3-5 and TM7-10 (c). Panel d gives a schematic description of the most prominent differences between the outward-facing crystal structure of LeuT (left) and the inward-facing crystal structure of vSGLT (right). TM helices are numbered according to LeuT.

Given that the inward-facing form of Mhp1 has been resolved only recently,[64] the structure of vSGLT in the inward-facing state, occluded form[28] served as a prototype for modeling the inward-facing conformation of NSS family members. In particular, a homology model for the inward-facing conformation of the eukaryotic serotonin transporter has been generated and experimentally validated.[63] Figure 12 summarizes the procedure for generating the inward-facing structural model for LeuT. As mentioned above, LeuT consists of two symmetrically related structural repeats: TM1-TM5 and TM6-TM10. The last three helices in each repeat may be superimposed onto each other as illustrated in Figure 12a–12c. In LeuT, the first two helices of each repeat, (TM1 and TM2) and (TM6 and TM7), contain the substrate binding site and form the transport machinery. They are referred to as the "bundle"; whereas the last three helices of each repeat are referred to as the "scaffold".[63] In the inward-facing model, the EC end of the bundle is closer to that of TM3 and TM8 (Figure 12d), whereas at the cytoplasmic side, the bundle is away from these two helices, exposing the substrate to the cytoplasm by opening an aqueous pathway from the binding site to the cytoplasm.[63] The generated model for the inward-facing state closely matches the crystal structure of vSGLT determined in the inward-facing state (Figure 11d).

5.2 *Results from elastic network models*

Although detailed all-atom simulations are an excellent source of information on local protein motions (e.g. the conformational changes along the horizontal arrows/steps in Figure 2), these methods have their limits. Global motions, such as the deformation between the inward- and outward-facing conformations, can alter a protein's structure by several Ångstroms RMSD and take microseconds or longer time scales to complete. Exploring these motions in large molecules like sodium-coupled transporters requires the use of coarse-grained models that use simplifying approximations to reduce the molecule's complexity. One such coarse-grained model that has been applied extensively to study the collective motions of biomolecular systems is the ENM. Computational analyses based on ENMs have proven in recent applications to shed light into conformational mechanisms relevant to the function (e.g. substrate binding, gating, etc.) of membrane proteins.[65]

In a typical ENM, each amino acid is represented by a single point particle, or "node", that is generally coincident with its C^α atom. Nodes within a certain interaction range (cutoff distance of about 7 Å or longer, depending on the particular ENM) are connected with springs, or "edges", each of which has a resting length equal to the equilibrium distance between the nodes it joins. The potential energy of the resulting network is a sum over pairwise harmonic potentials with uniform spring constants, and represented by a single well centered about the native state in the multidimensional description of the energy landscape. The equilibrium dynamics can be decomposed into a set of normal modes of oscillation under this potential, each mode representative of a collective direction of reconfiguration or a basis vector that spans conformational space. Global motions can be estimated from the lowest frequency (or softest) modes.

Normal modes are commonly utilized for generating alternative structures of a protein to be used in predicting its substrate/inhibitor-bound forms, in studying its transitions between different functional forms, and in fitting high-resolution structures into lower-resolution experimental data such as the density maps from cryo-electron microscopy. The growing size of the PDB enables benchmarking ENMs and understanding its limitations. Recent analysis of large ensembles of soluble protein structures showed that, in general regardless of the size of the protein, the experimentally observed variations in NMR models closely relate to different functional forms determined by X-ray crystallography,[66] and these functional forms in turn conform to the intrinsic global dynamics of the proteins as predicted by ENM-NMA.[67,68]

Here, we evaluate the ability of ENMs to predict the conformational transition from the outward-facing conformations of sodium-coupled transporters to inward-facing ones, or vice versa. Global modes of motion are calculated using the Anisotropic Network Model (ANM) with the standard cutoff distance of 15 Å.[69,70] We compare the low-frequency spectrum of ANM modes to the deformation vectors inferred from experiments for (i) Glt_{Ph} outward-to-inward transition (using the coordinates of the 1215 commonly resolved residues) and (ii) Mhp1 outward-to-inward transition (using 463 commonly resolved residues). As a metric of the correlation between predictions and experiments, we examine the level of cumulative overlap $(\Sigma_k \cos^2(d, u_k))^{1/2}$ between the structural change d inferred from experiments (upon suitable superposition of known structures), and the modal changes (eigenvectors, u_k) predicted by the ANM. Here suitable superimposition means the elimination of the difference in six external degrees of freedom using Kabsch algorithm.[71]

Results are presented in Figure 13. The labels refer to the "starting structure" in each case, for which ANM calculations have been performed. As seen in panel A, the first 80 modes (or 2% of the complete set of ANM modes) can satisfactorily account for about 80% of the structural change between the inward-facing and outward-facing forms of Glt_{Ph}. In the case of Mhp1, 30 modes (or 2%) account for about 60–70% of the change in the structure. These numbers are slightly lower than those obtained in studies of soluble globular proteins, for which around 80% correspondence is achieved by using 1–10 softest (non-degenerate) modes. Yet, they support

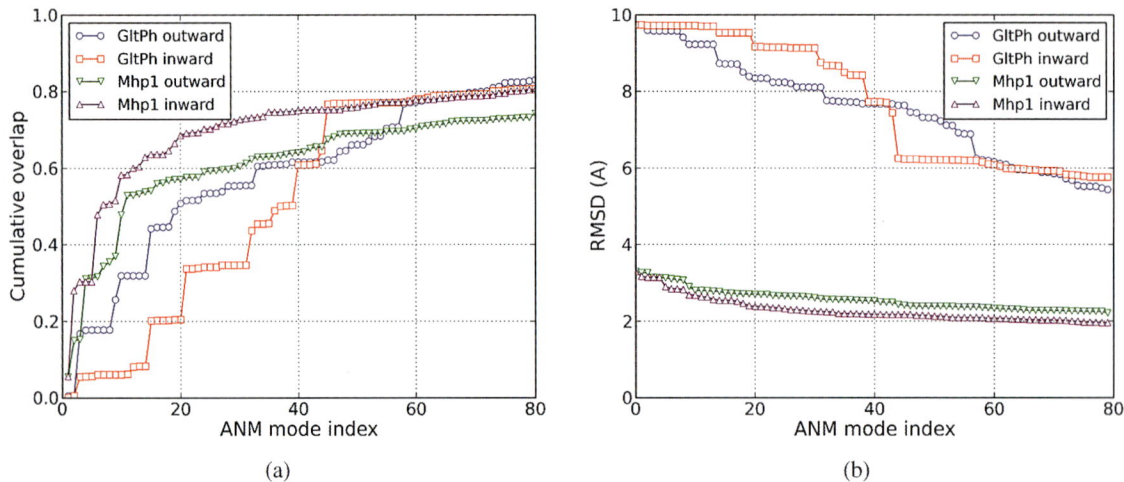

(a)

(b)

Figure 13. Examination of global transitions between the outward-facing and inward-facing structures of transporters using the ANM. (a) Cumulative overlap between the ANM modes of motion predicted for the indicated structures (see labels) and the experimentally observed structural changes between the alternative states of Glt_{Ph} and Mhp1. The structural changes from experiments are described in terms of 3N-dimensional deformation vectors d, which are compared with the 3N-dimensional eigenvectors predicted by the ANM. The deformation vector for Glt_{Ph} is evaluated from the comparison of the PDB entries 1xfh and 3kbc for the respective outward- and inward-facing states. That of Mhp1 is evaluated using the PDB entries 1jln and 2x79. (b) Gradual reconfiguration of a given structure (labeled) during stepwise movements along the ANM soft modes. The RMSDs between the instantaneous conformations and the target structures are shown.

the view that a small subset of modes in the lowest frequency end of the mode spectrum can be advantageously exploited for exploring the conformational subspace where the functional transitions take place.

We also note the stepwise increases in cumulative overlap at particular modes of Glt$_{Ph}$. Consistent with earlier studies, these are non-degenerate modes: they induce three-fold symmetric changes, equally affecting all three subunits. For example, the outward-facing state of Glt$_{Ph}$ favors a global opening/closing of the aqueous basin (mode 3), which alone contributes by ~0.2 to the cumulative overlap between predictions and experiments.[65] A similar feature is observed for the second ANM mode accessible to the Mhp1, suggesting that this mode plays a significant role in establishing the transition from the inward-facing state of Mhp1 to its outward-facing state.

In addition, the RMSD between the ANM-predicted instantaneous conformation and the target structure may be probed, as the starting conformation gradually reconfigures along the soft modes. Figure 13b displays the decrease in RMSD obtained for the two test cases' forward and reverse transitions. For Glt$_{Ph}$, the movement in the subspace of 70 ANM modes decreases the RMSD from 9.75 Å to about 5.5 Å upon starting from either conformation. For Mhp1, the decrease in RMSD is from 3.3 Å to 2 Å. These results suggest that the ANM can partially account for the structural changes observed in sodium-coupled transporters, and a few soft modes (about 2%) provide a reasonable estimate of the initial steps along the transition pathway. Further improvement in the methodology (and agreement with experiments) may presumably be achieved (i) upon more accurate description of the potentials (e.g. force constants) between interacting pair of residues to account for chain connectivity or other sequence-specific effects, and/or (ii) by explicitly taking the effect of the lipid bilayer into consideration in the analysis. The growing number of membrane protein structures will certainly guide us in developing more suitable ENMs or other coarse-grained models and methods.

6. Conclusion

Knowledge of transporter structure is a significant step towards gaining insights into the critical interactions that mediate substrate binding and translocation, or the structural determinants of the particular biological function. Structural data become especially useful when multiple structures along the translocation cycle are resolved. But, the information deduced from these structures on the molecular mechanisms of substrate transfer is only indirect: the detailed mechanisms and time evolution of the role of Na$^+$ ions or the conformational changes that control substrate translocation cannot be inferred from static structures. The resolved structures are, as we have shown, snapshots of a continuum of processes. Simulations conducted with resolved structures reveal the detailed (atomic scale) mechanisms of gate opening and closing events and substrate/cation release. Likewise, the "occluded" vSGLT conformer has been revealed by simulations to readily provide access to Na$^+$ migration to the cytoplasm. Simulations with Glt$_{Ph}$ outward-facing structure also revealed the role of Na$^+$ ions in stabilizing the closed form of the EC gate after substrate binding. Finally, coarse-grained simulations provide insights into evolutionarily optimized patterns/mechanisms that enable the transition between inward-facing and outward-facing states of these transporters. The pathway of transition between alternative functional states may now be efficiently explored by taking steps in the subspace spanned by soft modes. While there are significant advances in single-molecule measurements and detection of time-resolved data or transition events, for the most part, the collection of such data at atomic resolution is very hard, if not impossible. Likewise, while a number of structures along the allosteric cycle are known, a complete description of the sequence of events and their driving mechanisms at the microscopic scale is still lacking. Structure-based computations serve as important tools for improving our understanding of the molecular mechanisms of biological function, and are expected to become even more useful in the future with increasing data on the membrane proteins structures and function that will allow for optimizing computational models and methods.

References

1. C. J. Law, P. C. Maloney and D. N. Wang. 2008. Ins and outs of major facilitator superfamily antiporters. *Annu Rev Microbiol* **62**, 289–305.
2. B. Poolman. 1990. Precursor/product antiport in bacteria. *Mol Microbiol* **4**, 1629–1636.
3. B. I. Kanner and E. Zomot. 2008. Sodium-coupled neurotransmitter transporters. *Chem Rev* **108**, 1654–1668.
4. G. E. Torres and S. G. Amara. 2007. Glutamate and monoamine transporters: New visions of form and function. *Curr Opin Neurobiol* **17**, 304–312.
5. C. I. Wang and R. J. Lewis. 2010. Emerging structure-function relationships defining monoamine NSS transporter substrate and ligand affinity. *Biochem Pharmacol* **79**, 1083–1091.
6. T. S. Ikeda, E. S. Hwang, M. J. Coady, B. A. Hirayama, M. A. Hediger and E. M. Wright. 1989. Characterization of a Na^+/glucose cotransporter cloned from rabbit small intestine. *J Membr Biol* **110**, 87–95.
7. E. Padan, M. Venturi, Y. Gerchman and N. Dover. 2001. Na^+/H^+ antiporters. *Biochim Biophys Acta* **1505**, 144–157.
8. J. Orlowski and S. Grinstein. 2004. Diversity of the mammalian sodium/proton exchanger SLC9 gene family. *Pflugers Arch* **447**, 549–565.
9. J. M. Wood. 2007. Bacterial osmosensing transporters. *Methods Enzymol* **428**, 77–107.
10. S. Ressl, A. C. Terwisscha van Scheltinga, C. Vonrhein, V. Ott and C. Ziegler. 2009. Molecular basis of transport and regulation in the Na^+/betaine symporter BetP. *Nature* **458**, 47–52.
11. K. Tanaka, K. Watase, T. Manabe, K. Yamada, M. Watanabe, K. Takahashi, H. Iwama, T. Nishikawa, N. Ichihara, T. Kikuchi, S. Okuyama, N. Kawashima, S. Hori, M. Takimoto and K. Wada. 1997. Epilepsy and exacerbation of brain injury in mice lacking the glutamate transporter GLT-1. *Science* **276**, 1699–1702.
12. N. Zerangue and M. P. Kavanaugh. 1996. Flux coupling in a neuronal glutamate transporter. *Nature* **383**, 634–637.
13. S. G. Amara and A. C. Fontana. 2002. Excitatory amino acid transporters: Keeping up with glutamate. *Neurochem Int* **41**, 313–318.
14. N. C. Danbolt. 2001. Glutamate uptake. *Prog Neurobiol* **65**, 1–105.
15. B. I. Kanner. 1994. Sodium-coupled neurotransmitter transport: Structure, function and regulation. *J Exp Biol* **196**, 237–249.
16. A. L. Lomize, I. D. Pogozheva, M. A. Lomize and H. I. Mosberg. 2006. Positioning of proteins in membranes: A computational approach. *Protein Sci* **15**, 1318–1333.

17. M. H. Saier, Jr. and Q. Ren. 2006. The bioinformatic study of transmembrane molecular transport. *J Mol Microbiol Biotechnol* **11**, 289–290.

18. O. Boudker, R. M. Ryan, D. Yernool, K. Shimamoto and E. Gouaux. 2007. Coupling substrate and ion binding to extracellular gate of a sodium-dependent aspartate transporter. *Nature* **445**, 387–393.

19. N. Reyes, C. Ginter and O. Boudker. 2009. Transport mechanism of a bacterial homologue of glutamate transporters. *Nature* **462**, 880–885.

20. D. Yernool, O. Boudker, Y. Jin and E. Gouaux. 2004. Structure of a glutamate transporter homologue from *Pyrococcus horikoshii*. *Nature* **431**, 811–818.

21. A. Yamashita, S. K. Singh, T. Kawate, Y. Jin and E. Gouaux. 2005. Crystal structure of a bacterial homologue of Na$^+$/Cl$^-$ dependent neurotransmitter transporters. *Nature* **437**, 215–223.

22. T. L.Whitworth, L. C. Herndon and M. W. Quick. 2002. Psychostimulants differentially regulate serotonin transporter expression in thalamocortical neurons. *J Neurosci* **22**, RC192.

23. T. Beuming, J. Kniazeff, M. L. Bergmann, L. Shi, L. Gracia, K. Raniszewska, A. H. Newman, J. A. Javitch, H. Weinstein, U. Gether and C. J. Loland. 2008. The binding sites for cocaine and dopamine in the dopamine transporter overlap. *Nat Neurosci* **11**, 780–789.

24. E. M. Wright, D. D. Loo, B. A. Hirayama and E. Turk. 2004. Surprising versatility of Na$^+$-glucose cotransporters: SLC5. Physiology (Bethesda) **19**, 370–376.

25. E. M. Wright, B. A. Hirayama and D. F. Loo. 2007. Active sugar transport in health and disease. *J Intern Med* **261**, 32–43.

26. J. Francis, J. Zhang, A. Farhi, H. Carey and D. S. Geller. 2004. A novel SGLT2 mutation in a patient with autosomal recessive renal glucosuria. *Nephrol Dial Transplant* **19**, 2893–2895.

27. E. M. Wright, E. Turk and M. G. Martin. 2002. Molecular basis for glucose-galactose malabsorption. *Cell Biochem Biophys* **36**, 115–121.

28. S. Faham, A. Watanabe, G. M. Besserer, D. Cascio, A. Specht, B. A. Hirayama, E. M. Wright and J. Abramson. 2008. The crystal structure of a sodium galactose transporter reveals mechanistic insights into Na$^+$/sugar symport. *Science* **321**, 810–814.

29. J. Li and E. Tajkhorshid. 2009. Ion-releasing state of a secondary membrane transporter. *Biophys J* **97**, L29–L31.

30. E. Zomot and I. Bahar. 2010. The sodium/galactose symporter crystal structure is a dynamic, not so occluded state. *Mol Biosyst* **6**, 1040–1046.

31. P. Mitchell. 1957. A general theory of membrane transport from studies of bacteria. *Nature* **180**, 134–136.

32. I. Smirnova, V. Kasho, J. Sugihara and H. R. Kaback. 2009. Probing of the rates of alternating access in LacY with Trp fluorescence. *Proc Natl Acad Sci U S A* **106**, 21561–21566.

33. I. Smirnova, V. Kasho, J. Y. Choe, C. Altenbach, W. L. Hubbell and H. R. Kaback. 2007. Sugar binding induces an outward facing conformation of LacY. *Proc Natl Acad Sci U S A* **104**, 16504–16509.

34. L. Guan and H. R. Kaback. 2006. Lessons from lactose permease. *Annu Rev Biophys Biomol Struct* **35**, 67–91.

35. L. R. Forrest, S. Tavoulari, Y. W. Zhang, G. Rudnick and B. Honig. 2007. Identification of a chloride ion binding site in Na$^+$/Cl -dependent transporters. *Proc Natl Acad Sci U S A* **104**, 12761–12766.

36. E. Zomot, A. Bendahan, M. Quick, Y. Zhao, J. A. Javitch and B. I. Kanner. 2007. Mechanism of chloride interaction with neurotransmitter:sodium symporters. *Nature* **449**, 726–730.

37. S. K. Singh, A. Yamashita and E. Gouaux. 2007. Antidepressant binding site in a bacterial homologue of neurotransmitter transporters. *Nature* **448**, 952–956.

38. S. K. Singh, C. L. Piscitelli, A. Yamashita and E. Gouaux. 2008. A competitive inhibitor traps LeuT in an open-to-out conformation. *Science* **322**, 1655–1661.

39. L. Tang, L. Bai, W. H. Wang and T. Jiang. 2010. Crystal structure of the carnitine transporter and insights into the antiport mechanism. *Nat Struct Mol Biol* **17**, 492–496.

40. S. Weyand, T. Shimamura, S. Yajima, S. Suzuki, O. Mirza, K. Krusong, E. P. Carpenter, N. G. Rutherford, J. M. Hadden, J. O'Reilly, P. Ma, M. Saidijam, S. G. Patching, R. J. Hope, H. T. Norbertczak, P. C. Roach, S. Iwata, P. J. Henderson and A. D. Cameron. 2008. Structure and molecular mechanism of a nucleobase-cation-symport-1 family transporter. *Science* **322**, 709–713.

41. W. Humphrey, A. Dalke and K. Schulten. 1996. VMD: Visual molecular dynamics. *J Mol Graph* **14**, 33–38.

42. E. Gouaux. 2009. Review. The molecular logic of sodium-coupled neurotransmitter transporters. *Philos Trans R Soc Lond B Biol Sci* **364**, 149–154.

43. H. Krishnamurthy, C. L. Piscitelli and E. Gouaux. 2009. Unlocking the molecular secrets of sodium-coupled transporters. *Nature* **459**, 347–355.

44. D. A. Caplan, J. O. Subbotina and S. Y. Noskov. 2008. Molecular mechanism of ion-ion and ion-substrate coupling in the Na^+-dependent leucine transporter LeuT. *Biophys J* **95**, 4613–4621.

45. S. Y. Noskov and B. Roux. 2008. Control of ion selectivity in LeuT: Two Na+ binding sites with two different mechanisms. *J Mol Biol* **377**, 804–818.

46. I. Bahar, T. R. Lezon, L. W. Yang and E. Eyal. 2010. Global dynamics of proteins: Bridging between structure and function. *Annu Rev Biophys* **39**, 23–42.

47. I. H. Shrivastava, J. Jiang, S. G. Amara and I. Bahar. 2008. Time-resolved mechanism of extracellular gate opening and substrate binding in a glutamate transporter. *J Biol Chem* **283**, 28680–28690.

48. Z. Huang and E. Tajkhorshid. 2008. Dynamics of the extracellular gate and ion-substrate coupling in the glutamate transporter. *Biophys J* **95**, 2292–2300.

49. J. Dechancie and I. Bahar. 2011. The mechanism of substrate release by the aspartate transporter Glt(Ph): Insights from simulations. *Mol Biosyst* **7**, 832–842.

50. X. Gao, L. Zhou, X. Jiao, F. Lu, C. Yan, X. Zeng, J. Wang and Y. Shi. 2010. Mechanism of substrate recognition and transport by an amino acid antiporter. *Nature* **463**, 828–832.

51. S. Y. Noskov and B. Roux. 2008. Control of ion selectivity in LeuT: Two Na^+ binding sites with two different mechanisms. *J Mol Biol* **377**, 804–818.

52. D. A. Caplan, J. O. Subbotina and S. Y. Noskov. 2008. Molecular mechanism of ion-ion and ion-substrate coupling in the Na^+-dependent leucine transporter *LeuT Biophys J* **95**, 4613–4621.

53. T. Wein and K. T. Wanner. 2009. Generation of a 3D model for human GABA transporter hGAT-1 using molecular modeling and investigation of the binding of GABA. *J Mol Model* **16**, 155–161.

54. X. Huang and C. G. Zhan. 2007. How dopamine transporter interacts with dopamine: Insights from molecular modeling and simulation. *Biophys J* **93**, 3627–3639.

55. P. C. Gedeon, M. Indarte, C. K. Surratt and J. D. Madura. 2010. Molecular dynamics of leucine and dopamine transporter proteins in a model cell membrane lipid bilayer. *Proteins* **78**, 797–811.

56. L. Celik, B. Schiott and E. Tajkhorshid. 2008. Substrate binding and formation of an occluded state in the leucine transporter. *Biophys J* **94**, 1600–1612.

57. Y. Gu, I. H. Shrivastava, S. G. Amara and I. Bahar. 2009. Molecular simulations elucidate the substrate translocation pathway in a glutamate transporter. *Proc Natl Acad Sci U S A* **106**, 2589–2594.

58. L. Shi, M. Quick, Y. Zhao, H. Weinstein and J. A. Javitch. 2008. The mechanism of a neurotransmitter:sodium symporter — inward release of Na⁺ and substrate is triggered by substrate in a second binding site. *Mol. Cell* **30**, 667–677.

59. Z. Zhou, J. Zhen, N. K. Karpowich, R. M. Goetz, C. J. Law, M. E. Reith and D. N. Wang. 2007. LeuT-desipramine structure reveals how antidepressants block neurotransmitter reuptake. *Science* **317**, 1390–1393.

60. M. Quick, A. M. Winther, L. Shi, P. Nissen, H. Weinstein and J. A. Javitch. 2009. Binding of an octyl-glucoside detergent molecule in the second substrate (S2) site of LeuT establishes an inhibitor-bound conformation. *Proc Natl Acad Sci U S A* **106**, 5563–5568.

61. E. Turk, O. Kim, C. J. le, J. P. Whitelegge, S. Eskandari, J. T. Lam, M. Kreman, G. Zampighi, K. F. Faull and E. M. Wright. 2000. Molecular characterization of *Vibrio parahaemolyticus* vSGLT: A model for sodium-coupled sugar cotransporters. *J Biol Chem* **275**, 25711–25716.

62. T. J. Crisman, S. Qu, B. I. Kanner and L. R. Forrest. 2009. Inward-facing conformation of glutamate transporters as revealed by their inverted-topology structural repeats. *Proc Natl Acad Sci U S A* **106**, 20752–20757.

63. L. R. Forrest, Y. W. Zhang, M. T. Jacobs, J. Gesmonde, L. Xie, B. H. Honig and G. Rudnick. 2008. Mechanism for alternating access in neurotransmitter transporters. *Proc Natl Acad Sci U S A* **105**, 10338–10343.

64. T. Shimamura, S. Weyand, O. Beckstein, N. G. Rutherford, J. M. Hadden, D. Sharples, M. S. Sansom, S. Iwata, P. J. Henderson and A. D. Cameron. 2010. Molecular basis of alternating access membrane transport by the sodium-hydantoin transporter Mhp1. *Science* **328**, 470–473.

65. I. Bahar, T. R. Lezon, A. Bakan and I. H. Shrivastava. 2010. Normal mode analysis of biomolecular structures: functional mechanisms of membrane proteins. *Chem Rev* **110**, 1463–1497.

66. O. F. Lange, N. A. Lakomek, C. Fares, G. F. Schroder, K. F. Walter, S. Becker, J. Meiler, H. Grubmuller, C. Griesinger and B. L. de Groot. 2008. Recognition dynamics up to microseconds revealed from an RDC-derived ubiquitin ensemble in solution. *Science* **320**, 1471–1475.

67. A. Bakan and I. Bahar. 2009. The intrinsic dynamics of enzymes plays a dominant role in determining the structural changes induced upon inhibitor binding. *Proc Natl Acad Sci U S A* **106**, 14349–14354.

68. L. Yang, G. Song, A. Carriquiry and R. L. Jernigan. 2008. Close correspondence between the motions from principal component analysis of multiple HIV-1 protease structures and elastic network modes. *Structure* **16**, 321–330.

69. A. R. Atilgan, S. R. Durell, R. L. Jernigan, M. C. Demirel, O. Keskin and I. Bahar. 2001. Anisotropy of fluctuation dynamics of proteins with an elastic network model. *Biophys J* **80**, 505–515.

70. E. Eyal, L. W. Yang and I. Bahar. 2006. Anisotropic network model: Systematic evaluation and a new web interface. *Bioinformatics* **22**, 2619–2627.

71. W. Kabsch. 1976. A solution for the best rotation to relate two sets of vectors. *Acta Crystallographica* Section A **32**, 922–923.

Voltage-Gated Ion Channels

The Machines Responsible for the Nerve Impulse

12

Benoît Roux and Francisco Bezanilla

1. Introduction

The action potential is a transient change in the membrane electrical potential that nerve cells, skeletal and heart muscle and other excitable tissues use as a communication signal. The initial membrane depolarization, the rising phase of the action potential, usually develops in less than a millisecond and spans a voltage change from the resting potential of about −70 mV (negative inside) to a peak of about +40mV. The return of the membrane to its resting potential, the falling phase, is usually slower. At the heart of the action potential generation, there are ionic conductances, which are ion-selective and voltage-dependent.

The ionic basis of the generation and propagation of the action potential was put forward by the elegant experiments of Hodgkin and Huxley that were summarized in a mathematical formulation of the voltage dependence of the ion-selective conductances in 1952.[1] It was subsequently understood that the molecular basis of those ion conductances were specialized membrane-embedded proteins called "ion channels". For much of the following decades, scientists further refined knowledge of these systems, probing the size and shape of channels with various artificial blockers,[2,3] the gating "current" corresponding to the movements of protein-bound charges within the membrane potential,[4,5] and the microscopic manifestations of ion selectivity.[6-8] The advent of single-channel recordings in the late 1970's allowed one to measure the activity of isolated channels in live membranes for the first time.[9] In the 1980's, the cloning of the Na^+ channel,[10,11] and of the K^+ channel,[12] made it possible to analyze the primary amino-acid sequence of those channels, and predict the existence of the transmembrane (TM) segments of the protein. Further progress in molecular biology then allowed the introduction of artificial site-directed mutations into the protein.[13-15] The study of ion channels made an even bigger leap with the availability of high-resolution X-ray crystallographic structures in the last decade or so.[16-21]

Today, many ion channels have been cloned and expressed, allowing their functional characterization with electrophysiological techniques and in some cases, their structure has been obtained and major progress has been done in correlating structure with function. The combination of

231

atomic-level information from experiments together with the advances in molecular dynamics methods and the increasing power of modern computers now make it possible to simulate "virtual" models of these proteins.[21–29]

In the large family of ion channels, voltage-dependent ion channels are the molecular machines underlying the generation and propagation of the action potential. In this chapter, we will give a brief overview of elementary physical aspects of membrane excitability and the generation of the action potential. We will then describe the basic properties of these channels. The description will attempt to give a modern view of the structural features that give origin to the two basic properties, selectivity and voltage dependence, as have been obtained by a combination of electrophysiology, molecular biology, spectroscopy, crystallography and computational chemistry.

2. Biological Membranes and Channels

The membranes of living cells are complicated extensive molecular structures, about 30–40 Å thick, held together by many cooperative noncovalent interactions.[30] Lipid molecules, which represent the main constituents of membranes, possess long nonpolar hydrocarbon chains attached to a polar moiety. The nonpolar "hydrophobic" chains exclude water, while the polar "hydrophilic" headgroup of the molecule is strongly associated with water molecules.[30,31] By virtue of their physical properties, lipid molecules spontaneously self-assemble to form an extensive supramolecular structure called a bilayer, with the polar headgroup of the lipids exposed to bulk water and the hydrocarbon chains segregated from it (Figure 1). The cohesion of the bilayer structure arises from its ability to simultaneously satisfy the contrasting solvation requirements of the hydrophobic and hydrophilic moieties of the lipid molecules.

The viability of living cells depends upon the ability of the membranes to isolate and protect the integrity of its content and to provide regulatory, selective, permeation mechanisms for the

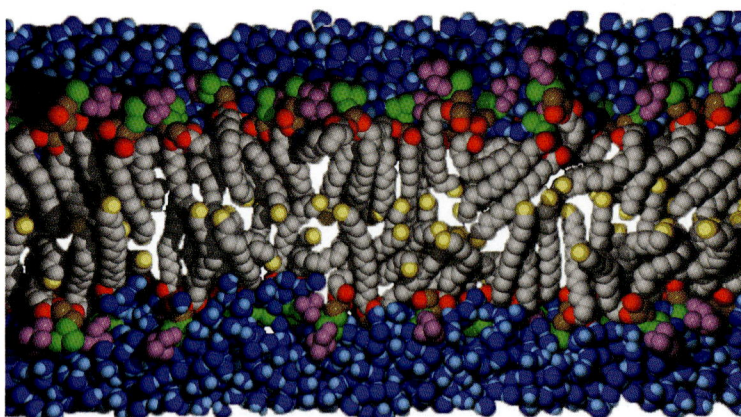

Figure 1. Molecular graphics representation of a phospholipid membrane bilayer of dipalmitoyl phosphatidylcholine (DPPC). The nonpolar hydrocarbon chains of the lipids are represented in gray with the last carbon as a yellow sphere. The hydrophilic headgroup is represented in red, green, purple and brown. The surrounded water molecules are represented in blue. (Figure adapted/reproduced with permission/modified from Feller *et al.*)[31]

transport of material in and out of the numerous cellular compartments. The hydrocarbon chains form a nonpolar (low dielectric) core at the center of the bilayer. Small ions such as Na^+ and K^+ are strongly bound to water molecules in bulk solution. Highly specific membrane-spanning macromolecular structures, ion channels and transporters serve to facilitate and control the passage of selected ions across the lipid barrier.[32] Without such specialized proteins, the hydrocarbon region of the lipid membrane would be essentially impermeable by presenting a prohibitively high energy barrier to the passage of any ion. It is the role of an ion channel to control and facilitate the passage of selected ions across the membrane by providing coordinating groups to help compensate the loss of hydration of the ions while they traverse following their electrochemical gradient.[32,33] An important aspect of ion channels is their selectivity, i.e. the channel provides a favorable passage across the membrane in a manner that is more favorable to one type of ion over another type of ion. Selectivity is so important that channels are typically named after the ion that they favor, hence, K^+ and Na^+ channels are selective for K^+ and Na^+ ions, respectively. In the nerve membrane, there are several types of channels, each of which is selective to a specific ion. Therefore, the situation of zero net flow across the membrane does not depend on one particular ion concentration gradient but it involves the concentration of the other permeant ions and their relative permeabilities. In this situation, we have to consider the individual fluxes and the stationary state solution.

In response to different stimuli, such as a change in the membrane potential, the conformation of the channel molecules can change and the ionic pathway may be open or closed. This is the gating process. Because of thermal motion, the protein undergoes conformational transitions between the closed and open states. Such transitions occur as random events but in voltage-gated channels, the probability of being in the closed state as opposed to the open state is biased by the membrane potential. Na^+- and K^+-channels in neurons are "voltage-dependent": their open probability is low at negative (hyperpolarized) potentials and is high at positive (depolarized) potentials.[34] This means that it is not possible to predict at any given time whether the channel will be open or closed. The laws of statistical probability, however, allow us to make certain predictions of the average behavior of the channel, which naturally emerges from large populations of channels.

Voltage-gated K^+ (Kv) channels are transmembrane proteins that control and regulate the flow of K^+ ions across cell membranes. In response to changes in the membrane potential, they undergo conformational changes, thereby allowing or blocking the passage of selected ions. Structurally, Kv channels are formed by four subunits surrounding a central aqueous pore for K^+ permeation (Figure 2). Each subunit comprises six transmembrane α-helical segments called S1 to S6. The first four segments, S1–S4, constitute the voltage sensor domain (VSD). The discovery of similar domains in unrelated membrane proteins lacking a conduction pore have established the concept of the VSD as an independent functional module. The last two transmembrane segments, S5 and S6, from each of the four subunits join around a common axis to form a selective ion-conducting pore domain (PD). Some K^+-selective channel structures possess only an ion-conducting PD, without the four VSDs. For example, the bacterial channel KcsA from *Streptomyces lividans* is formed by four subunits, each comprising two transmembrane segments with a high sequence similarity to the S5–S6 segments of Kv channels.[16] Although there is no atomic resolution information about Na^+ channels yet, their overall structure is believed to closely parallel that of K^+ channels, with a single long polypeptide chain forming four homologous domains, each with six putative membrane spanning

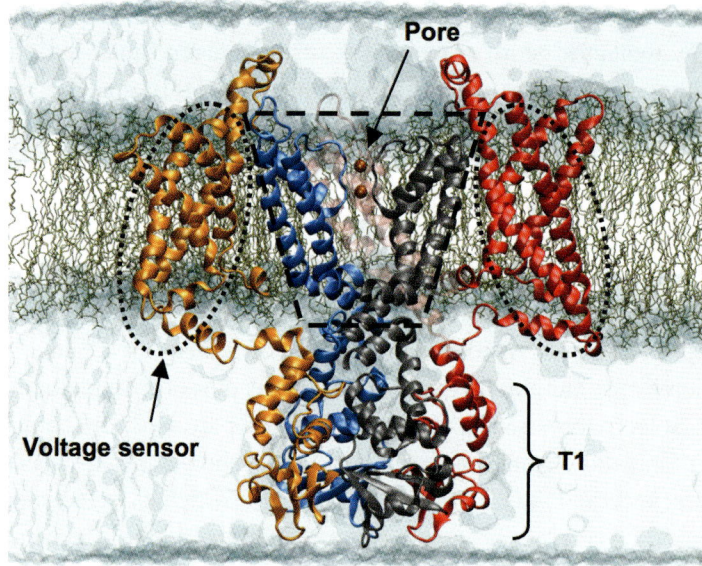

Figure 2. Schematic view of the Kv1.2 channel in a membrane. The channel is a tetramer but only two subunits of the pore domain are shown (blue and gray) and only two voltage sensor domains (VSD) are shown (orange and red). The intracellular T1 tetramerization domain at the bottom is shown with all four subunits. (Figure adapted/reproduced with permission/modified from Khalili *et al.*)[29]

segments forming a PD surrounded by four VSDs.[32] Such *functional modularity*, with different protein domains playing distinct functional roles, appears to be an important concept in how the molecular machinery of ion channels is constructed. From a functional point of view there are important differences between Na$^+$ and K$^+$ channels. First, K$^+$ channels are highly selective for K$^+$ ions over Na$^+$ by almost a factor of 10^3–10^4 whereas, Na$^+$ channels are selective for Na$^+$ by only a factor of about 15.[32] Another striking difference is the rate of activation. The latency to first opening is about 10 times shorter in Na$^+$ channels than in K$^+$ channels, yielding an overall activation time about 10 times faster.[34] We will return to ion channels below, but first we should explore an important consequence of a selective membrane: the Nernst membrane potential.

3. The Nernst Potential

An electrical potential difference exists across the cell membrane. This potential difference, or *membrane potential*, is due to the differences in ionic concentration in the cytoplasm and the extracellular medium.[35] The membrane potential underlies numerous important physiological processes including the propagation of the nerve impulse, cell excitability, and signal transduction. By virtue of the long-range nature of the Coulombic interactions, variations in the membrane potential insure the synchronization of various molecular processes taking place at distances far larger than typical short range intermolecular forces. It constitutes a unified communication system for physiological information throughout the cell.

The existence of the potential largely depends on the ability of a membrane to act as an insulator by presenting a selective barrier to the passive diffusion of small ions. As first explained by

Nernst in 1889,[36] a potential difference appears spontaneously across a semi-permeable membrane separating two ionic solutions of different concentrations. The potential results from a balance between the entropic tendency to homogenize the ion concentrations in the system, and the necessity to maintain local charge neutrality as much as possible. The struggle between these two opposing forces takes place in the region of the semi-permeable membrane, and the potential difference is the result of this interfacial phenomena. In this situation, which corresponds to thermodynamic equilibrium, the bulk ionic solutions remain electrically neutral and the potential difference across the membrane arises from a very small imbalance of net charges on each side of the membrane. For a monovalent salt, the membrane potential difference (traditionally called E) between the intracellular (i) and outside (o) side is equal to $E = V_i - V_o = -(k_B T/q) \ln([C_i]/[C_o])$, where q is the charge of the ion that is permeable through the membrane, k_B is the Boltzmann constant, T is the temperature, and $[C_i]$ and $[C_o]$ are the ionic concentration inside and outside the cell. For example, the concentration of K^+ inside the cell is about 160 mM, while it is about 10 mM outside the cell. This yields a potential $E_K = -70$ mV inside the cell relative to the outside. This value reflects the membrane potential under resting conditions, indicating that the basal polarized state of the cell membrane is to be largely impermeable to all ions except K^+. On the other hand, the concentration of Na^+ inside the cell is about 10 mM, while it is about 150 mM outside the cell, which yields a potential $E_{Na} = +70$ mV inside the cell relative to outside. The situation where the membrane potential is dominated by the Na^+ concentration gradient corresponds to a depolarized state of the cell membrane.

The origin of the membrane potential can be illustrated with a simple example. Assume that we have a membrane separating two compartments (Figure 3) that has channels only permeable to K^+ such that no other ions can permeate (note that the channels are not shown in Figure 3 for the sake of simplicity). Initially the channels are closed and we add 160 mM of KCl to the lower compartment (say the interior of the cell) and 10 mM of KCl to the upper compartment (the outside). As we have added a neutral salt, there will be the same number of cations and anions in the lower compartment; the same will be true for the upper compartment (even though the total number of

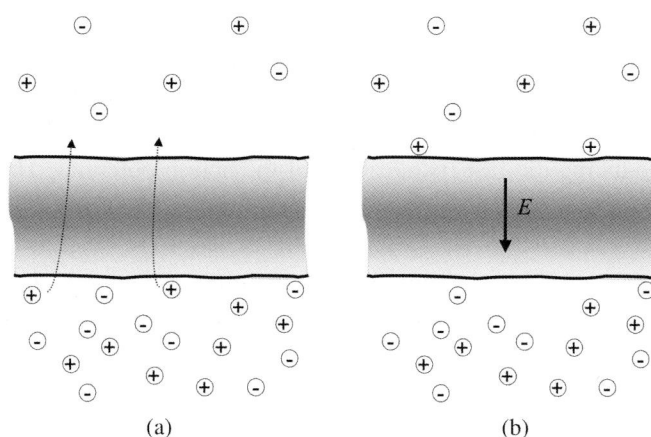

(a)　　　　　　　　　　　　(b)

Figure 3. The origin of the Nernst membrane potential for a semi-permeable membrane that allows the passage of the cations (+). In (a), the number of anions (−) and cations (+) is equal on both side of the membrane. Initially, more cations will diffuse from the high concentration to the low concentration. In the equilibrium state, a net imbalance of charge (shown in b) creates an electric field in the opposite direction.

ions is 10 times lower in the upper compartment). The consequence of the electroneutrality in each side will be a zero charge difference across the membrane and consequently the membrane potential difference will be zero. The permanent thermal motion of the ions will make them move randomly but they will not be able to cross the membrane because they are poorly permeant through the bilayer and the channels are closed. Suppose that at one point we open the K^+ channels. Then, as there are 10 times more K^+ ions in the bottom compartment than in the top compartment, there will be 10 times more chances of an ion crossing up than down. This initial situation is schematically pictured in Figure 1a, where each K^+ ion crossing in the upward direction leaves a Cl^- ion behind. This diffusive flow, which is proportional to the concentration gradient, increases the top compartment by one positive charge and the lower compartment by one negative charge, producing a charge separation. This charge separation introduces a non-random net electrostatic force acting on the ions, which tends to drive the cations from the top compartment back into the bottom compartment. The final result is that the charge separation will build up a voltage across the membrane, reached when the electrostatic force balances the diffusive flow produced by the concentration gradient. When that happens, any ion that crosses in one direction will be counterbalanced on average by another crossing in the opposite direction, maintaining an equilibrium situation.

4. Selective Channels and the Action Potential

The electric processes underlying the action potential are illustrated in Figure 4. Under resting conditions, the cell membrane is polarized (negative inside the cell), and most of the voltage-gated channels are closed.[34] At this voltage, the probability that the K^+ channels are open is higher than that of the Na^+ channels, yielding a semi-permeable membrane that supports a potential that is very near the K^+ Nernst potential, dominated by the concentrations of K^+ in (high) and out (low) of the cell.[a] This is the polarized or resting state of the cellular membrane. To initiate the action potential, a perturbation occurs (normally a slight depolarization of the membrane caused by the opening of receptors activated by neurotransmitters), which increases the opening probability of voltage-gated Na^+ channels. This suddenly allows an inward Na^+ current across the membrane, which tends to support a Nernst potential E_{Na} dominated by the concentrations of Na^+ in (low) and out (high) of the cell. Consequently, the membrane potential becomes more positive, which in turn, opens even more Na channels, producing the fast rising phase of the action potential. This is the depolarized state of the cellular membrane. At this point, the membrane depolarization starts to trigger the opening probability of voltage-sensitive K^+ channels. The latter are kinetically slower to activate, therefore, their influence starts to be important only after the initial depolarizing phase has occurred. At this point, the K^+ channels allow an outward K^+ current across the membrane, which tends to return toward the resting Nernst potential E_K (repolarization), which is dominated by the K^+ concentrations. The rising phase of the action potential usually develops in less than a

---

[a] The net ionic current (Itot) across the membrane is zero but there is a small inward leak of Na^+ (I_{Na}) and a larger outward leak of K^+. As the permeability to K^+ is larger than that of Na^+, the resting membrane potential V is typically close to the Nernst equation given by the intracellular and extracellular K^+ concentrations.

millisecond and spans a voltage change from the resting membrane potential of about −70mV (negative inside) to a peak of about +40 mV, while the falling phase is usually slower in returning the membrane to its resting level, either directly or sometimes by transiently overshooting its resting value. The time-dependence and the kinetics of the action potential is more complicated than this simple explanation due to the delays caused by the charging of the effective capacitance of the membrane and the voltage and time dependence of the kinetic activation and inactivation of the Na^+ channels, but this is the overall mechanism. The transient depolarization that we just described is generated locally in a circumscribed region of the cell membrane, but it acts as a trigger to initiate other action potentials in contiguous regions of the membrane, thus propagating the impulse as a wave along the neuron.

With all these ionic currents, one might wonder how the cell does not run out of ions! Actually, the amount of ions flowing across the membrane needed to reset the potential is tiny. This can be understood by picturing the lipid bilayer as an insulator separating two conducting media: the ionic solutions on each side of the membrane. This constitutes an electric capacitor with two conducting plates (the ionic solutions) separated by a low dielectric insulator (the membrane).[35] The capacitance C increases with the area of the plates and decreases with the separation between the plates according to the relation $C = \varepsilon_m A/d$, where A is the membrane area, d is the membrane thickness, and ε_m is the dielectric constant of the membrane. As the thickness d is only on the order of 25 Å, the capacitance per unit of area of the membrane ($C = c/A$) is very high, close to 1 $\mu F/cm^2$. In a capacitor, the potential difference V is related to the excess charge by $Q = CmV$. In the case of a cell membrane, a small amount of charge separation is able to generate a large potential difference. For example, to obtain a membrane potential of 100 mV, it is necessary to separate the product of $C_m = 1(\mu F/cm^2)$ times $V_m = 0.1$ (Volts), that is, $Q = 0.1$ μCoulombs/cm^2. To get an idea, this corresponds to only one elementary charge per surface of 130×130 Å^2. Since a physiological salt concentration of 150 mM approximately corresponds to one cation-anion pair per volume of $22 \times 22 \times 22$ Å^3 of solution, the membrane potential arises from a strikingly small accumulation of net charge relative to the bulk concentration. Thus, in the presence of a membrane potential, the bulk

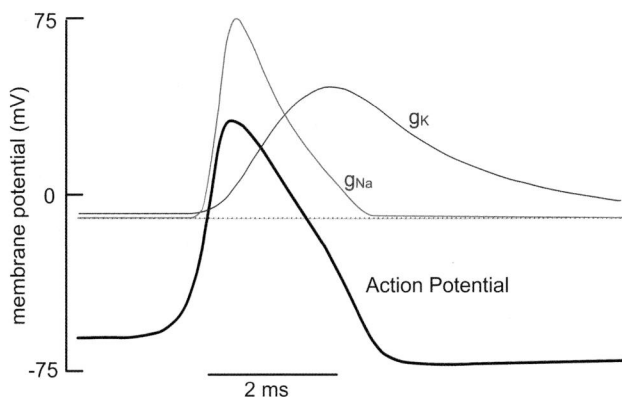

Figure 4. Propagated action potential as computed from the Hodgkin and Huxley equations.[1] These equations allow the computation of the time course of the ionic conductances g_K and g_{Na} that give origin to the action potential. The time course of g_K and g_{Na} during a non-propagated action potential has been verified in direct measurements of conductances by interrupting the action potential with an imposed voltage.[37]

solution remains electrically neutral and only a small charge imbalance is distributed in the neighborhood of the interfaces. This also means that there is not a major change in the bulk salt concentrations caused by the ionic currents that are associated with one action potential. After a train of action potentials, the intracellular ionic concentrations are restored by the ATP-driven Na/K pump, this action is particularly important in small cells that have a large surface to volume ratio.[32] In the following section, we return to survey the functional features of K^+ channels at the atomic level.

5. Channel Function at the Atomic Level

5.1 *Fast ion conduction and selectivity*

A remarkable feature of K^+ channels is their ability to discriminate very effectively for K^+ over Na^+ while being able to maintain a very large throughput rate. K^+ channels can discriminate between K^+ and Na^+ by approximately a factor of 10^3–10^4. As we have seen above, a high selectivity is critical to establish the Nernst resting potential of the living cell. At the same time, conduction through K^+ channels is near the diffusion limit. A question of central interest for biophysicists is thus: What is the molecular mechanism allowing K^+ channels to be so selective while maintaining such a large conductance?

The determination of the structure of the KcsA channel using protein X-ray crystallography provided the first view of the general architecture of the central pore domain — the key molecular structure responsible for controlling ion conduction and selectivity.[16] The KcsA channel is made of four identical subunits, closely homologous to the S5–S6 transmembrane segments of Kv channels, disposed symmetrically around a common axis corresponding to the ion-conducting pore. The "selectivity filter," which corresponds to the narrowest part of the pore, is located near the extracellular side. It is formed by the backbone carbonyl oxygens from a highly conserved sequence of amino acids: TTVGYGD.[13] The narrow filter is only 12 Å long, comprising five specific binding sites called S_0 at the extracellular end to S_4, at the intracellular end. The selectivity filter is surrounded by four short α-helices, the "pore helices," which point their C-terminal end toward the center of a wide water-filled vestibular cavity leading to the intracellular side. The long-range electric field from the dipole of the pore helices helps stabilize and attract incoming K^+ ions on the intracellular side.[22]

One of the most fascinating mechanistic questions about K^+ channels is how they are able to achieve a fast throughput rate and yet remain highly selectivity for K^+ over Na^+. Because of its small radius, small cations such as K^+ are very strongly associated with water molecules in the bulk phase. The hydration free energy of a single K^+ ion is on the order of −80 kcal/mol.[32] Dehydrating a K^+ ion, which is energetically very unfavorable, is nonetheless necessary for the selectivity filter to recognize whether the proper ion is attempting to go through the membrane (see below). To enable fast conduction, the channel must thus provide coordinating ligands that favorably help compensate the loss of hydration. In the K^+ channel, those coordinating ligands are the backbone carbonyl groups along the selectivity filter. This is schematically illustrated in Figure 5a.

But could the strongly attractive interactions between the K^+ ions and the selectivity filter actually slow down the rapid conduction? The reason why this does not happen is because the five

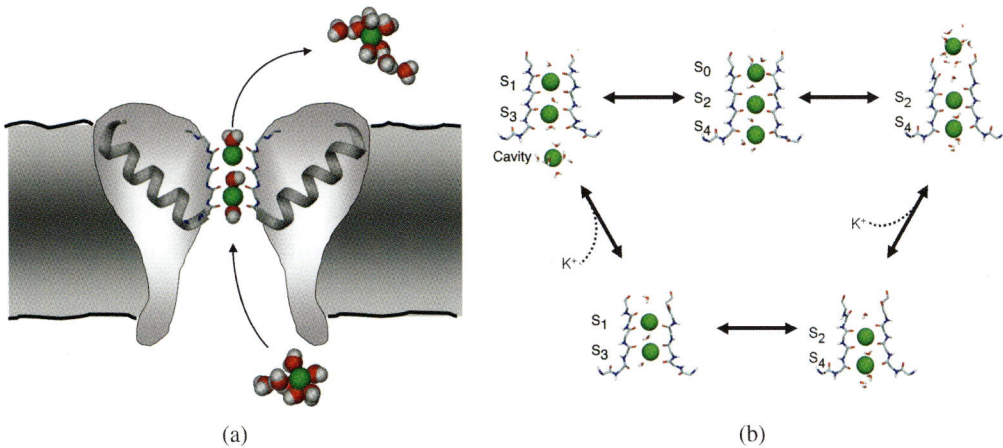

(a)   (b)

Figure 5. Schematic representation of the ion conduction process in K$^+$ channels. (a) A K$^+$ ion on the intracellular bulk solution is completely hydrated. Its hydration free energy is on the order of −80 kcal/mol. To enter into the narrow selectivity filter, the ion must first become almost completely dehydrated. At this stage, the strongly favorable interactions with the carbonyl groups lining the selectivity filter must compensate for the loss of ion solvation. Then, the K$^+$ ion exits on the extracellular side where it can become hydrated once again. (b) The multi-ion translocation process relies on the balance of ion-filter attraction and ion-ion repulsion to enable fast conduction. The transport cycle involves a transition K$^+$···[K$^+$, K$^+$] ↔ [K$^+$, K$^+$]···K$^+$, similar to the knock-on mechanism proposed more than 50 years ago by Hodgkin and Keynes.

binding sites S_0–S_4 along the selectivity filter can be occupied by more than one K$^+$ ion simultaneously. The electrostatic repulsion between the closely spaced K$^+$ ions helps overcome the otherwise very strongly attractive interactions with the carbonyl groups, thus permitting rapid conduction. This rapid multi-ion conduction along a narrow single file pore is called "knock-on," a mechanism proposed more than 50 years ago by Hodgkin and Keynes.[38] The mechanism assumes that ion-channel attraction and ion-ion repulsion play compensating effects, as several ions move simultaneously in single file through the narrow pore: the approach of one ion from one side of the selectivity filter is coupled to the simultaneous exit of another ion on the opposite side: according to the process, K$^+$···[K$^+$, K$^+$] ↔ [K$^+$, K$^+$]···K$^+$. Though plausible, the concept of knock-on relies on a strikingly delicate energy balance. To allow rapid ion conduction, the strong attraction between the ions and the channel must be exquisitely counterbalanced by the electrostatic repulsive forces between the ions.

The multi-ion free energy surface governing the conduction process computed using molecular dynamics (MD) simulation studies considerably clarified the basis for the knock-on mechanism.[24] The computations showed that the ion-ion electrostatic repulsion applies mainly at short distance, although the presence of an incoming K$^+$ at one end of the pore is nonetheless sufficient to destabilize and promote the unbinding of a K$^+$ at the opposite end of the pore. The dominant elementary events underlying the ion conduction process are best visualized with the transport cycle between multi-ion states shown in Figure 5b. At physiological concentration, the selectivity filter is predominantly occupied by two K$^+$ ions, in the configurations [S_1, S_3] or [S_2, S_4].[17,23,39] During an outward conduction event, a third ion hops from the intracellular vestibular cavity into the site S_4

while two ions in the selectivity filter are located in the sites S_1 and S_3. The incoming ion induces a concerted transition to a state in which the three K^+ ions occupy the sites S_4, S_2 and S_0, which is then followed by the rapid dissociation and departure of the outermost ion in S_0 on the extracellular side, yielding the conduction of one K^+.

Fast ion conduction can thus be understood at the atomic level, but what about selectivity? In simple terms, selectivity reflects the fact that a "wrong" ion (Na^+) must encounter more difficulty than a "correct" ion (K^+) when it attempts to go through the pore. The molecular mechanism underlying the rapid discrimination between K^+ and Na^+ is, in particular, fascinating because these two monovalent cations are very similar, differing only slightly in their atomic radius (by ~0.38 Å).[32] Because of its smaller radius, the hydration free energy of Na^+ is ~18 kcal/mol more negative than that of K^+, i.e. $G_{bulk}(Na^+) \approx G_{bulk}(K^+) - 18$ kcal/mol. A channel provides coordinating groups that help compensate the loss of hydration. From a physical point of view, selectivity arises when this process is more unfavorable for one type of ion than for another, i.e. the wrong ion must experience an environment that is energetically unfavorable (relative to the bulk phase). In this sense, ion selectivity is first and foremost about free energy. While non-equilibrium kinetic aspects may ultimately be taken into consideration, a robust selectivity arises because the entry of K^+ into the narrow pore is thermodynamically more favorable than for Na^+.[40] In terms of thermodynamics free energy, this can be stated as,

$$[G_{pore}(Na^+) - G_{bulk}(Na^+)] - [G_{pore}(K^+) - G_{bulk}(K^+)] = \Delta\Delta G, \qquad (3)$$

and to produce a channel that is selective for K^+, the free energy difference $\Delta\Delta G$ must be a positive number. According to electrophysiological measurements, $\Delta\Delta G$ is on the order of 4–6 kcal/mol for K^+ channels.[40] The key issue about the selectivity is to identify the physical origin of the unfavorable free energy $\Delta\Delta G$. On the basis the X-ray structure, it was initially suggested that the filter is constrained in an optimal geometry by a network of aromatic residues surrounding the selectivity filter so that a dehydrated K^+ ion fits snugly with proper coordination by the backbone carbonyl oxygens, but it cannot distort sufficiently to coordinate a smaller cation such as Na^+.[16,17] However, as the atomic radius of K^+ and Na^+ differs only by 0.38 Å, the snug-fit mechanism assumes that the selectivity filter is able to rigidly retain its geometry with sub-ångstrom precision in order to discriminate between these two cations. Proteins, like most biological macromolecular assemblies, are "soft materials" displaying significant structural flexibility at room temperature. This suggests that, at room temperature, the channel should be able to distort easily to cradle Na^+ with very little energetic cost. This is indeed seen in many MD simulations with Na^+ in KcsA.[41,42] Based on the estimated $\Delta\Delta G$ (~6 Kcal/mol), one may note that $G_{pore}(Na^+) \approx G_{pore}(K^+) - 12$ kcal/mol, which implies that — in absolute terms — Na^+ in the pore is more strongly solvated than K^+ in the pore. This is generally indicative that the selectivity filter is flexible and that the backbone carbonyl oxygens coordinate a Na^+ transiently crossing the pore.

All such arguments are qualitative, but can they be made more quantitative? Free energy perturbation (FEP) based on all-atom MD simulations represents a powerful approach to investigate the microscopic origin of thermodynamic factors in biological systems.[43,44] By carrying FEP/MD simulations, it is possible to incorporate the effect of thermal fluctuations and the contributions from all the atomic coordinates. When this method is applied to the KcsA, the results show that the narrow pore can indeed be selective for K^+ over Na^+, despite atomic fluctuations of the selectivity filter on the order of 0.5 to 1.0 Å RMS.[24] Analysis of the FEP/MD calculations shows that it is the interplay of the attractive ion-ligand (favoring the smaller cation) and repulsive ligand-ligand interactions (favoring the larger cation) that govern size selectivity in a flexible protein binding site.[26] Because such interactions can be directly modulated by the number and the type of ligands involved in ion coordination, altering the composition of the molecular groups forming a binding site provides a very potent molecular mechanism to achieve and maintain a high selectivity in flexible proteins.

5.2 *Channel gating and membrane voltage*

Any channel (even a very selective one) would only be able to act as a passive device facilitating the passage of ions down their electrochemical gradients without gating. However, additional complexity emerges when there are mechanisms by which channels can open and close in response to cellular signals.[1,2] There are many different kind of signals that can affect the gating of ion channels, including ligand binding (protons, Ca^{2+} ions, neurotransmittors, nucleotides, etc.), changes in the mechanical and physical states of the membrane bilayer, changes in temperature, etc. In the case of voltage-gated channels, the probability of opening the pore, P_o, changes abruptly in response to the membrane potential V.[34] Upon membrane depolarization, the voltage sensor in each subunit undergoes a voltage-dependent transition from a resting to an activated state, resulting in a conformation that allows the opening of the pore domain (Figure 2). The change in the configuration of the four VSDs then increases the probability of opening of the main pore gate, thus causing the activation of the channel (Figure 6).

To fully understand how voltage-gated channels function requires knowledge of the structure of the channel in its various states to atomic resolution, and knowledge of how the different parts of the protein change their conformation during voltage-gating. But first, let us make the concept of voltage-gating a little more quantitative by considering a channel with only two conformational states, open (o) and closed (c). The probability of the open state can be expressed in terms of the total free energy as,

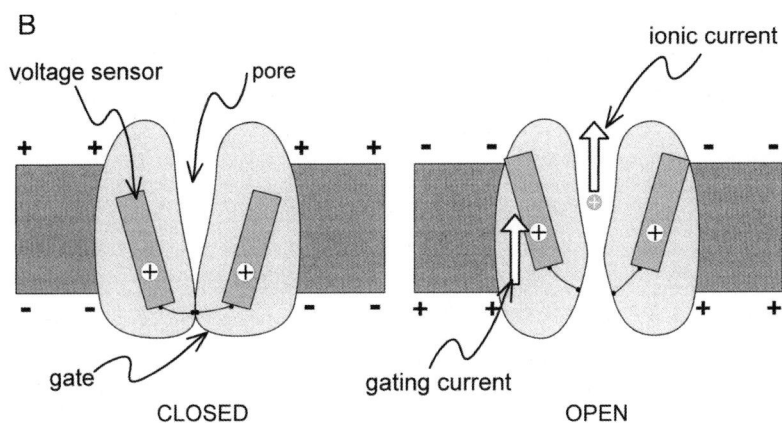

Figure 6. Schematic effect of membrane potential on channel gating.

$$P_{\mathrm{o}} = \frac{\exp[-G_{\mathrm{o}}^{\mathrm{tot}}/k_{\mathrm{B}}T]}{\exp[-G_{\mathrm{o}}^{\mathrm{tot}}/k_{\mathrm{B}}T] + \exp[-G_{\mathrm{c}}^{\mathrm{tot}}/k_{\mathrm{B}}T]}. \tag{1}$$

The total free energy of the protein in the open or closed states can be written as the sum of two contributions. The first contribution represents the intrinsic energy of the protein, which is independent of the membrane potential. The second contribution represents the coupling between the protein and the membrane potential V. This coupling takes the form of an effective state-dependent charge Q multiplying the membrane potential V. The total free energy of the open state is thus,

$$G_{\mathrm{o}}^{\mathrm{tot}} = G_{\mathrm{o}} + Q_{\mathrm{o}}V, \tag{2}$$

and likewise for the closed state. A simple rearrangement allows to rewrite P_{o} as,

$$P_{\mathrm{o}}(V) = \frac{\exp[\Delta Q(V - V_{1/2})/k_{\mathrm{B}}T]}{1 + \exp[\Delta Q(V - V_{1/2})/k_{\mathrm{B}}T]}. \tag{3}$$

where $\Delta Q = (Q_{\mathrm{c}} - Q_{\mathrm{o}})$ and $V_{1/2} = (G_{\mathrm{o}} - G_{\mathrm{c}})/\Delta Q$. The quantity $V_{1/2}$ corresponds to the voltage at which one half of the population of proteins are in the open state, and one half are in the closed state. It is directly related to the relative free energy of these two states in the absence of an applied voltage. The quantity ΔQ is called the "gating charge".[1,4,5,34,35] Kv channels display a large gating charge, on the order of about 12 to 14 elementary charge.[32,34] The large positive charge ΔQ is responsible for the strong coupling of the conformation of the channel to the transmembrane voltage. As illustrated in Figure 7, if ΔQ were small, then the open probability of the channel, P_{o}, would only be weakly affected by changes in the transmembrane potential ($\Delta V = V - V_{1/2}$).

In simple terms, channel opening corresponds to the outward translocation of a large positive charge ΔQ equal to 12–14 elementary charges. It is understood that the four positively charged arginine residues along S4, referred to as R1 to R4, provide the dominant contributions to the total "gating charge" ΔQ and are, thus, mainly responsible for the coupling to the membrane voltage.[14,15] Other important charged residues in the VSD include K5 and R6 along S4, E0 along S1,

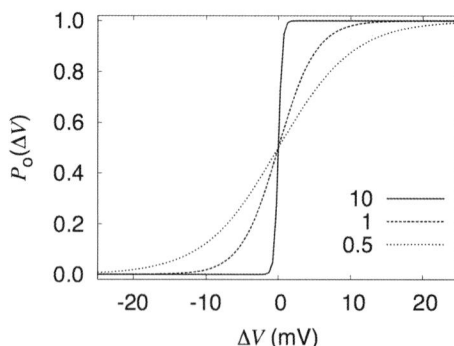

Figure 7. Probability of activation as a function of the gating charge for a simple voltage-gated channel represented as a two-state model based on Eq. (3). P_{o} for values of ΔQ equal to 0.5, 1.0, and 10 are shown.

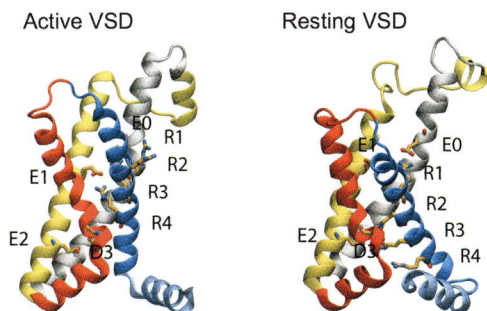

Figure 8. The VSD, where the voltage function sensing lies, is a small bundle of four antiparallel helices (S1, gray; S2, yellow; S3, red; S4, blue) packed in a counterclockwise fashion (seen from the extracellular side). The main charged residues in the VSD include Arg294 (R1), Arg297 (R2), Arg300 (R3), Arg303 (R4), Lys305 (K5), and Arg309 (R6) along S4, Glu183 (E0) along S1, Glu226 (E1) and Glu236 (E2) along S2, and Asp259 (D3) along S3. Those charges provide the dominant contributions to the total "gating charge" ΔQ. Adapted from Khalili *et al.*[9] The residue numbers correspond to the Kvl.2 channel.

E1 and E2 along S2, and D3 along S3 (Figure 8). It is important to rigorously link the gating charge ΔQ, which is responsible for the strong coupling to the transmembrane potential, to the atomic movements in structural models of the channel. In the simplest case of a perfectly planar membrane, the atomic charges of the VSD are expected to interact directly with the (constant) transmembrane electric field. In such a case, the total gating charge ΔQ would simply be related to the displacement z_i of each of the residues carrying a charge q_i in the direction perpendicular to the membrane surface, i.e. $\Delta Q = \sum_i q_i(z_i/d)$, where d is the thickness of the membrane (the region where the field is non-zero). This is the familiar electrostatic formula that is derived from the displacement current arising from a charge moving between the two conducting surfaces of a parallel plate capacitor. However, the situation is far more complicated because the surface of the protein and of the VSD is irregular, with multiple complex aqueous high dielectric regions. A fundamental consequence of such high dielectric regions is to focus the membrane potential, i.e. as with a parallel plate capacitor, the electric field tends to be more intense when the low dielectric insulating region is thin.[35,45]

To quantitatively treat those effects, it is necessary to know the structure of the channel in its various states at atomic resolution. Crystallographic studies have provided atomic resolution structures of a number ion channels in different conformations, most importantly of the Kv1.2 channel from rat brain.[18,19] However, no atomic-resolution crystal structure of a voltage-gated K+ channel in the closed/resting state is currently available. Experimental studies in combination with computer modeling have sought to complement the missing structural information. One strategy is to carefully construct atomic models of both active and resting states that are consistent with all available experimental information using protein structure prediction algorithms.[46] This approach has produced an atomic model of the closed/resting state that is broadly constrained by a wide range of experimental data (Figure 8). Using MD simulations initiated from such models, it has been found that the calculated gating charge of the channel is equal to 12–13 elementary charges,[29] in good accord with experimental estimates. This suggests that the open and closed state models offer a realistic depiction of voltage-gating.

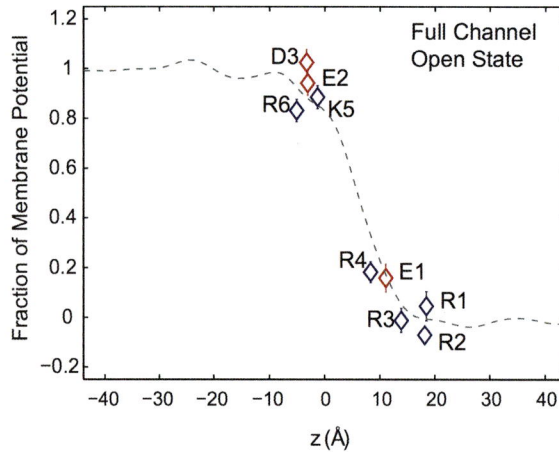

Figure 9. Fraction of the membrane potential calculated from free energy MD simulations carried out on the open/active conformational of the channel. Note that the membrane field is focused over roughly 10 Å on the outer-leaflet of the bilayer (0 Å > Z > 10 Å). For example, R4 and K5 along S4 experience 0.16 and 0.89 of the potential, respectively, but are separated only by one helix turn. (Figure adapted/reproduced with permission/modified from Khalili *et al.*)[29]

Additional FEP/MD computations also indicate that the membrane potential sensed by the charged residues in the VSD varies abruptly over the extracellular half of the membrane in the arginine-rich region of S4. The results are shown in Figure 9.[29] In the active state of the channel the three outermost arginines (R1, R2, and R3) are positioned near the extracellular solution at ~0.1 of the membrane potential. Deeper inside the VSDs, E1 on S2 and R4 on S4 are located at ~0.2 of the membrane potential. Further along S4, K5 is located at ~0.6 of the potential in the isolated VSD, and at ~0.9 of the potential in the full tetrameric channel. Near the intracellular membrane-solution interface are E2 on S2, D3 on S3, and R6 on S4, which are positioned within 0.8 of the membrane potential. Therefore, rather than being spread over the width of the entire bilayer, the membrane field is focused over a fairly narrow region corresponding roughly to the outer leaflet of the bilayer. For example, R4 and K5 experience 0.16 and 0.89 of the potential, respectively, even though they are separated by less than one helix turn along S4. Thus, a large gating charge ΔQ can be reproduced without a full translocation of S4 across the bilayer. Accordingly, the voltage gradient (the field) is focused, rather than being roughly distributed over the entire thickness of the membrane. This is also in qualitative agreement with previous experimental studies and computations based on simple structural models.[25,47–49]

5.3 *The functional implications of a focused field*

What are the implications of a focused membrane field? Principally, it means that a relatively modest movement of the charges within the VSD is strongly coupled to the membrane potential. From a kinetic point of view, a focused field concentrates the effect of the voltage drop across the membrane over a relatively short distance, resulting in a more intense driving force that can be utilized to rapidly and efficiently induce the essential conformational changes within the VSD. This has important implications, particularly from the point of view of timescales and kinetics.

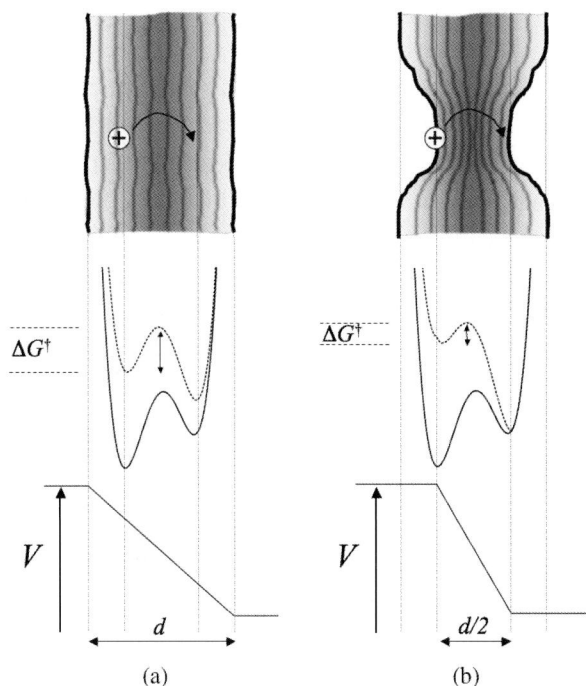

Figure 10. Schematic representation of the effect of a focused field on a simple two-state model of voltage activation. (a) The voltage difference V drops over the distance d, and the transition rate from the closed (c) to open (o) state is accelerated weakly by the applied membrane field. (b) The voltage difference drops over a distance $d/2$, and the transition rate is greatly accelerated by the applied membrane field.

To illustrate this, let us consider a simple example where the free energy profile of the VSD along a reaction coordinate describing the process of voltage activation, $G(z)$, can be written as a sum of a voltage-independent component, $G_0(z)$, and a component corresponding to the transmembrane field driving a gating charge ΔQ. As shown in Figure 10, the applied membrane potential shifts the equilibrium from the left free energy well (closed state) to the right free energy well (open state). The membrane potential also decreases the activation free energy $\Delta G^\dagger$ for crossing the free energy barrier from the closed (left) to the open (right) state. If the voltage drop V across the membrane is spread over a large distance d as shown in (a), then the intensity of the field does not give rise to a fast conformational change within the VSD. In contrast, the transition rate can be greatly accelerated if the voltage drops over a shorter distance $d/2$ as shown in (b) due to the more intense membrane field. This is important because the activation by voltage-gated channels must occur sufficiently rapidly, within milliseconds, to produce a proper action potential able to carry useful information. An important caveat with the schematic illustration shown in Figure 10 is the assumption that each component, i.e. the intrinsic voltage-independent free energy profile, $G_0(z)$, the gating charge ΔQ and the spatial-dependence of the voltage drop are assumed to be independent from one another. For instance, the overall gating charge "measured externally" would be different, as it reflects the coupling to the electromotive force. In reality, all these quantities are set by the physics of the microscopic situation, and they would all vary if the thickness of the membrane changes as in Figures 10a and 10b.

6. Future Outlook

A complete understanding of the function of ion channels implicated in the generation and propagation of the nerve impulse will require a characterization of the all microscopic features responsible for ion conduction, selectivity, and voltage-gating. Another complex process that we have not discussed in detail is called inactivation, by which the PD of ion channels converts to a nonconductive conformation. At the time when this chapter was written, important but limited information at the atomic level was available about voltage-gated K^+ channels and the behavior of the ion-selective PD and the VSDs. In large part, the game-changer in the field of K^+ channels has been the availability of high-resolution structure from X-ray crystallography. This combined with additional information from a broad range of experiments together with sophisticated computational methods made it possible to rationally understand critical functional features at the molecular level, including fast ion conduction, high selectivity, and the origin of the gating charge with the focused membrane field. Much less is currently known about Na^+ channels, and conceptual ideas about these are essentially developed by analogy with what is known about K^+ channels. Particularly intriguing are the microscopic organization of the selectivity filter and the origin of the fast activation kinetics and large coupling between the four non-identical VSD surrounding the PD of Na^+ channels.[50] Undoubtedly, there will be additional surprises and unexpected twists, as more atomic-resolution information about Na^+ channels will become available. Ultimately, one would like to know all the conformational states of K^+ and Na^+ channel and *visualize*, atom-by-atom, how these macromolecules move as a function of time in response to the changes in the membrane potential. Then, with the use of sophisticated computational methods, it may be possible to predict the various effects of site-directed mutations and of pharmacological compounds. Little by little, progress is being made toward this goal.

Suggested Additional Reading Materials

F. Bezanilla. 2005. Voltage-gated ion channels. *IEEE Trans Nanobioscience* **4**, 34–48.

B. Roux, T. Allen, S. Bernèche and W. Im. 2004. Theoretical and computational models of biological ion channels. *Q Rev Biophys* **37**, 15–103.

B. Roux. 2008. The membrane potential and its representation by a constant electric field in computer simulations. *Biophys J* **95**, 4205–4216.

References

1. A. L. Hodgkin and A. F. Huxley. 1952. A quantitative description of membrane current and its application to conduction and excitation of nerve. *J Physiol (Lond.)* **117**, 500–544.

2. C. Armstrong. 1969. Inactivation of the potassium conductance and related phenomena caused by quaternary ammonium ion injection in squid axons. *J Gen Physiol* **54**, 553–575.

3. C. Armstrong. 1971. Interaction of tetraethylammonium ion derivatives with the potassium channels of giant axons. *J Gen Physiol* **58**, 413–437.

4. C. M. Armstrong and F. Bezanilla. 1973. Currents related to movement of the gating particles of the sodium channels. *Nature* **242**, 459–461.

5. C. M. Armstrong and F. Bezanilla. 1974. Charge movement associated with the opening and closing of the activation gates of the Na channels. *J Gen Physiol* **63**, 533–552.

6. G. Eisenman and S. Krasne. 1973. In *IV International Biophysics Congress Symposium on Membrane Structure and Function*. (Moscow).

7. F. Bezanilla and C. M Armstrong. 1972. Negative conductance caused by entry of sodium and cesium ions into the K^+ channels of squid axon. *J Gen Physiol* **53**, 342–347.

8. B. Hille. 1973. Potassium channels in myelinated nerve-selective permeability to small cations. *J Gen Physiol* **61**, 599.

9. E. Neher and B. Sackmann. 1976. Single-channel currents recorded from membrane of denervated frog muscle fibers. *Nature (Lond.)* **260**, 779–802.

10. M. Noda, *et al*. 1984. Primary structure of electrophorus-electricus sodium-channel deduced from cDNA sequence *Nature* **312**, 121–127.

11. M. Noda, *et al*. 1986. Expression of functional sodium-channels from cloned cDNA. *Nature* **322**, 826–828.

12. B. L. Tempel, D. M. Papazian, T. L. Schwarz, Y. N. Jan and L. Y. Jan 1987. Sequence of a probable potassium channel component encoded at Shaker locus of Drosophila. *Science* **237**, 770–775.

13. L. Heginbotham, Z. Lu, T. Abramson and R. Mackinnon. 1994. Mutations in the K^+ channel signature sequence. *Biophys J* **66**, 1061–1067.

14. S. Aggarwal and R. MacKinnon. 1996. Contribution of the S4 segment to gating charge in the Shaker K^+ channel. *Neuron* **16**, 1169–1177.

15. S. Seoh, D. Sigg, D. Papazian and F. Bezanilla. 1996. Voltage-sensing residues in the S2 and S4 segments of the Shaker K^+ channel. *Neuron* **16**, 1159–1167.

16. D. A. Doyle, *et al*. 1998. The structure of the potassium channel: Molecular basis of K^+ conduction and selectivity. *Science* **280**, 69–77.

17. M. Zhou, J. H. Morais-Cabral, S. Mann and R. MacKinnon. 2001. Chemistry of ion coordination and hydration revealed by a K^+ channel-Fab complex at 2.0 A resolution. *Nature* **414**, 43–48.

18. S. B. Long, E. B.Campbell and R. Mackinnon. 2005. Crystal structure of a mammalian voltage-dependent Shaker family K^+ channel. *Science* **309**, 897–903.

19. S. B. Long, E. B.Campbell and R. Mackinnon. 2005. Voltage sensor of Kv1.2: Structural basis of electromechanical coupling. *Science* **309**, 903–908.

20. L. G. Cuello, V. Jogini, D. M. Cortes and E. Perozo. 2010. Structural mechanism of C-type inactivation in K^+ channels. *Nature* **466**, 203–208.

21. L. G. Cuello, *et al*. 2010. Structural basis for the coupling between activation and inactivation gates in K^+ channels. *Nature* **466**, 272–275.

22. B. Roux and R. MacKinnon. 1999. The cavity and pore helices the KcsA K^+ channel: Electrostatic stabilization of monovalent cations. *Science* **285**, 100–102.

23. J. Åqvist, V. Luzhkov. 2000. Ion permeation mechanism of the potassium channel. *Nature* **404**, 881–884.

24. S. Bernèche and B. Roux. 2001. Energetics of ion conduction through the K^+ channel. *Nature* **414**, 73–77.

25. B. Chanda, O. K. Asamoah, R. Blunck, B. Roux and F. Bezanilla. 2005. Gating charge displacement in voltage-gated ion channels involves limited transmembrane movement. *Nature* **436**, 852–856.

26. S. Noskov, S. Bernèche and B. Roux. 2004. Control of ion selectivity in potassium channels by electrostatic and dynamic properties of carbonyl ligands. *Nature* **431**, 830–834.

27. J. F. Cordero-Morales, *et al.* 2006. Molecular determinants of gating at the potassium-channel selectivity filter. *Nature Structural & Molecular Biology* **13**, 311–318.

28. J. F. Cordero-Morales, *et al.* 2007. Molecular driving forces determining potassium channel slow inactivation. *Nature Structure and Molecular Biology* **14**, 1062–1069.

29. F. Khalili-Araghi, *et al.* 2010. Calculation of the gating charge for the Kv1.2 voltage-activated potassium channel. *Biophys J* **98**, 2189–2198.

30. R. W. Pastor, R. M. Venable and S. E. Feller. 2002. Lipid bilayers, NMR relaxation, and computer simulations. *Acc Chem Res* **35**, 438–446.

31. S. E. Feller, R. M. Venable and R. W. Pastor. 1997. Computer simulation of a DPPC phospholipid bilayer: structural changes as a function of molecular surface area. *Langmuir* **13**, 6555–6561.

32. B. Hille. 2001. *Ion Channels of Excitable Membranes*. Sinauer, Sunderland, MA, Third edition.

33. B. Roux, S. Bernèche and W. Im. 2000. Ion channels, permeation and electrostatics: Insight into the function of KcsA. *Biochemistry* **39**, 13295–13306.

34. F. Bezanilla. 2008. How membrane proteins sense voltage. *Nat Rev Mol Cell Biol* **9**, 323–332.

35. B. Roux. 1997. The influence of the membrane potential on the free energy of an intrinsic protein. *Biophys J* **73**, 2980–2989.

36. W. Nernst. 1889. Die elektromotorische Wirksamkeit der Ionen. *Z Phys Chem* **4**, 129–181.

37. E. Rojas, F. Bezanilla and R. E. Taylor. 1970. Demonstration of sodium and potassium conductance changes during a nerve action potential. *Nature* **225**, 747–748.

38. A. L. Hodgkin and R. D. Keynes. 1955. The potassium permeability of a giant nerve fibre. *J Physiol (Lond.)* **128**, 61–88.

39. M. Zhou and R. A. MacKinnon. 2004. mutant KcsA K(+) channel with altered conduction properties and selectivity filter ion distribution. *J Mol Biol* **338**, 839–846.

40. J. Neyton and C. Miller. 1988. Discrete Ba2+ block as a probe of ion occupancy and pore structure in the high-conductance Ca^{2+}-activated K^+ channel. *J Gen Physiol* **92**, 569–586.

41. L. Guidoni, V. Torre and P. Carloni. 1999. Potassium and sodium binding to the outer mouth of the K^+ channel. *Biochem.* **38**, 8599–8604.

42. I. H. Shrivastava, D. P. Tieleman, P. C. Biggin and M. S. Sansom. 2002. K(+) versus Na(+) ions in a K channel selectivity filter: A simulation study. *Biophys J* **83**, 633–645.

43. J. McCammon and T. Straatsma. 1992. Alchemical free energy simulation. *Ann Rev Phys Chem* **43**, 407.

44. P. A. Kollman. 1993. Free energy calculations: Applications to chemical and biochemical phenomena. *Chem Rev* **93**, 2395–2417.

45. V. Jogini and B. Roux. 2007. Dynamics of the Kv1.2 voltage-gated K^+ channel in a membrane environment. *Biophysical Journal* **93**, 3070–3082.

46. M. M. Pathak, *et al.* 2007. Closing in on the resting state of the shaker K^+ channel. *Neuron* **56**, 124–140.

47. L. D. Islas and F. J. Sigworth. 2001. Electrostatics and the gating pore of Shaker potassium channels. *J Gen Physiol* **117**, 69–89.

48. O. K. Asamoah, J. P. Wuskell, L. M. Loew and F. Bezanilla. 2003. A fluorometric approach to local electric field measurements in a voltage-gated ion channel. *Neuron* **37**, 85–97.

49. D. M. Starace and F. Bezanilla. 2004. A proton pore in a potassium channel voltage sensor reveals a focused electric field. *Nature* **427**, 548–553.

50. B. Chanda, O. K. Asamoah and F. Bezanilla. 2004. Coupling interactions between voltage sensors of the sodium channel as revealed by site-specific measurements. *J Gen Physiol* **123**, 217–230.

Voltage-Gated Channels and the Heart

13

Jonathan R. Silva and Yoram Rudy

1. Introduction

Modern devices, such as radios and televisions are controlled by electrical circuits that determine the frequency they are tuned to, their volume and whether they are turned on or off. Similarly, the heart contains circuitry that controls how fast it beats, how hard it beats and what part of it is contracting at any given time. If this heart circuitry is somehow damaged, a dangerous heart rhythm, called an arrhythmia, can arise in which electrical excitation becomes disorganized and prevents the heart from being an efficient pump. The lack of blood flow to vital organs can result in death. Arrhythmias have a variety of causes including damage caused by lack of blood flow to a region of the heart (a heart attack/myocardial infarction), inheritance of a genetic defect, or an inadvertent drug effect. In the United States, every day an estimated 1200 people suffer from sudden cardiac death,[1] making it a major focus of medical research.

This chapter aims to introduce the reader to cardiac electrophysiology and provide a description of how a particular molecular machine, the ion channel, fits into the larger picture. We begin with a description of the electrical activity of the heart and then explain how ion channels in individual cells generate electrical activity. To conclude, a detailed account is provided as to how a specific ion channel, the slow delayed rectifier K^+ channel (I_{Ks}), functions and how it participates in generating the heart rhythm.

2. Cardiac Electrophysiology Background

2.1 *Organ level physiology*

The primary function of the heart is to collect the blood returning from the body, send it to the lungs to be oxygenated, collect it from the lungs, and then send it full of oxygen to the body. This task is accomplished by four chambers which are joined together by unidirectional valves. The right atrium collects the blood from the body and then passes it through the tricuspid valve to the

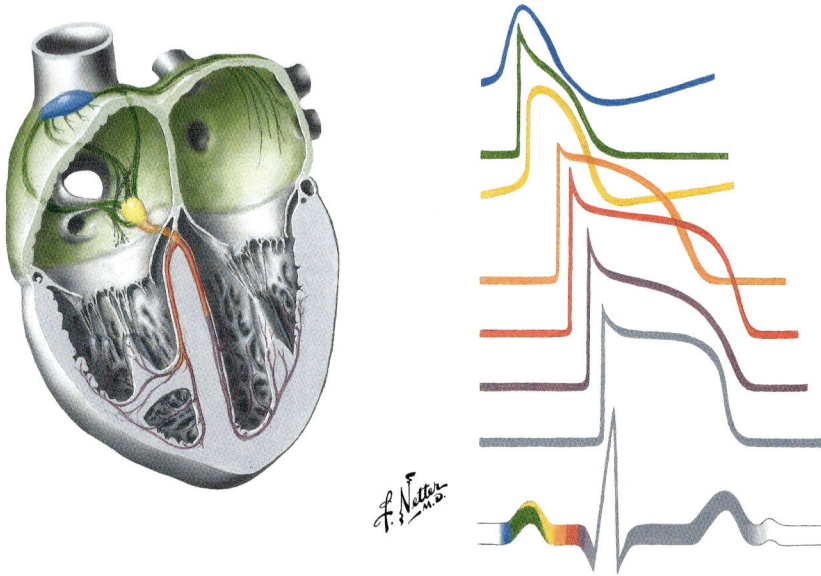

Figure 1. The heart chambers and the cardiac conduction system. Colored regions generate the corresponding AP waveforms shown on the right, which in turn correlate to different intervals of the ECG (bottom right). Prominent deflections observed in the ECG are labeled: P (atrial activation), QRS (ventricular activation), and T (ventricular repolarization). Netter illustration from www.netterimages.com © Elsevier Inc. All rights reserved.

right ventricle by contracting. After a delay, the right ventricle contracts and sends the blood to the lungs through the pulmonary valve. At the same time, the left atrium is collecting oxygenated blood from the lungs and passes it through the bicuspid mitral valve to the left ventricle, which sends the blood out to the body through the aortic valve (Figure 1).

The timing and contractile force of each chamber is controlled by an electrical impulse that originates from a specialized group of cells located in the right atrium, called the sino-atrial (SA) node. After initiation, the signal spreads throughout the atria via intercellular connections, called gap junctions, which are analogous to pipes that allow ions and even small metabolites to pass between myocytes. This initial spread of the impulse signals contraction in the right and left atria, allowing them to deliver the blood that they contain to the ventricles. Atrial excitation then brings the electrical signal to the atrio-ventricular (AV) node, which carries it from the atria to the ventricles. Signal conduction through the AV node is slow, which allows time for the blood to pass from the atria to the ventricles. However, once the signal emerges from the AV node, conduction is rapidly spread throughout the ventricles with the help of specialized cells (called Purkinje cells) that form the right and left bundles of the ventricular conduction system (Figure 1).

The heart signal can be observed on the body surface by measuring the difference in electrical potential between two points. The archetypal signal, often shown on television, comes from the potential difference between the left and right arms (Lead I in the 12-lead electrocardiogram, ECG). Figure 1 shows the correspondence between activation of different regions of the heart and the Lead I ECG signal. The first deflection, termed the P-wave, correlates to excitation of the atria. After a delay that reflects the slow conduction through the AV node, a strong impulse is observed that corresponds to ventricular excitation, called the QRS complex. The magnitude of the signal generated by the ventricles is much stronger than the P-wave due to a much larger mass of tissue generating the electrical activity. Ventricular relaxation is observed in the final deflection, the T-wave. The interval between the QRS spike and the T-wave is called the QT interval and its prolongation is often observed in patients that are pre-disposed to arrhythmias.

2.2 Cellular level electrophysiology

Each individual cardiac cell has the ability to generate an electrical impulse, known as an action potential (AP). The shape of the AP in a given cell is uniquely suited to its function (Figure 1). As initiators of the heart rhythm, SA nodal cells generate an oscillatory AP that determines the rapidity of the heart beat. In stark contrast are the ventricular cells that we will focus on. These cells (called myocytes) cannot initiate an AP on their own, but wait for a sig-nal from a neighboring cell to be excited. The ventricular AP is also quite long, which is important to its function — allowing sufficient calcium flow into the myocyte to signal contraction (described below).

2.2.1 The Hodgkin-Huxley model

The ventricular AP is generated by electrical current through ion channels that open and close to generate its unique waveform. Before the discovery of ion channels, a similar current in neu-rons was described by Hodgkin and Huxley using a circuit model.[2] This model has two primary components, a capacitor and a parallel set of variable resistors that are each in series with a voltage generator (battery) (Figure 2a). The capacitor represents a separation of charge by the cell membrane, which is composed of a low dielectric lipid bilayer. Each resistor represents a current that can cross the cell membrane that carries a particular ion such as Na^+ or K^+. We now know that each branch represents a population of ion channels that is selective for a given ionic species. The relationship between the current passing through the resistors and the voltage across the capacitor is described by the following equation, which can be derived from Kirchoff's current law:

$$\frac{dV_m}{dt} = -\frac{1}{C_m} \cdot I_{ion},$$ (1)

where V_m is the transmembrane potential (intracellular space with respect to the extracellular space that is ground), t is time, C_m is the membrane capacitance and I_{ion} is the total current through all the branches with resistors.

Referring to Figure 2a, in terms of the cellular entities this equation describes the relationship between the potential across the cell membrane (V_m) and the current through the branches (I_X and I_Y). The batteries describe the potential caused by the difference in concentration of a particular ion species between the inside and outside of the cell across the corresponding channel conductance

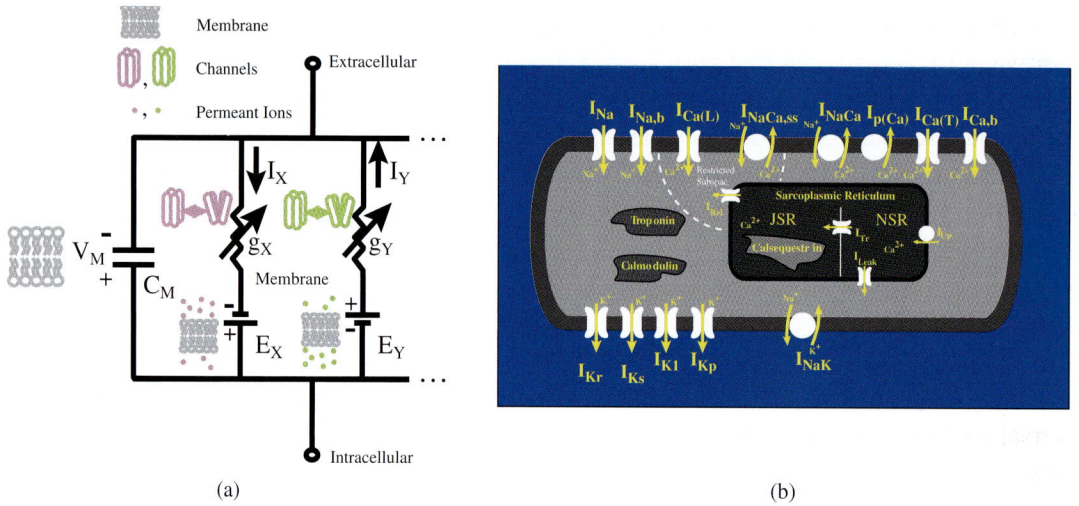

Figure 2. Electrophysiological Models. (a) Generic Circuit Model for Excitable Tissue. Channel opening and closing is modeled by resistors with variable conductance (g_X and g_Y). These are positioned in series with the reversal potential (E_X and E_Y) typically calculated via the Nernst equation, which depends on the difference between intra and extracellular ion concentrations. The currents in each of the branches (I_X and I_Y) are in parallel with a capacitor C_M, which represents charge separation by the membrane. (b) Schematic of a ventricular cell model. The restricted subspace represents a small volume near the interface of the $I_{Ca(L)}$ channels and the ryanodine receptors release intracellular Ca^{2+}. Definitions: I_{Na}, fast sodium current; $I_{Ca(L)}$, calcium current through L-type calcium channels; $I_{Ca(T)}$, calcium current through T-type calcium channels; I_{Kr}, rapid delayed rectifier potassium current; I_{Ks}, slow delayed rectifier potassium current; I_{K1}, inward rectifier potassium current; I_{Kp}, plateau potassium current; $I_{Na,b}$, sodium background current; $I_{Ca,b}$, calcium background current; I_{NaK}, sodium-potassium pump current; $I_{NaCa,ss}$ sodium-calcium exchange current in the restricted space; I_{NaCa}, sodium-calcium exchange current outside the restricted space; $I_{p(Ca)}$ calcium pump in the sarcolemma; I_{up}, calcium uptake from the myoplasm to network sarcoplasmic reticulum (NSR); I_{rel}, calcium release from junctional sarcoplasmic reticulum (JSR); I_{leak}, calcium leakage from NSR to myoplasm; I_{tr}, calcium translocation from NSR to JSR. Calmodulin and troponin represent calcium buffers in the myoplasm. Calsequestrin is a calcium buffer in the JSR. Detailed description of cardiac cell models and computer code are available at http://rudylab.wustl.edu. (Figure modified with permission from Zipes *et al.*)[35]

(g_X or g_Y in the circuit branches of Figure 2a). The equation typically used to calculate this potential is the Nernst equation:

$$E_X = \frac{RT}{F} \cdot ln\frac{[X]_e}{[X]_i}, \tag{2}$$

where X is the ion that the channel is selective for, R is the natural gas constant, F is the Faraday constant, T is the temperature, and $[X]_e$ and $[X]_i$ are the extracellular and intracellular ion concentrations. The current through a particular branch in Figure 2a is then, according to Ohm's law:

$$I_X = g_x \cdot (V_m - E_X), \tag{3}$$

where I_X is the current through and g_x is the conductance of the channels. Since E_X is the potential when the current switches signs (positive to negative or vice versa), it is often referred to as the reversal potential. By convention, the flow of current into the cell is considered negative.

In neurons and myocytes, ionic pumps and exchangers work to keep the concentration of K^+ ions high and the concentration of Na^+ and Ca^{2+} ions low inside the cell, relative to the extracellular concentration. According to Eqs. (2) and (3), in the absence of a transmembrane potential, K^+ current flows outward (positive current), and Na^+ and Ca^{2+} currents flow inward (negative current). The branches in the circuit model of Figure 2a can then be set to correspond to a certain type of ion-channel that passes one type of ion. For example, we could label one branch I_{Na}, which passes inward Na^+ current, and another branch I_{Ca} which passes inward Ca^{2+} current and the third branch I_K, which passes outward K^+ current (when V_m is 0). If V_m overwhelms the potential specified by the battery, then the current will go in the opposite direction. This value of V_m is known as the reversal potential. According to Eq (1), the change in membrane potential is negative when current is positive and vice versa. So, Na^+ currents increase (depolarize) V_m and K^+ currents tend to make it more negative (a process called repolarization).

The final component of the circuit model is the variability of the conductance, which depends on V_m. The total conductance (the inverse of resistance) accounts for the amount of current that can pass through a single channel, how many channels there are, and the likelihood that the channel is open via the following equation:

$$g_X = g_{sc} \cdot n \cdot P_o, \tag{4}$$

where g_{sc} is the single channel conductance, n is the number of channels and P_o is the probability that a channel is open. Describing P_o is the primary challenge in constructing a channel model. The simplest approach is to employ equations that describe independent gates that open and close depending on voltage. If we assign the variable m to the probability that our gate is open, we can describe its dependence on voltage with the following equation:

$$\frac{dm}{dt} = \alpha(V_m) \cdot (1 - m) - \beta(V_m) \cdot m \tag{5}$$

where $\alpha(V_m)$ is a voltage-dependent rate constant that determines how likely a channel is to transition from a closed to open state and $\beta(V_m)$ determines how likely channels are to transition from open to closed in a given amount of time. Since the quantity m is the probability that channels are open, the quantity $(1 - m)$ is the probability that the channels are closed. Multiple independent gates can be incorporated by multiplying them together. For example, $P_o = m^3$ would imply three gates that would have to be open at the same time for current to pass through. How the rate constants are determined is discussed in detail below in the section on I_{Ks} modeling.

In a cardiac cell, many types of ion channels contribute to the current. To account for each different population of channels, additional branches are added to the circuit diagram of Figure 2a. Each branch has its own variable conductance that reflects how much current a given channel is allowing to pass in response to a given V_m, and the battery in the branch reflects the propensity of ions (that the channel is selective for) to cross the membrane due to concentration differences. For the I_K branch, we can specify several channel populations (sub-branches) that carry K^+ current. In cardiac myocytes these include: the fast delayed rectifier (I_{Kr}), the slow delayed rectifier (I_{Ks}), the inward rectifying K$^+$ channel (I_{K1}) and the plateau K$^+$ current (I_{Kp}). As can be seen in Figure 2b, many channels can carry each type of ion. Description of all the types of ion channels shown is beyond the scope of this chapter. The most important channels are the voltage-gated sodium channel (I_{Na}), the L-type calcium channel ($I_{Ca,L}$) and the K$^+$ channels (I_{Ks}, I_{Kr} and I_{K1}).[3,4] In the next subsection, we describe how these channels interact with each other to form the cardiac AP.

2.2.2 *Modeling the cardiac ventricular action potential*

At the resting state, before a cell is excited by an impulse from a neighboring myocyte, V_m resides near the K$^+$ Nernst potential (approximately −90 mV). In this state, I_{K1} channels are open, allowing K$^+$ to pass freely through the membrane, explaining why V_m is near E_K. However, once a neighboring cell is excited, it depolarizes the resting myocyte, raising its V_m. This process occurs because of a difference in V_m between the excited cell and the resting cell that causes current to cross through gap junctions to the resting cell. Once V_m reaches −60 mV due to this depolarization, fast Na$^+$ channels begin to open.

Since Na$^+$ channels are highly selective, I_{Na} is an inward current (Na$^+$ flows from outside to inside) and depolarizes the cell further. This depolarization further activates the channel, creating a positive feedback (more depolarization causes more opening which causes more depolarization) which results in a rapid AP upstroke (Figure 3a). Shortly after I_{Na} opens, it quickly closes via a process termed inactivation. However, the rapid depolarization caused by I_{Na} activation affects all other major currents. $I_{Ca,L}$

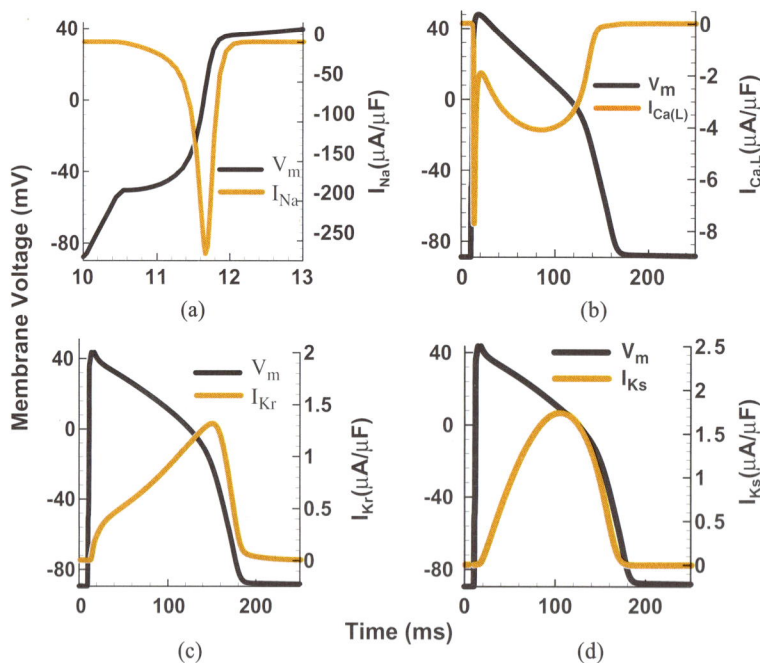

Figure 3. Major currents (orange) during the AP (black). (a) I_{Na} and the AP upstroke (note the shortened time scale). (b) $I_{Ca,L}$. (c) I_{Kr} (d) I_{Ks}. (Figure modified with permission from Zipes *et al.*)[35]

responds to depolarization by activating rapidly, but in contrast to I_{Na}, its inactivation is only partial (Figure 3b). I_{Kr} opens quickly, but just as rapidly inactivates (Figure 3c), and I_{Ks} begins to open, albeit rather slowly (Figure 3d). I_{K1} closes quickly and does not reopen until the end of the AP (not shown).

Following depolarization a plateau is observed, which is primarily sustained by a persistent inward current carried by $I_{Ca,L}$ (see V_m traces in Figure 3). The delivery of Ca^{2+} into the cell by this current causes further Ca^{2+} release from intracellular stores (Ca^{2+}-induced Ca^{2+} release, CICR). This large intracellular Ca^{2+} release signals the contractile machinery in the cell to begin its mechanical work, thereby accomplishing the primary task of the AP — to initiate myocyte contraction. The next task is to restore V_m to its resting state, so that the next heartbeat can occur. This task is carried out by the opening of the K^+ channels, I_{Ks} and I_{Kr}. Depending on species, the role of each of these channels varies. In human and canine ventricular myocytes, I_{Ks} plays a marginal role during a single AP opening slowly and contributing a modest amount of outward current to repolarization.[5] In contrast, I_{Kr} opens rapidly and mostly inactivates, however, the moderate amount of current passing through the channel counters what is left of the partially inactivated $I_{Ca,L}$ and begins to lower V_m. As V_m decreases, I_{Kr} further recovers from its inactivation and more current is available to repolarize the cell to its resting state. At the end of the AP, I_{K1} opens and the AP finishes with I_{K1} channels remaining open while the rest of the channels deactivate (close) due to the negative resting V_m.

2.2.3 *Rate dependence*

While the standing AP (a single AP initiated from the resting state) provides a controlled experimental and theoretical platform for model creation, the heart rhythm is a continuous process with cells generating APs at least once per second (60 beats per minute, bpm). The heart rate is determined by the rate of firing of the SA node. When the body needs higher cardiac output — more blood and therefore more oxygen — the cells of the SA node oscillate at a higher rate. This SA node frequency increase is typically a response to increased adrenergic tone (i.e. the blood adrenaline concentration increases).

From the perspective of a ventricular myocyte, an increased heart rate requires a shortened AP. Physiologically, this shortening occurs for two reasons. The first is that a typical human QT interval (a reflection of ventricular AP duration, APD) at slow rate nominally can last up to 430 ms. At 180 bpm (3 beats per second — a heart rate achieved during strenuous exercise) the APs would not fit into the allotted time (3 QT intervals would last 1.3 seconds). QT interval shortening down to 200 ms allows the tissue to recover between beats so that it is electrically ready for subsequent stimulation.[6] The second reason for shortening has to do with the contraction signaled by Ca^{2+} coming into the cell during the AP. As discussed above, during each heartbeat, the P-wave signals contraction of the atria. When the atria contract, the ventricles should be relaxed and ready to receive blood. However, during an AP, Ca^{2+} is entering the myocyte, enhancing contraction. By shortening the AP, Ca^{2+} enters for a shorter duration of time and the myocyte has more time to recover for the next beat.

2.2.4 β-*adrenergic regulation*

An increase in adrenergic tone is a signal for the heart to beat faster and to beat harder, both of which increase cardiac output (the amount of blood the heart pumps through over time). While the

SA node is responsible for increased heart rate, individual cells play a significant role in increasing the strength of contraction. One mechanism for increasing contractile force is to increase the rate of Ca^{2+} entry into the cell, which is accomplished through an increase in $I_{Ca,L}$ conductance. However, at the same time that $I_{Ca,L}$ is increasing the rate of Ca^{2+} entry, it is also bringing in more current and raising V_m. As mentioned in the previous section, a shorter APD is desirable at a fast heart rate. To compensate for this inward current, an outward current must increase. This role is filled by I_{Ks}, which is directly regulated by β-adrenergic stimulation. In fact, several key proteins in the β-adrenergic cascade are incorporated directly into the I_{Ks} channel complex.[7] The mechanism whereby I_{Ks} counters $I_{Ca,L}$ is described in detail below.

2.2.5 *The long QT syndrome*

Just as increasing outward repolarizing currents and decreasing inward depolarizing currents causes AP shortening, decreasing outward currents and increasing inward currents can prolong the AP. There are many potential causes for changes in current magnitude such as drugs and inherited genetic mutations.

Drugs can often interact with ion channels, and sometimes they are given intentionally for this purpose. For example, lidocaine is often prescribed for pain, and works by inhibiting Na^+ channels in nerves to inhibit the pain signal. Lidocaine can also be used therapeutically to block Na^+ channels in the heart and prevent arrhythmias. However, drugs can also interact with certain populations of channels unintentionally. A commonly found interaction is the blockade of the I_{Kr} current, which can be an effect of many well known drugs such as methadone, erythromycin, and cocaine.[8] In fact, drug block of I_{Kr} is so common that most drugs today must be screened for this effect before they come to market. Clinically, when a drug blocks an outward current such as I_{Kr}, the outward manifestation is a prolonged QT interval, which reflects an increase in ventricular APD. This is known as drug-induced Long QT (LQT) Syndrome.

Genetic defects can also cause LQT Syndrome. The most prevalent type is LQT1 which causes a reduction in I_{Ks} current. Two other less common types of LQT are LQT2, which causes a reduction in I_{Kr}, and LQT3, which causes an increase in late I_{Na}. Because of the role I_{Ks} plays in responding to β-adrenergic stimulation, the effects of I_{Ks} mutations (LQT1) are typically observed during strenuous exercise- one common trigger is swimming.[9] Patients with LQT1 can be treated with β-blockers, which inhibit the β-adrenergic cascade, thereby preventing situations where an increase in I_{Ks} is necessary.

The rest of this chapter will focus on I_{Ks} as a molecular machine, showing how details of its molecular structure and motions determine its ability to affect AP repolarization.

3. I_{Ks} as a Molecular Machine

3.1 *Structure and kinetics*

Like many K^+ channels, I_{Ks} is formed by two types of subunits, α and β.[10] The α-subunit is KCNQ1 and was discovered because mutations to it were linked to patients with the LQT Syndrome. In fact, it was initially called K_vLQT1 (i.e. a voltage gated K^+ channel linked to

LQT). Four KCNQ1 subunits can come together to form a functional, homomeric (all subunits are the same), K$^+$ channel. Each of the subunits has six hydrophobic segments that span the cell membrane — segments S1–S6. The fourth segment contains a series of positive charges that confer voltage dependence to the channel opening and closing. The fifth and sixth segments contain the channel pore that allows current to pass through and determines the ion selectivity of the channel. One controversy surrounding I$_{Ks}$ since the discovery of its constitutive subunits has been how many β subunits assemble with the four α subunits and where in the channel they reside. The most recent data indicate that there are two KCNE1 subunits per channel[11] and that they reside at the inter-VSD (Voltage Sensing Domain) S4–S1 interface between two adjacent subunits.[12]

Channels such as KCNQ1 are often studied by injecting the message RNA (mRNA) that encodes the protein sequence into an expression system such as a frog oocyte. Once the mRNA is injected, the oocyte's protein synthesizing machinery works to make the channel and incorporate it into the cell membrane (channel expression). As opposed to the cardiac myocyte with all its branches in the circuit model, an expression system has ideally only one circuit model branch — the channel of interest. This approach simplifies the study of ion channels by isolating them.

After expression, the effect of voltage on the channel can be carefully studied by applying voltage protocols. The simplest voltage protocol is to step from a negative voltage (such as −80 mV) — where all the channels are closed — to a positive voltage where channels begin to open. This stepping is similar to what an AP does, although the AP voltage waveform is more complex. By stepping to various voltage levels, we can see how readily the channel opens in response to the voltage change. As can be seen in Figure 4a, homomeric KCNQ1 channels open quite quickly (activation) in response to a step to +40 mV. By stepping back to a negative potential (such as −70 mV), we can watch the channel closing process (deactivation). However, careful examination of the current during deactivation shows that there is a small hook in the current traces. This hook reflects channels that entered a closed state that was only accessible after the channel opened (inactivation). When we step back down to the negative potential, these channels come back from inactivation to the open state (recovery from inactivation) causing a small rise in the current. As time progresses, the channels close.

Comparing the KCNQ1 channel kinetics from this protocol to what we see during the AP shows a discrepancy. While the KCNQ1 channel opened quite quickly, I$_{Ks}$ opens slowly. This difference is the result of the second type of subunit that forms I$_{Ks}$, the β subunit KCNE1. When both α and β subunits are expressed in oocytes, the current changes quite dramatically. The activation is much slower, the conductance increases, and the inactivation that we observed through the hook in the deactivation current is no longer present (Figure 4c).[13] This new current, which includes the effects of KCNE1, behaves much more closely to the native I$_{Ks}$ current, especially with its slow activation.

3.2 *Kinetic modeling of I$_{Ks}$*

Quantitative assessment of how a current will behave in an AP, based on its response to voltage protocols in an oocyte, can be accomplished via computer modeling. As with the Hodgkin-Huxley

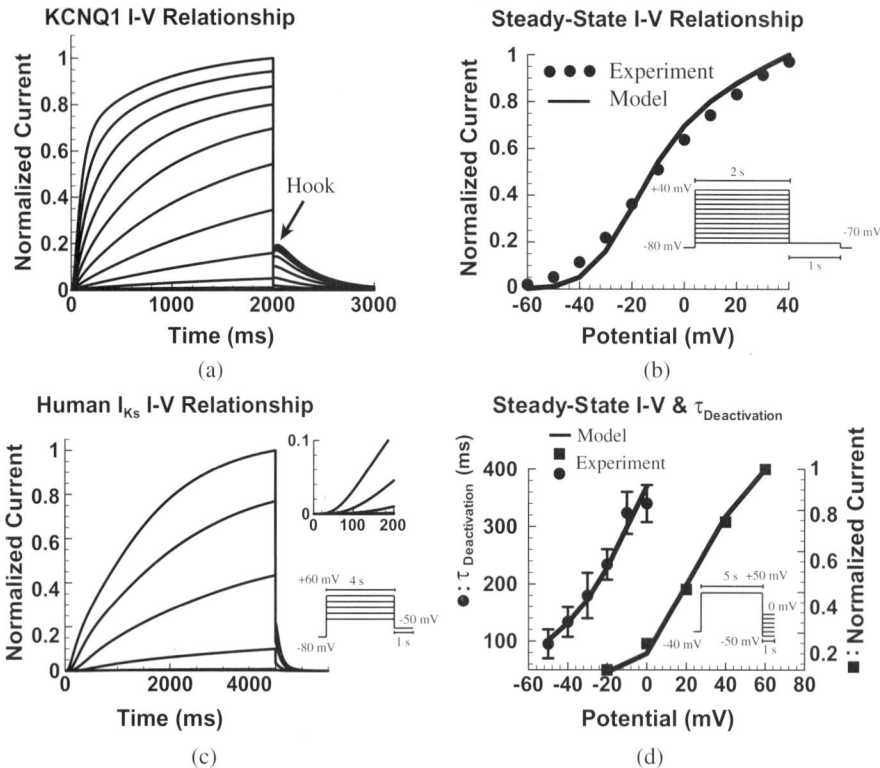

Figure 4. (a) Simulated time dependent KCNQ1 current resulting from steps from −80 mV to various potentials (−70 to 40 mV) followed by a step to −70 mV (protocol shown in Panel B). After stepping down to −70 mV, a hook is observed in the tail current indicating the presence of an inactivated state. (b) Simulated KCNQ1 steady-state I-V curve (solid line) is compared to experiment[13] (circles). (c) Time dependent human I_{Ks} current resulting from the protocol shown to the right. Note much slower activation compared to KCNQ1 and disappearance of the hook. A significant initial delay of activation (20 ms) is also reproduced by the model (inset). (d) Steady state I-V relationship for human I_{Ks} is compared to experiment[36] (circles, protocol shown in C) and time constant of deactivation is also compared to experiment[5] (squares, protocol shown in D inset).

model of the neuronal K$^+$ channel, the first model of I_{Ks} used independent gates to model the channel activation. The equation for the current took the following form:

$$I_{Ks} = \overline{g_{Ks}} \cdot xs_1 \cdot xs_2 \cdot (V_m - E_{Ks}),\tag{6}$$

where $\overline{g_{Ks}}$ is the maximum conductance that I_{Ks} can achieve, xs_1 and xs_2 are gates with a voltage dependence of opening, and E_{Ks} is the I_{Ks} reversal potential. Note that the two probabilities multiplied ($xs_1 \cdot xs_2$) together implies two independent transitions with different rates required to activate the channel. While the two gates (xs_1 and xs_2) in this model accurately reproduce channel activation, it is difficult (if not impossible) to assign these opening processes to any specific molecular motion.[4,14] The difficulty arises because the channel is tetrameric with four identical subunits, each with its own set of positive charges (the same in all subunits) that confer voltage dependence. The model in Eq. (6) reflects only two voltage gates, each with different voltage-dependence characteristics.

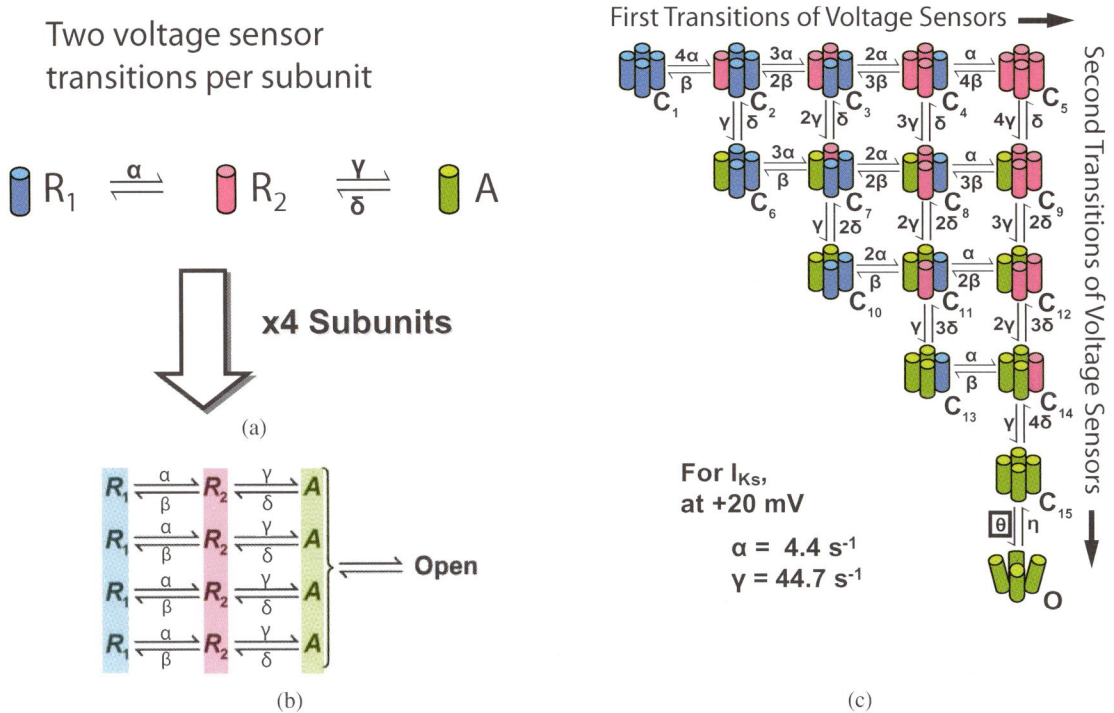

Figure 5. Conformational changes of K$^+$ channels during activation. (a) Two voltage sensor transitions before channel opening. (b) All four α-subunits that form the channel undergo a first transition from a resting state (R1) to an intermediate state (R2) and a second transition from R2 to an activated state (A). Once all voltage sensors are in the activated state, the channel can open. (c) Total number of combinations of voltage-sensor positions in the four subunits is 15 and can be represented by 15 closed states before channel opening. Blue, red, green indicate a voltage sensor in position R1, R2 or A, respectively. (Figure reproduced with permission from Rudy *et al.*)[37]

One way to formulate the model to represent the tetrameric structure, as well as two distinct rates of opening, is to employ dependent transitions.[15] In the case of the model in Eq. (6) with independent gates, one gate does not "feel" what another gate does. That is, the rate of gate opening and closing does not depend on what position the other gate is in. With dependent transitions, we can include event order. The model in Figure 5 shows a schematic that represents each of the four channel voltage sensors undertaking two sequential steps to activation. Each state in the model represents a permutation of voltage sensors. For example, the first represents all four voltage sensors in the resting state. The second state corresponds to one voltage sensor in the intermediate position and all other three in the resting conformation. When all four voltage sensors are in the activated position — they have each undergone two transitions — the channel can make a cooperative transition to the open state. (A cooperative transition requires the participation of all 4 voltage sensors, i.e. all voltages sensors are activated conformation). The model transition rates are derived from the number of voltage sensors that can make the transition. From the first state, C_1, four voltage sensors can move from the resting (R_1) to the intermediate state (R_2), so the forward rate is 4α. Since, only one voltage sensor can move from intermediate (R_2) to resting (R_1) between state C_2 and C_1, the transition rate is simply β.[16] The equations describing the occupancy of each state rely on the occupancy of adjacent states and the transition rates between them, for example:

$$\frac{dC_1}{dt} = \beta C_2 - 4\alpha C_1. \tag{7}$$

$$\frac{dC_2}{dt} = 4\alpha C_1 + \delta C_6 + 2\beta C_3 - (\beta + 3\alpha + \gamma)C_2. \tag{8}$$

Once the model structure is defined, the rate constants must be fit to describe their dependence on voltage. Typically, the equations for the rate constants take the following form, which reflects exponential dependence on voltage:

$$\alpha = \alpha_0 \cdot e^{\frac{z \cdot V_m \cdot F}{R \cdot T}}. \tag{9}$$

$$\beta = \beta_0 \cdot e^{-\frac{z \cdot V_m \cdot F}{R \cdot T}}, \tag{10}$$

where α_0 and β_0 are the rates for the transitions in the absence V_m, and z reflects the effective charge transported through the electric field imparted by V_m and confers voltage dependence.[17] In practice, the parameters α_0 and z are set so that the macroscopic current predicted by the model behaves similarly in response to a voltage protocol to the experimentally recorded current (Figures 4b and 4d). The model can then be used to study the consequences of two-stage voltage sensor activation in the context of the AP. Above, we discussed two situations where additional outward, repolarizing current is required to shorten the AP — at fast heart rates and in the presence of β-adrenergic stimulation. Figure 6 shows APs at fast and slow rates along with the occupancy of I_{Ks} channels in different "Zones" of the model. Zone 2 corresponds to channels with at least one voltage sensor that has not yet made a first activation transition. Zone 1 represents channels that only have to make second transition to open. Since the second transition is faster than the first, channels in Zone 1 are able to open much more quickly than channels in Zone 2. The consequence of the two zones is that at fast rates, while channels can still close to Zone 1 between APs, they do not have sufficient time to transition to the deep closed states of Zone 2. Therefore, at fast rates there is accumulation of channels in Zone 1 from where they can open readily. Consequently, I_{Ks} increases faster at fast rate, fulfilling its role of shortening the AP when the heart is beating quickly.

As discussed, a second role of I_{Ks} is to provide outward current to counter inward current carried by $I_{Ca,L}$ in the presence of β-adrenergic stimulation. Recently, the two-stage model of I_{Ks} has been expanded to describe the I_{Ks} response in this situation.[18] β-adrenergic stimulation causes I_{Ks} to activate more quickly. Careful fitting of the model to experimentally recorded I_{Ks} currents shows that this increased rate of activation is partially the result of channels shifting to closed states that are near to the open state. In this case, phosphorylation by the β-adrenergic effector Protein Kinase A (PKA) is responsible for this accumulation, instead of a rapid sequence of AP driven voltage depolarizations at a fast heart rate.

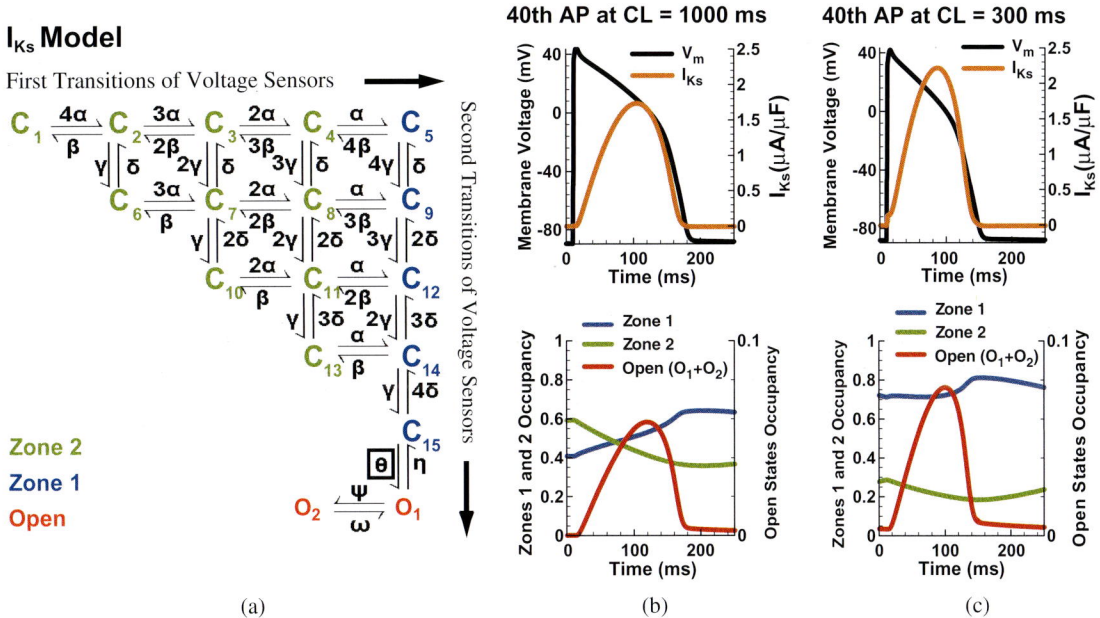

Figure 6. Kinetic Transitions of I_{Ks} Channels During the AP at Slow and Fast Rate. (a) Markov model of the I_{Ks} channel[15]. States are color coded according to their type: Zone 2; closed states for which not all voltage sensors have completed the first transition (light green). Zone 1; closed states for which all four voltage sensors have completed the first transition (blue). Open (red). (b) I_{Ks}, V_m and channel state occupancies during the 40th AP at slow rate, CL = 1000 ms. I_{Ks} rises slowly, resulting in peak current at the end of the AP where it most efficiently contributes to repolarization. Only 40% of channels reside in zone 1 at AP onset and can activate rapidly. While V_m remains depolarized, channels continue to transition from zone 2 to zone 1. (c) I_{Ks}, V_m, and channel state occupancies during the 40th AP at fast rate, CL = 300 ms. Since the diastolic interval is shorter at CL = 300 ms, V_m stays at depolarized potentials for a greater percentage of time, which causes channel accumulation in zone 1 of closed states. At AP onset 75% of channels reside in zone 1, facilitating rapid transitions to the open state. This results in increased I_{Ks} late during the AP and APD shortening. Note that the mechanism for I_{Ks} increase is accumulation in closed states near the open state (zone 1) as opposed to open state accumulation. The accumulation in zone 1 creates a reserve of channels that are ready to open rapidly, "on demand" to generate a greater repolarizing current; we call this pool of channels "available reserve". (Figure modified from Zipes *et al.*)[35]

The above example demonstrates the ability of kinetic models to predict the role of a given channel transition in the cardiac AP. Here, the presence of multiple activation transitions confers on I_{Ks} the ability of increasing its opening rate when more current is needed. However, one question that still remains is what molecular motions these transitions correspond to. In the next section, we describe a framework to directly incorporate what we know about the molecular structure into the model of channel function.

3.3 *Molecular modeling of I_{Ks}*

The first step in creating a molecularly based model is to propose a hypothetical structure. Ideally, the model will be based on a crystal structure of the channel, but to date, a structure of the KCNQ1 channel has not been published. Fortunately, the structure of a similar channel, $K_v1.2$, has recently become available[19] and can be used as a starting point for an I_{Ks} model. This approach is known as homology modeling and relies on similarity between proteins with similar sequences to infer structure. The first step in homology modeling is to align the residues of the protein of interest (KCNQ1) with similar residues in the template ($K_v1.2$). For our purposes, the charged residues provide the most useful alignment information. Since the discovery of ion channels, it has been hypothesized that positively charged residues move through the lipid membrane, where their solvation is energetically unfavorable, by interacting with negatively charged residues also lying within the membrane spanning domains of the channel protein. The crystal structure of $K_v1.2$, in addition to experimental evidence, indicates that the positively charged residues in S4 interact with negatively charged residues in S2 and S3. Given that it is likely that these residues are critical to enabling S4 motion, it is plausible that the charged residues occupy a similar position in KCNQ1. The alignment for the molecular KCNQ1 model is shown in Figure 7a. After the alignment is determined, a homology model can be improved by minimizing a function that quantitatively accounts for the alignment with the template protein in addition to the intrinsic molecular energy that comes from molecular forces such as van der Waal's interactions, Coulombic forces, and specific residue conformations.[20]

Crystal structures are often obtained in nonphysiological conditions to facilitate crystal formation. Therefore, even models based directly on a crystal structure must be refined to account for the physiological environment. In a cell, the channel is situated in the membrane surrounded on all sides by lipid molecules and on top and bottom by water molecules. This environment can be simulated using molecular dynamics on a full system that includes the atoms in the channel, the lipid membrane and the water molecules (Figures 7b and 7c).[21] If the crystal structure is near a physiological state, the channel will quickly relax into this conformation. However, since the system typically contains over 100 000 atoms and their interactions, the time that can be simulated, on the order of tens of microseconds at best, is not sufficient to simulate a channel transition during gating, which has time constants that last milliseconds to seconds.

To simulate a gating channel transition with current computing technology, several approximations are required. One approximation is to simulate the macroscopic behavior of the lipid and water solvents (implicit solvent) instead of every atomic interaction (explicit solvent). Implicit solvents require specification of several parameters: the dielectric constant, ion accessibility and the protein charge density. For our purposes these parameters have two important effects. First, a solvent with a high dielectric constant and ion accessibility (such as the water on each side of the membrane) will solvate charged residues with less energy than the low dielectric, ion- inaccessible membrane. These implicit solvent properties correctly favor charged residues on the outside of the membrane and nonpolar residues within the membrane. Once the implicit membrane has been specified (Figure 7e), the energy of a given channel conformation can be approximated with the Poisson-Boltzmann equation:

$$-\nabla \cdot [\varepsilon(\vec{r})\nabla\varphi(\vec{r})] + \bar{\kappa}^2(\vec{r})\varphi(\vec{r}) = \frac{e_c}{k_B T}4\pi\rho(\vec{r}), \tag{11}$$

where $\varphi = \frac{e\Phi}{k_B T}$ is the reduced electrostatic potential and Φ is the electrostatic potential (statvolt); ε is the inhomogeneous dielectric constant (statcoul² erg⁻¹ cm⁻¹); ρ is the density of charge within the protein (stacoul cm⁻³); e_c is the electron charge (statcoul); k_B is the Boltzmann constant

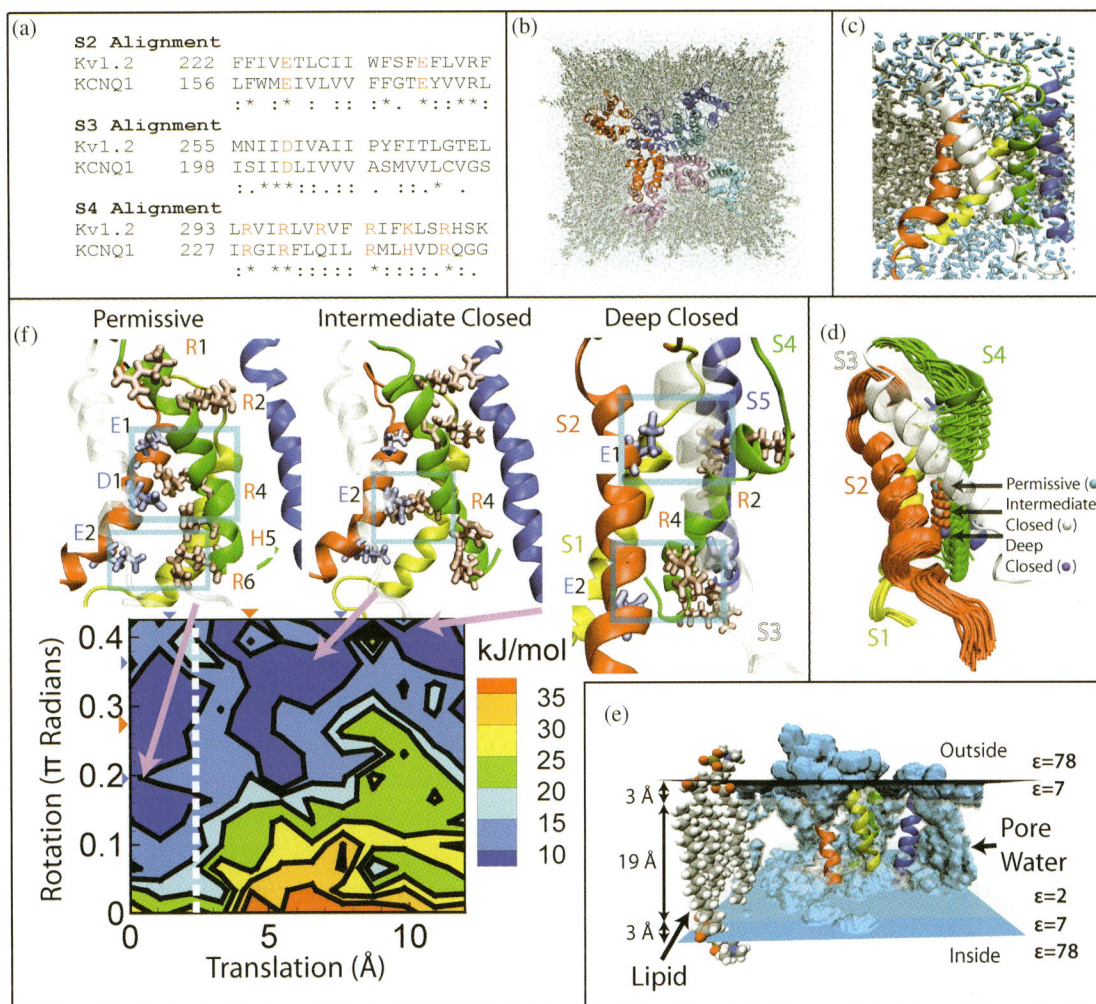

Figure 7. KCNQ1 Model and Energy Landscape. (a) Kv1.2/KCNQ1 Alignment. Symbols: (*) identical; (:) and (.) conserved and semi-conserved substitutions. Red residues are charged. (b) Top-down (extracellular) view of the all atom system. Each subunit (S1-S6) is color coded. Pore regions (S5-S6) of one subunit interact with the adjacent subunit voltage-sensing region (S1-S4). (c) Voltage-sensing region with lipid (gray) and water (light blue) solvent molecules. (d) S4 translation and rotation. β-carbon of R4 labeled with red beads shows motion. Arrows indicate stable configurations (labeled). S1-Yellow, S2-Red, S3-White, S4-Green, S5-Blue. (e) Implicit membrane. Isocontours at dielectric constant ε = 78 (blue) and ε = 2 (transparent) show the transition region representing the lipid esters. Water (blue) can penetrate the protein as in Panel C and is represented by ε = 78. (f) Energy landscape at V_m = 0 mV and associated conformations. Increasing translation corresponds to movement toward the intracellular space; increasing rotation is counter-clockwise as viewed from extracellular space. Left of dotted white line is permissive state. Positively/negatively charged residues are labeled with red/blue letters. Energy landscape minima (deep blue) correspond to stable configurations; right to left: deep closed state, intermediate closed state, permissive state. Boxes on deep closed state conformation indicate interaction between R2 and E1, and between R4 and E2. Intermediate closed state shows interaction between R4 and D1 and E2. In the permissive state, interaction is observed between R6 and E2, and between R4 and E1. (Figure adapted with permission from Silva *et al.*)[28]

(erg K^{-1}); T is the temperature (K). $\bar{\kappa}^2 = \varepsilon_w \kappa^2$, with $\kappa^2 = \frac{4\pi}{k_B T} \Sigma_\alpha q_\alpha \rho_\alpha / \varepsilon_w$ while κ^{-1} is the Debye-Hückel screening parameter (cm), q_α is the charge (statcoul), ρ_α is the bulk ion density of each ion and ε_w is the dielectric constant of water (statcoul^2 erg^{-1} cm^{-1}). This equation accounts for solvent effects and charged residue interactions, which play a critical role in channel gating. However, other forces may also be just as important, including van der Waal's force and the bending and twisting of the protein residues into favorable and unfavorable conformations. In the future, it may be necessary to additionally include these forces to accurately simulate channel gating.

With a method for computing the energy (by integrating the potentials obtained via the Poisson-Boltzmann equation over space) of a given conformation, we can proceed to think about the motion of S4, which has been a topic of discussion for decades. Several recent experiments point to translation of the S4 up and down through the membrane approximately 12 Å while at the same time rotating counter-clockwise (Figure 7d).[22] As the channel moves from a closed to open state, this motion will alter interactions between the positively and negatively charged residues on S2, S3 (negative charges) and S4 (positive charges). A rigorous sampling of this motion is currently not computationally feasible due to the number of degrees of freedom involved (many for each residue). As a first approximation, we can move S4 through this predicted range, calculating the energy for several conformations along the way. Note that this approach assumes that S4 remains rigid and that the other helices do not move, which is not necessarily true. With the putative motion decided upon, a two dimensional energy landscape can be constructed, representing with each axis a degree of freedom (translation and rotation; Figure 7f). Once the energy landscape is constructed, stable points (associated with energy minima) can be examined as potential conformations along the activation pathway. As can be seen in Figure 7f, each stable point corresponds to different interactions between the charged residues of the channel.

The next challenge we face is incorporating the effect of V_m on the channel gating. Utilizing our implicit membrane, the effect of the membrane potential by modifying the PB equation as follows:[23,24]

$$-\nabla \cdot [\varepsilon(\vec{r}) \nabla \varphi(\vec{r})] + \bar{\kappa}^2(\vec{r}) \varphi(\vec{r}) = \frac{e_c}{k_B T} 4\pi \left(\rho(\vec{r}) + \frac{\bar{\kappa}^2(\vec{r}) V_m \Theta(\vec{r})}{4\pi} \right), \qquad (12)$$

where $\Theta(\vec{r})$ is a Heaviside step function whose value is 1 for $\vec{r}$ within the intracellular space and 0 elsewhere. This new equation, the Poisson-Boltzmann Voltage (PB-V) equation, accounts for voltage as a function of charge on either side of the membrane. The potential across a slice of the membrane predicted by this equation is shown in Figure 8a. The effect on the energy landscape of applying a positive or negative voltage can be seen in Figure 8c. As expected, negative voltages favor the closed state, allowing positive charges on S4 to move closer to the negatively charged interior of the cell. Similarly, positive voltages favor the open state.

Experimentally, charges moving across the membrane are observed as a capacitive gating current. Integrating this current gives a quantitative assessment of how many charges are moving across the field as the channels transition from the closed to the open state, if the total number of channels is known. Theoretically, we can calculate how many effective charges we expect. This is accomplished by multiplying the charges in the channel by the fraction of the

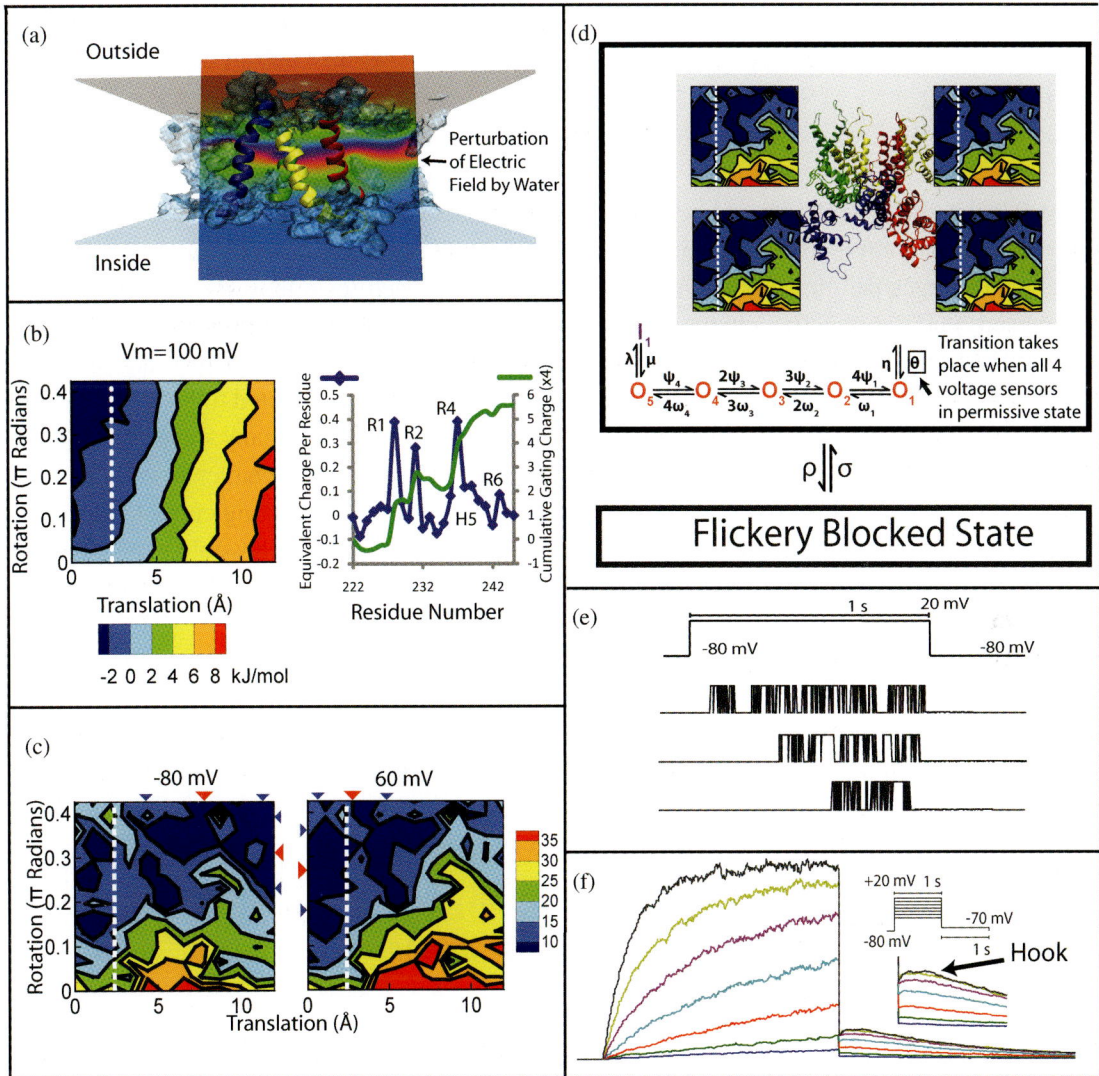

Figure 8. V_m Effect and Kinetic Model of KCNQ1. (a) Normalized transmembrane potential (a.u.). Water penetration causes significant perturbation of the electric field (ripples in the potential lines, arrow). (b) Left: Membrane potential ($V_m = 100$ mV) causes stabilization of the open state and introduces a net difference of ~10 kJ/mol between closed and open configurations. Right: Gating charge contribution for each residue in S4 caused by moving across energy landscape (blue, left axis), and cumulative gating charge for whole channel (green, right axis). (c) Energy landscape at positive and negative V_m. Triangles are mean S4 position (red) and standard deviation (blue). (d) KCNQ1 Markov model. Energy landscapes of 4 subunits determine residency in permissive state. All 4 subunits in the permissive state (left of dotted white line) enables open state (O1–O5) transitions. Inactivated state, I, is only accessible from O5. Flickery blocked state is available from any conformation.[38] (e) Single channel currents; traces show random nature of opening. (f) Macroscopic KCNQ1 current is the sum of 1000 single channels. A hook (arrow), characteristic of KCNQ1, is observed upon repolarization and is caused by channel recovery from inactivation. (Figure adapted with permission from Silva *et al.*)[28]

electric field at their location in space at the closed and the open conformations according to the following equation:[23]

$$Q = \Sigma_p q_p \varphi_{mp}(\vec{r}) \tag{13}$$

where Q is the effective charge, φ_{mp} represents the fraction of the membrane potential traversed, q_p is the protein charge. Figure 8b shows the total effective charge as 2.5 charges for each subunit for a total of 6 charges per channel. The charge contributed by each residue is also plotted showing, as expected, that the effective charge is due to motion of the positive charges through the field, most of the charge being carried by the first, second and fourth positively charged arginines (R1, R2, R4) on S4. (KCNQ1 has a glutamine where the third arginine is expected to be, so R4 actually refers to the third KCNQ1 arginine.) While the gating current from many channels has been observed, the KCNQ1 gating current has yet to be recorded. In part, this challenge is due to difficulty in growing large numbers of channels in expression systems (the gating current per channel is quite small, so large numbers of channels are required). Additionally, we would expect less gating current per channel for KCNQ1 compared to similar channels because it is missing a charge (R3) in S4.

Since we cannot validate our model with experimental gating current, we turn to the ionic current going through the channel pore. The ionic current reflects the opening and closing transitions of the channel and the dependence of its kinetics on voltage. Our task then is to correlate our energy landscape to the kinetics of channel activation and deactivation. The assumption that we make is that S4 motion is similar to the motion of a particle governed by diffusion and the energy landscape. The Smoluchowski equation describes such a scenario:

$$\frac{\partial p(x,t)}{\partial t} = \frac{\partial}{\partial x} D\left(\frac{\partial}{\partial x} p(x,t) - \beta F(x)p(x,t)\right). \tag{14}$$

$p(x, t)$ is the probability of finding a partical at a position x (Å) at a time t (ms). D(Å²ms⁻¹) is the diffusion constant, $F(x)$ is the force on the particle (kJ·mol⁻¹·Å⁻¹), and β is $1/(k_B \cdot T)$ (kJ·mol⁻¹)⁻¹. The result from this equation is the probability that a particle will move from one point to an adjacent point as a function of its concentration and the force introduced by the energy landscape.[25,26] If we assume that our particle has no mass ($F = -\nabla W$, where ∇W is the gradient of the potential energy W in kJ·mol⁻¹), we can write the Smoluchowski equation in the adjoint form:

$$\frac{\partial p(x,t)}{\partial t} = \frac{\partial}{\partial x} D(x)e^{-\beta W(x)}p(x,t)\frac{\partial}{\partial x}(e^{\beta W(x)}p(x,t)). \tag{15}$$

This form has the advantage of leading to easily definable transition rates between points on the energy landscape and the conformations that they correspond to.[27] With these transition rates, we essentially have a large kinetic model similar to the one shown in Figures 5 and 6a. The states of the model correspond to conformations of the channel and the transition rates reflect the probability that the channel will transition from one conformation to a neighboring conformation.

Since we have only simulated the voltage sensor motion, we need to incorporate other potential channel movements into the model. These motions can include activation transitions from other parts of the channel like the hinged gates in S6 or pore transitions that close the channel after it has been open for some time (inactivation). To include these transitions, we couple them to the kinetic model that describes the voltage sensor motion. One difficulty that arises is that there are four voltage sensors that need to be simulated. In Figure 5, we lumped similar states together, but in that case we only had two transitions per voltage sensor, which led to 15 closed states. If we have 10 rotation steps and 10 translation steps, we will have 100 states for each of the four voltage sensors resulting in a computationally intractable model. One way to overcome this difficulty is to simulate each voltage sensor separately with a random particle.[28] An alternative way to overcome this challenge has been described recently.[29] Once all four voltage sensors have translated a certain distance (past the dotted line, Figure 8D), the channel can move to the open states, and from there to inactivated states, which are described in much less detail.

At this stage, the model reveals several aspects of I_{Ks} channel gating that are not well understood. One challenge is how to define the permissive state, which determines when the channel can transition to the open (O) and inactivated (I) states. After what amount of voltage sensor translation can the channel enter the O and I states? Do these transitions depend on the amount of rotation as well? Once the channel has made it into the O and I states, can the voltage sensors move back into the resting conformations? What part of the channel is responsible for the transition to the O states and the transitions between these states? These questions and others remain unanswered experimentally. As this experimental data become available the model could be refined, improving its accuracy.

The model in this form can be used to simulate the open and closing of a single channel (Figure 8e). If the channel starts from the closed state (all four voltage sensors in the resting state), the random openings begin to occur when all four voltage sensors are in the permissive conformation and the channel can transition to the O states. Channels then close in one of two ways, they move into the inactivated state or they transition back into the closed state. By lumping many (~1000) of these single channel constructs together, we can find the channel open probability as a function of time. As explained above [Eq. (4)], the open probability can be used to calculate the macroscopic current (Figure 8f).

The ability to simulate the macroscopic current enables validation against whole cell currents recorded from native cells or expression systems. As for the kinetic model, the first step is to reproduce the current response to voltage steps (Figure 9b). However, since we have a model whose kinetics are based on molecular structure, we can go one step further and simulate how mutations will affect the channel. One intriguing residue is the extracellular negative charge on S2, E1. As seen above, this negative charge interacts with positive charges on S4, affecting its gating motion. To test the model, three mutations were simulated that progressively reduced the amount of negative charge at E1 (a polar glutamine, a neutral alanine, and a positively charged lysine). The energy landscapes that result from these mutations are shown in Figure 9a. Applying the same procedure that is used to obtain the wild type current, the predicted current from the mutant channel can be simulated. Just as the wild-type current can be measured in an expression system, the current from a mutant channel can also be measured and compared to the model prediction, providing model validation (Figure 9).

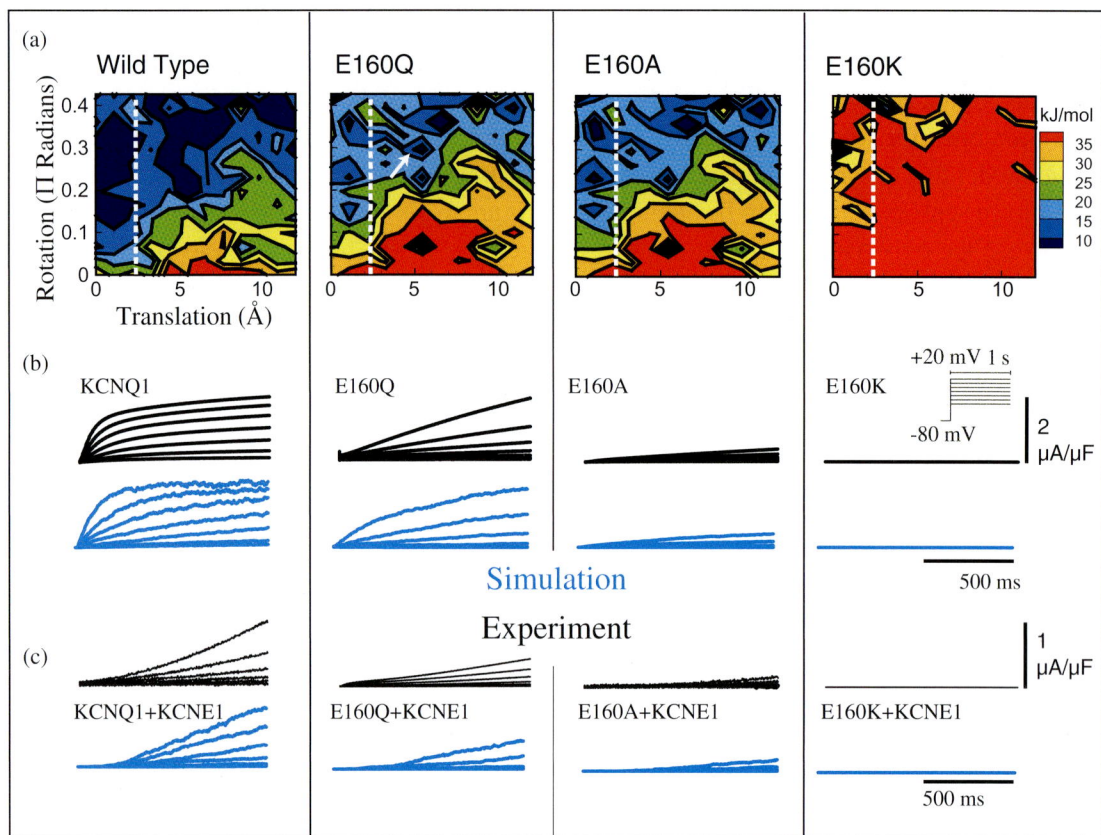

Figure 9. Mutation effects. (a) Energy landscapes for wild-type KCNQ1 and mutants E160Q, E160A and E160K at $V_m = 0$ mV. (b) Comparison of macroscopic current between experiment (black) and simulation (cyan) for homomeric wild-type KCNQ1 and mutants (Protocol in inset). (c) I_{Ks} (KCNQ1+KCNE1) is accurately simulated by slowing S4 movement. Mutations cause slowing of activation, in addition to KCNE1 effect. (Figure adapted with permission from Silva *et al.*).[28]

To reproduce the native I_{Ks} current, we also need to account for the effect of the β-subunit, KCNE1. As discussed above, its location and how many subunits are in a native channel have been a matter of contention for some time. Since I_{Ks} displays slower activation compared to the homomeric channel, it was simply assumed that KCNE1 slows down the motion of the voltage sensor. The effect was implemented in the model by reducing the magnitude of the diffusion coefficient in the Smoluchowski equation, D. As can be seen from Figure 9c this simple manipulation is able to reproduce the experimentally observed kinetics of the heterologously expressed channel. However, this step raises further questions as to the function of I_{Ks}. For example, does the voltage sensor move along the same path albeit more slowly or does the path it takes change? Since the current literature indicates that two KCNE1 subunits are likely to be present in the native channel, do two voltage sensors activate with one set of kinetics while the other two activate differently? Or, do the voltage sensors move in the same way that they do in KCNQ1 and a downstream step in activation is affected by KCNE1? As before, experimental data providing insight into these questions are crucial to improving our understanding of and ability to model I_{Ks}.

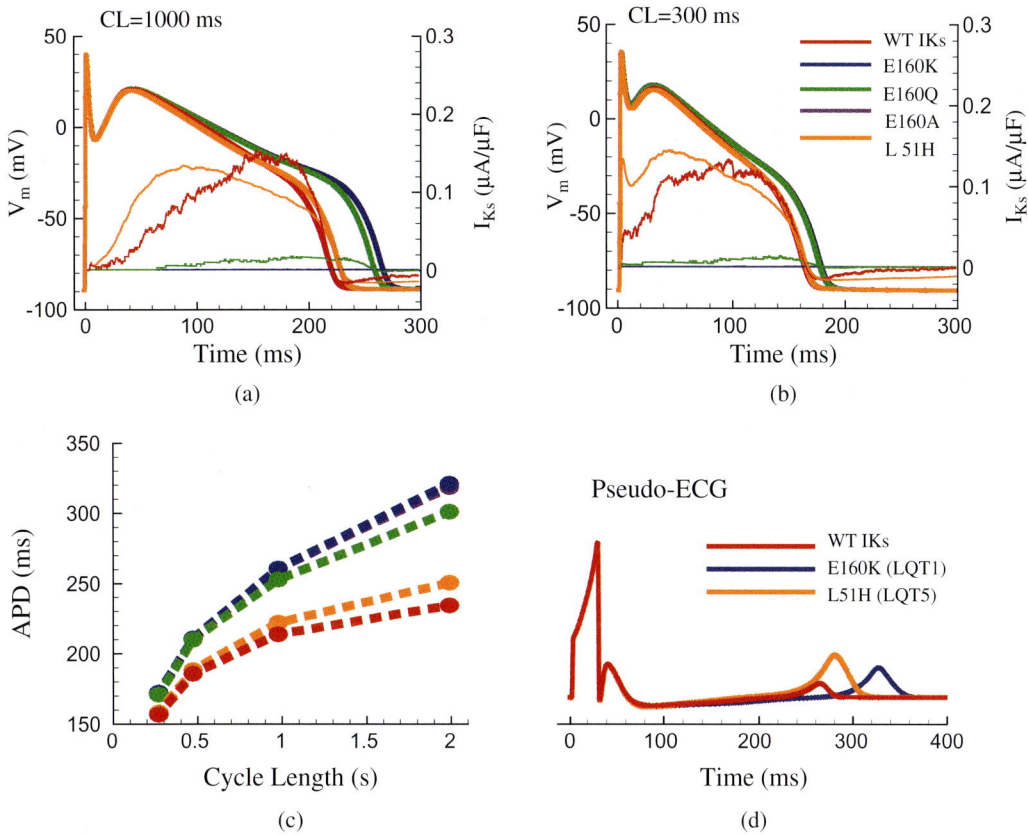

Figure 10. AP and ECG. V_m and I_{Ks} currents are color coded. All currents, with the exception of L51H, are in the presence of KCNE1. (a) AP and I_{Ks} at slow rate (CL = 1000 ms). (b) AP and I_{Ks} at fast rate (CL = 300 ms). (c) Adaptation curves, which plot AP duration (APD) dependence on CL. (d) Simulated pseudo-ECGs. LQT1 mutation E160K shows significant QT interval prolongation, compared to mild prolongation by LQT5 mutation L51H, consistent with clinical phenotypes,[34] (Figure adapted with permission from Silva *et al.*)[28]

Within the assumptions that we have made, we can examine the cellular level consequences of KCNQ1 mutations. To simulate the myocyte, we expand the model to include all the currents incorporated in the Hund-Rudy dynamic model of the canine ventricular myocyte.[30,31] In this model, the I_{Ks} branch of the circuit contains the molecularly based model, and all the other branches contain previously validated Hodgkin-Huxley type models of the other currents that participate in generating the AP. Since a proposed function of I_{Ks} is to shorten the AP at fast rates, we compare its role in the AP at cycle lengths (the interval between stimuli) of 300 ms and 1000 ms. As can be seen from Figure 10, I_{Ks} is able to reduce AP duration by increasing at fast rates. The mechanism is similar to what was demonstrated previously in the simpler 15-state kinetic model of the channel; at fast rates voltage sensors do not return to closed states that are far from the open state.

An I_{Ks} mutation that replaces the negative charge E1 at residue 160 on S2 by a positive charge K (E160K), is known clinically to cause LQT1. As shown in Figure 9 (right column), the mutation results in essentially complete loss of current and consequently, severe APD prolongation

(Figure 10). Also of interest is the behavior of KCNQ1 during the AP in the absence of KCNE1. In the presence of the KCNE1 LQT5 mutation L51H, KCNE1 does not interact with functional channels and reduces the number of KCNQ1 channels that make it to the membrane by 30%.[32] As noted above, homomeric KCNQ1 expression level is much smaller than that of heteromeric I_{Ks}. However, the channel activates much more quickly than I_{Ks}, which nearly compensates for the reduced number of channels.

Clinically, physicians do not have access to recordings from patient's ventricular myocytes. Instead they rely on the surface ECG to assess how long it takes for the ventricles to repolarize by using the QT interval as a marker. To link our molecular model to the ECG, we utilize a pseudo-ECG, which simulates the potential generated by AP propagation across a segment of the ventricular wall.[33] While the excitation wavefront in the intact ventricles is much more complex, it is reasonable to expect that the pseudo-ECG would accurately reflect the effect on the QT interval of a mutant channel. Three cases are shown in Figure 10d, LQT1 (E160K), LQT5 (L51H) and the normal wild-type channel. As is the case in the clinical phenotype, LQT1 prolongs the QT interval more severely than LQT5.[34]

Conclusion

The model example presented in this chapter is intended to illustrate how an ion channel operates as a molecular machine in order to participate effectively in the generation and shaping of the AP. A modeling approach can help to elucidate the role of molecular transitions in electrophysiology of the heart. At the same time, this exercise also highlights the immense amount of work that could be accomplished in the future to attain a better understanding of how the molecular motions of ion channels work together to form the cardiac AP. Here we have highlighted one channel, but similar work can be done for the other major channels as well. Eventually, it is hoped that such work will result in better understanding of how damage from heart attacks, inherited diseases and adverse drug interactions predispose patients to arrhythmias in order to develop better treatments.

Further Reading

Y. Rudy and J. R. Silva. 2006. Computational biology in the study of cardiac ion channels and cell electro-physiology. *Q Rev Biophys* **39**, 57–116.

J. R. Silva and Y. Rudy. 2010. Multi-scale electrophysiology modeling: From atom to organ. *J Gen Physiol* **135**, 575–581.

References

1. Z. J. Zheng, J. B. Croft, W. H. Giles and G. A. Mensah. 2001. Sudden cardiac death in the United States, 1989 to 1998. *Circulation.* **104**, 2158–2163.
2. A. L. Hodgkin and A. F. Huxley. 1952. A quantitative description of membrane current and its application to conduction and excitation in nerve. *J Physiol* **117**, 500–544.
3. C. H. Luo and Y. Rudy. 1994. A dynamic model of the cardiac ventricular action potential. I. Simulations of ionic currents and concentration changes. *Circ Res* **74**, 1071–1096.

4. J. Zeng, K. R. Laurita, D. S. Rosenbaum and Y. Rudy. 1995. Two components of the delayed rectifier K+ current in ventricular myocytes of the guinea pig type. Theoretical formulation and their role in repolarization. *Circ Res* **77**, 140–152.

5. L. Virag, N. Iost, M. Opincariu, J. Szolnoky, J. Szecsi, G. Bogats, P. Szenohradszky, A. Varro and J.G. Papp. 2001. The slow component of the delayed rectifier potassium current in undiseased human ventricular myocytes. *Cardiovasc Res* **49**, 790–797.

6. S. Genovesi, D. Zaccaria, E. Rossi, M. G. Valsecchi, A. Stella and M. Stramba-Badiale. 2007. Effects of exercise training on heart rate and QT interval in healthy young individuals: Are there gender differences? *Europace* **9**, 55–60.

7. S. O. Marx, J. Kurokawa, S. Reiken, H. Motoike, J. D'Armiento, A. R. Marks and R. S. Kass. 2002. Requirement of a macromolecular signaling complex for beta adrenergic receptor modulation of the KCNQ1-KCNE1 potassium channel. *Science* **295**, 496–499.

8. D. M. Roden. 2004. Drug-induced prolongation of the QT interval. *N Engl J Med* **350**, 1013–1022.

9. P. Ott, F. I. Marcus and A. J. Moss. 2002. Images in cardiovascular medicine. Ventricular fibrillation during swimming in a patient with long-QT syndrome. *Circulation* **106**, 521–522.

10. M. C. Sanguinetti, M. E. Curran, A. Zou, J. Shen, P. S. Spector, D. L. Atkinson and M. T. Keating. 1996. Coassembly of K(V)LQT1 and minK (IsK) proteins to form cardiac I(Ks) potassium channel. *Nature* **384**, 80–83.

11. H. Chen, L. A. Kim, S. Rajan, S. Xu and S. A. Goldstein. 2003. Charybdotoxin binding in the I(Ks) pore demonstrates two MinK subunits in each channel complex. *Neuron* **40**, 15–23.

12. D. Y. Chung, P. J. Chan, J. R. Bankston, L. Yang, G. Liu, S. O. Marx, A. Karlin and R. S. Kass. 2009. Location of KCNE1 relative to KCNQ1 in the I(KS) potassium channel by disulfide cross-linking of substituted cysteines. *Proc Natl Acad Sci U S A* **106**, 743–748.

13. M. Tristani-Firouzi and M. C. Sanguinetti. 1998. Voltage-dependent inactivation of the human K+ channel KvLQT1 is eliminated by association with minimal K+ channel (minK) subunits. *J Physiol* **510** **(Pt 1)**, 37–45.

14. P. C. Viswanathan and Y. Rudy. 1999. Pause induced early after depolarizations in the long QT syndrome: A simulation study. *Cardiovasc Res* **42**, 530–542.

15. J. Silva and Y. Rudy. 2005. Subunit interaction determines IKs participation in cardiac repolarization and repolarization reserve. *Circulation* **112**, 1384–1391.

16. J. P. Keener. 2009. Invariant manifold reductions for Markovian ion channel dynamics. *J Math Biol* **58**, 447–457.

17. W. N. Zagotta, T. Hoshi and R. W. Aldrich. 1994. Shaker potassium channel gating. III: Evaluation of kinetic models for activation. *J Gen Physiol* **103**, 321–362.

18. S. Severi, C. Corsi, M. Rocchetti and A. Zaza. 2009. Mechanisms of beta-adrenergic modulation of I(Ks) in the guinea-pig ventricle: Insights from experimental and model-based analysis. *Biophys J* **96**, 3862–3872.

19. S. B. Long, E. B. Campbell and R. Mackinnon. 2005. Crystal structure of a mammalian voltage-dependent Shaker family K+ channel. *Science* **309**, 897–903.

20. A. Sali and T. L. Blundell. 1993. Comparative protein modelling by satisfaction of spatial restraints. *J Mol Biol* **234**, 779–815.

21. J. C. Phillips, R. Braun, W. Wang, J. Gumbart, E. Tajkhorshid, E. Villa, C. Chipot, R. D. Skeel, L. Kale and K. Schulten. 2005. Scalable molecular dynamics with NAMD. *J Comput Chem* **26**, 1781–1802.

22. F. Tombola, M. M. Pathak and E. Y. Isacoff. 2005. How far will you go to sense voltage? *Neuron* **48**, 719–725.

23. B. Roux. 1997. Influence of the membrane potential on the free energy of an intrinsic protein. *Biophysical Journal* **73**, 2980–2989.

24. M. Grabe, H. Lecar, Y. N. Jan and L. Y. Jan. 2004. A quantitative assessment of models for voltage-dependent gating of ion channels. *Proc Natl Acad Sci U S A* **101**, 17640–17645.

25. D. Sigg and F. Bezanilla. 2003. A physical model of potassium channel activation: From energy landscape to gating kinetics. *Biophysical Journal* **84**, 3703–3716.

26. D. Sigg, H. Qian and F. Bezanilla. 1999. Kramers' diffusion theory applied to gating kinetics of voltage-dependent ion channels. *Biophysical Journal* **76**, 782–803.

27. A. Ansari. 2000. Mean first passage time solution of the Smoluchowski equation: Application to relaxation dynamics in myoglobin. *J Chem Phys* **112**, 2516–2522.

28. J. R. Silva, H. Pan, D. Wu, A. Nekouzadeh, K. F. Decker, J. Cui, N. A. Baker, D. Sept and Y. Rudy. 2009. A multiscale model linking ion-channel molecular dynamics and electrostatics to the cardiac action potential. *Proc Natl Acad Sci U S A* **106**, 11102–11106.

29. A. Nekouzadeh, J. R. Silva and Y. Rudy. 2008. Modeling subunit cooperativity in opening of tetrameric ion channels. *Biophysical Journal* **95**, 3510–3520.

30. K. F. Decker, J. Heijman, J. R. Silva, T. J. Hund and Y. Rudy. 2009. Properties and ionic mechanisms of action potential adaptation, restitution and accommodation in canine epicardium. *Am J Physiol Heart Circ Physiol* 1017–1026.

31. T. J. Hund and Y. Rudy. 2004. Rate dependence and regulation of action potential and calcium transient in a canine cardiac ventricular cell model. *Circulation* **110**, 3168–3174.

32. A. Krumerman, X. Gao, J. S. Bian, Y. F. Melman, A. Kagan and T. V. McDonald. 2004. An LQT mutant minK alters KvLQT1 trafficking. *Am J Physiol Cell Physiol* **286**, C1453–C1463.

33. K. Gima and Y. Rudy. 2002. Ionic current basis of electrocardiographic waveforms: A model study. *Circ Res* **90**, 889–896.

34. I. Splawski, J. Shen, K. W. Timothy, M. H. Lehmann, S. Priori, J. L. Robinson, A. J. Moss, P. J. Schwartz, J. A. Towbin, G. M. Vincent and M. T. Keating. 2000. Spectrum of mutations in long-QT syndrome genes. KVLQT1, HERG, SCN5A, KCNE1, and KCNE2. *Circulation* **102**, 1178–1185.

35. D. P. Zipes and J. Jalife. 2009. *Cardiac Electrophysiology: From Cell to Bedside.* Saunders/Elsevier, Philadelphia, Fifth Edition.

36. S. Kupershmidt, I. C. Yang, M. Sutherland, K. S. Wells, T. Yang, P. Yang, J. R. Balser and D. M. Roden. 2002. Cardiac-enriched LIM domain protein fhl2 is required to generate I(Ks) in a heterologous system. *Cardiovasc Res* **56**, 93–103.

37. Y. Rudy and J. R. Silva. 2006. Computational biology in the study of cardiac ion channels and cell electrophysiology. *Q Rev Biophys* **39**, 57–116.

38. M. Pusch, L. Bertorello and F. Conti. 2000. Gating and flickery block differentially affected by rubidium in homomeric KCNQ1 and heteromeric KCNQ1/KCNE1 potassium channels. *Biophysical Journal* **78**, 211–226.

Index